Advances in the Measurement, Utility and Evaluation of Precipitation Observations

Advances in the Measurement, Utility and Evaluation of Precipitation Observations

Guest Editors

Jiangjiang Zhang
Junliang Jin

MDPI

Basel • Beijing • Wuhan • Barcelona • Belgrade • Novi Sad • Cluj • Manchester

Guest Editors

Jiangjiang Zhang
Yangtze Institute
for Conservation
and Development
Hohai University
Nanjing
China

Junliang Jin
Yangtze Institute
for Conservation
and Development
Hohai University
Nanjing
China

Editorial Office
MDPI AG
Grosspeteranlage 5
4052 Basel, Switzerland

This is a reprint of the Special Issue, published open access by the journal *Hydrology* (ISSN 2306-5338), freely accessible at: https://www.mdpi.com/journal/hydrology/special_issues/DN07NH7D61.

For citation purposes, cite each article independently as indicated on the article page online and as indicated below:

Lastname, A.A.; Lastname, B.B. Article Title. *Journal Name* **Year**, *Volume Number*, Page Range.

ISBN 978-3-7258-5721-0 (Hbk)
ISBN 978-3-7258-5722-7 (PDF)
https://doi.org/10.3390/books978-3-7258-5722-7

Contents

About the Editors

Jiangjiang Zhang

Jiangjiang Zhang is a Professor at the Yangtze Institute for Conservation and Development, Hohai University, Nanjing, China. He received his Ph.D. from Zhejiang University in 2017. From 2015 to 2016, he was a visiting scholar at Pacific Northwest National Laboratory (PNNL) and Purdue University, and he conducted postdoctoral research at Zhejiang University from 2017 to 2020. His research focuses on the simulation of diverse hydrological and environmental systems, data assimilation, uncertainty quantification, and machine learning.

Junliang Jin

Junliang Jin is a Professor at the Yangtze Institute for Conservation and Development, Hohai University, Nanjing, China. His research focuses on extreme precipitation, climate change impacts, hydrological processes, and flood modeling. He has contributed to multiple projects investigating the dynamics of extreme rainfall events and their influence on riverine and urban flooding. His work integrates observational data, numerical modeling, and climate projections to improve the understanding and prediction of hydrologic extremes. In addition to serving as a Guest Editor for this Special Issue, he holds editorial responsibilities for leading journals in hydrology and water resources.

hydrology

MDPI

Article

Creation and Comparison of High-Resolution Daily Precipitation Gridded Datasets for Greece Using a Variety of Interpolation Techniques

Giorgos Ntagkounakis [1,*], Panagiotis Nastos [1], John Kapsomenakis [2] and Kostas Douvis [2]

1 Laboratory of Climatology and Atmospheric Environment, Department of Geology and Geoenvironment, National and Kapodistrian University of Athens, 15784 Athens, Greece; nastos@geol.uoa.gr
2 Research Center for Atmospheric Physics and Climatology, Academy of Athens, 11527 Athens, Greece; jkaps@academyofathens.gr (J.K.); kdouvis@academyofathens.gr (K.D.)
* Correspondence: georgedag@geol.uoa.gr

Abstract: This study investigates a range of precipitation interpolation techniques with the objective of generating high-resolution gridded daily precipitation datasets for the Greek region. The study utilizes a comprehensive station dataset, incorporating geographical variables derived from satellite-based elevation data and integrating precipitation data from the ERA5 reanalysis. A total of three different modeling approaches are developed. Firstly, we utilize a General Additive Model in conjunction with an Indicator Kriging model using only station data and limited geographical variables. In the second iteration of the model, we blend ERA5 reanalysis data in the interpolation methodology and incorporate more geographical variables. Finally, we developed a novel modeling framework that integrates ERA5 data, a variety of geographical data, and a multi-model interpolation process which utilizes different models to predict precipitation at distinct thresholds. Our results show that using the ERA5 data can increase the accuracy of the interpolated precipitation when the station dataset used is sparse. Additionally, the implementation of multi-model interpolation techniques which use distinct models for different precipitation thresholds can improve the accuracy of precipitation and extreme precipitation modeling, addressing important limitations of previous modeling approaches.

Keywords: interpolation; daily gridded precipitation totals; extreme precipitation; wet days; Greece; ERA5; general additive models; Indicator Kriging; regression kriging

1. Introduction

Climate datasets at a high spatial resolution are essential for evaluating climate risks and for designing and implementing adaptation strategies and mitigation measures [1]. High-resolution gridded datasets of meteorological variables can be produced by Global Climate Models (GCMs), Regional Climate Models (RCMs), and statistical interpolation of existing climate data. GCMs and RCMs can also simulate future scenarios for the atmosphere of the Earth and/or a given region. The different methodological approaches and the resulting datasets have different applications. GCMs and RCMs are utilized to forecast future climate scenarios. In contrast, gridded datasets generated using statistical interpolation are essential for understanding current climate conditions and for evaluating and refining the output from RCMs and GCMs.

 https://doi.org/10.3390/hydrology12020031

1.1. Global and Regional Climate Models

GCMs use quantitative analysis to simulate the Earth's climate and to make predictions about the future. These models are usually calculated on a coarse resolution and are not suited for regional or national climate analysis. To analyze climatic variables at a regional level, downscaling methodologies are employed. These methods are broadly classified into two categories: dynamical downscaling and statistical downscaling.

In the dynamical downscaling approach, the output from GCMs is used to develop RCMs with higher resolution, providing a more accurate representation of the region's climatic patterns. The most widely used regional climate models are the EURO-CORDEX RCMs which have a horizontal resolution of about 12 km [2]. These models also simulate climate parameters for a variety of different scenarios (RCP 2.6, RCP 4.5, RCP 8.5) and are widely used for assessing climate change and its impacts in different regions [3–5]. If the resolution of the EURO-CORDEX RCMs proves insufficient other dynamical downscaling techniques are also used by incorporating the data outputs of the EURO-CORDEX RCMs and developing more specialized models for specific areas [6,7]. The accuracy evaluation of dynamical downscaling and RCMs outputs involves a comparison with reference datasets, which commonly include observational data or high-resolution reanalysis products.

Statistical downscaling aims to leverage observational data to enhance the outputs of global or regional climate models. According to the methodology, a statistical model is used to establish some statistical relationship between observations of a climatic parameter and independent or predictor values that are usually available on a higher resolution. Once the statistical relationship has been established, predictions can be made for the climatic parameters that are more accurate and usually have a higher resolution. Additionally, to validate the results, the observation dataset is usually split into a dataset that is used for training and a dataset used for validation. Once the statistical model has established the mathematical relationships using the training dataset, predictions are made for the validation dataset and the predicted values are compared to the observed values to assess the model's performance. The statistical models and methodologies employed for this type of downscaling vary greatly depending on its intended use and the data available. Initial statistical models frequently employ regression analysis and regression kriging, whereas contemporary approaches encompass more intricate techniques such as Generalized Additive Models, Bias Adjustment techniques, Random Forests, and combinations of the different models using different approaches [8–12].

Both approaches are widely used to create high-resolution climate data for different uses, utilizing a variety of modeling approaches and data. Dynamical downscaling is based on GCM outputs and can be used to make predictions about the future without the need for robust observational data, but is usually computationally intense. Statistical downscaling approaches are less computationally intense and can be more easily applied in a variety of resolutions; however, they require long-term observational records in order to establish mathematical relationships between the climate variables and independent values, which may not be available in every region. Although such observational databases may not be needed for the dynamical downscaling approach, observational records are used for validation in both techniques. In the last three decades, RCMs have been used extensively for providing climate data due to the coarse resolution of the GCMs; however, in recent research, the need for RCMs has been questioned because the new generation of high-resolution GCMs seems to have managed to match the performance of RCMs [13,14].

1.2. Gridded Datasets Based on Observational Values

Gridded datasets based on observational values use complex statistical interpolation techniques with existing climate data from gauges or satellites in order to create high-resolution gridded datasets. These datasets can then be used to validate or downscale climate models providing data that are inherently closer to the observed climate parameters of a country or region. Additionally, databases covering reference periods between 1970 and 2010 are essential for analyzing how climate change has affected a region and for understanding how the spatial distribution of climate variables has changed.

With a spatial resolution of 0.1 degrees, the E-OBS dataset is widely recognized as one of the highest-quality gridded observational datasets for Europe, providing valuable insights into climate patterns [15]. The methodology employed by the E-OBS for creating the temperature dataset involves first interpolating monthly values to a high-resolution grid using station data. Then, a Generalized Additive Model (GAM) is fitted to the daily station values in order to interpolate to the rest of the grid. The formula employed by the E-OBS for the GAM uses longitude, latitude, altitude, and mean monthly background interpolated temperatures. For precipitation, a very similar methodology is used in the E-OBS dataset with the GAM formula utilizing longitude, latitude and the square root of monthly background interpolated totals predicting the square root of daily precipitation. In their documentation, they explain that altitude was not used because it did not improve the performance of the model. Additionally, they used the squared root of the precipitation and the precipitation totals in order to remove some of the skewness of the precipitation data.

In the past decade, there has been a proliferation of region-specific or country-specific climate databases. This development is primarily driven by the need to increase the accuracy of regional climate data and to provide higher temporal or spatial resolution, typically around 1–2 km [16–20]. In Greece, the only gridded dataset has been developed by Varotsos et al. [21,22] with a 1 km resolution. In their research for the temperature grid dataset, they applied a hybrid approach blending outputs from the non-hydrostatic Weather Research and Forecasting (WRF) meso-scale meteorological model with the available gauge data, while for precipitation they chose to use only station data.

Among all climatic variables studied, precipitation has consistently presented substantial challenges in accurately reproducing geographical variability and the occurrence of extreme events [13,21,22]. The main reason that precipitation has been so difficult to simulate is because the variable does not follow a normal distribution and it is not a continuous variable like temperature. Additionally, precipitation extremes are even more difficult to simulate [23–25] and current research shows that the frequency, intensity, and spatial distribution of precipitation extremes are changing [26,27]. In Greece, precipitation exhibits intense seasonal variability [28–30], while additionally the high degree of spatial variability of the Greek landscape influences precipitation [31] and creates different microclimates in the region, with the mountainous high-elevation regions of Greece exhibiting a higher number of wet days and annual precipitation totals while the islands in the Aegean Sea experience lower precipitation totals and droughts in the summer [32–34].

The goal of this study is to create a high-resolution daily gridded precipitation database at a $0.01^\circ \times 0.01^\circ$ resolution using novel statistical techniques which can be used for analyzing precipitation changes as well as for validation or bias-adjustment of RCMs or GCMs. In this research, we will focus on precipitation as there are additional challenges in creating a high-resolution precipitation dataset in Greece since the station network that is available is quite sparse and the spatial and seasonal variability of the variable in the region is very high. The aforementioned challenges hinder the accurate interpolation of precipitation in the region by traditional approaches necessitating the use of additional independent variables and data sources including reanalysis data. The methodologies employed in our research will be particularly useful for Greece since the country has experienced a significant number of extreme precipitation events in recent years [35] and higher-resolution precipitation data will be useful in analyzing precipitation climatology and improving the performance of climate models over the region, informing policy-makers about mitigation and adaptation measures that may be needed currently or in the future. Finally, the novel multi-model approach developed improves upon existing statistical interpolation techniques by blending ERA5 reanalysis data in the methodology which utilizes dynamical modeling techniques and the incorporation of extreme precipitation models which improve the distribution of precipitation in the studied region.

2. Materials and Methods

2.1. Data

In this research, we decided to use a hybrid method where gauge precipitation data are used as dependent variables, while ERA5 reanalysis data in combination with elevation, longitude, latitude, distance from sea coastal line, and the AUREHLY principal components are used as independent variables. Reanalysis data have been used for Greece in previous research [36] to supplement the sparse gauge dataset that is available in the Greek region for the reference period of 1980–2010. The time frame of 1980–2010 was selected for this study due to its widespread use as a reference period in both regional and global climate modeling, serving as a benchmark for comparison against future scenario simulations. Consequently, precise gridded climate datasets spanning this period are of critical importance for conducting a thorough analysis of regional climate change and variability.

2.1.1. Gauge Precipitation Dataset

In order to construct the high-resolution grid, daily quality-controlled precipitation gauge data for the Greek region were provided by the Hellenic National Meteorological Service (HNMS) for the period 1980–2010. A total of 97 stations were used, covering most of the region studied, while out of the total available data points, 23.86% were missing across the period studied. The dataset used has a density of one station per 1360 km^2 which is very sparse for interpolating precipitation, highlighting the challenges in creating an accurate and representative dataset for the region. The dataset represents the best available data for the Greek region for the period studied. To better examine the performance of the interpolation methods, the study area was divided into 13 different subregions that represent the different climates and precipitation regimes that exist in Greece. The stations were divided into Crete (C), Dodecanese (D), eastern Peloponnese (EP), western Peloponnese (WP), Cyclades (CY), eastern Aegean (EA), Ionian Islands (I), western Greece (WG), central-eastern Greece (CEG), northern Aegean (NAe), west-central Macedonia (WCM), east Macedonia and Thrace (EMT), and Cities (CT). The station density per area studied is also presented in Table S1 in the Supplementary Materials section. The regions were chosen based on published literature for the region [33,37], while for the purposes of this research, we decided to add Cities as a separate category. This category includes large cities that experience the urban heat island effect, a phenomenon that can influence precipitation regimes [38,39]. Additionally, the high population density within these urban areas underscores the critical need for accurate climate information to support evidence-based policy decisions that address the unique challenges and vulnerabilities of urban environments.

2.1.2. Geospatial Data

The elevation data employed in this study were generated by upscaling the 12 m resolution TanDEM-X Elevation Model, which is a product generated from the TerraSAR-X satellite mission [40]. The final Digital Elevation Model (DEM) along with the stations used and the different subregions studied are presented in Figure S1 in the Supplementary Materials. To calculate the distance from the sea coastal line that is used as an independent variable in the models developed, we used coordinates provided by the https://geodata.gov.gr/ (accessed on 1 June 2024) [41] government platform.

2.1.3. Reanalysis Data

For independent variables, we also utilize reanalysis precipitation data from the ERA5 reanalysis [42,43], which were provided at a 0.1° resolution in the WGS84 geographic coordinate system. The ERA5 precipitation dataset has been shown to overestimate the frequency and duration of precipitation while underestimating its intensity [44,45], and more specifically in Greece, there are also differences in the performance of the dataset between the island and continental regions [36]. Despite its known shortcomings, the ERA5 reanalysis dataset has been shown that it can be used for downscaling climate variables in previous research in a variety of different regions [36,45–47]. Additionally, the ERA5 dataset has demonstrated a capacity to effectively simulate precipitation within regions characterized by sparse gauge networks, in contrast to the E-OBS dataset, which may exhibit inaccurate precipitation estimates due to limited data availability in regions such as Southern Europe [48]. Additionally, the reanalysis dataset employs dynamical modeling techniques which will supplement the sparse station network of the region. Finally, while the E-OBS provides climate data for Europe, the global availability of ERA5 data enhances the transferability of research findings. Techniques and methodologies developed using ERA5 can be more readily applied to other regions worldwide.

2.2. Data Transformations and Preprocessing

To facilitate the analysis process, all source geospatial data were reprojected to the World Geodetic System 1984 geographic coordinate system using ESRI ArcGIS' Project function. Moreover, the grid presented in this study was generated by upscaling the TanDEM-X Elevation Model to a 0.01° resolution using ESRI ArcGIS' Resample function with bilinear interpolation. To map the coarser 0.1° resolution ERA5 grid to the 0.01° raster, a nearest-neighbor interpolation approach was used. Regarding the resolution chosen, prior research conducted in the Greek region [21,22,36,49] and other European countries [16–20] has established that a 0.01° spatial resolution is sufficient for the analysis of precipitation totals and extreme events.

Additionally, the principal components (PCs) generated by the Analyse Utilisant le RELief pour les besoins del' HYdrométéorologie (AURELHY) method [50] were also used as independent variables in the models developed. The AURELHY PCs are variables that are obtained through the application of principal component analysis on the elevation differences between a number of different grid points. This approach provides valuable insights into the underlying topographical structure of the region studied. The methodology employed for the calculation of principal components is elaborated upon in prior research conducted within the Greek region [36,51,52], where AURELHY PCs have been effectively used to generate high-resolution gridded datasets of temperature and precipitation. The AUREHLY PCs were calculated for the grid dataset used, as well as the ERA5 grid and the meteorological stations. On all occasions, the 0.01° elevation grid was used to calculate the elevation differences between the neighboring grid points, the ERA5 data grid, and the stations' positions.

Finally, the distance from the sea coastal line, which has been used to interpolate precipitation in Greece in previous research [36,51], was calculated using ESRI ArcGIS' Near function and the geospatial coordinates of the shoreline, the ERA5 grid and the final grid that was used for the analysis.

2.3. *Methodology*

2.3.1. Interpolation Models

In this research, we employed a number of different methodologies and interpolation techniques to construct daily gridded precipitation datasets in order to find the most optimal for the region and our station dataset. Initially, a monthly interpolation was performed to establish monthly background totals, which will subsequently be utilized in the daily interpolation process, as independent variables. We used a regression kriging model to interpolate the daily rain gauge data accumulated by month using longitude, latitude, and elevation as covariates on one instance (MoInt) and considering longitude, latitude, elevation, distance from sea coastal line, the ERA5 monthly sums and a number of AUREHLY PCs on another instance (MoIntERA5). The AUREHLY PCs are chosen based on their usage in previous research in Greece [36,51,52].

For the daily interpolation grid, Generalized Additive Models (GAMs) are used in combination with an Indicator Kriging (IK) model. Generalized Additive Models are an extension of Generalized Linear Models that assume the underlying functions are additive and that the predictors are smooth. Interpolation approaches that use GAMs have been used to generate the E-OBS dataset [15] and daily precipitation datasets in Greece [21,22]. In our research, we used a variety of different GAMs, the formula of the different models is described in Figure 1 and in the Supplementary Materials in Figures S2 and S3. The GAMs and the Indicator and regression kriging models are developed using the R programming language [53] and a variety of different libraries [54–57].

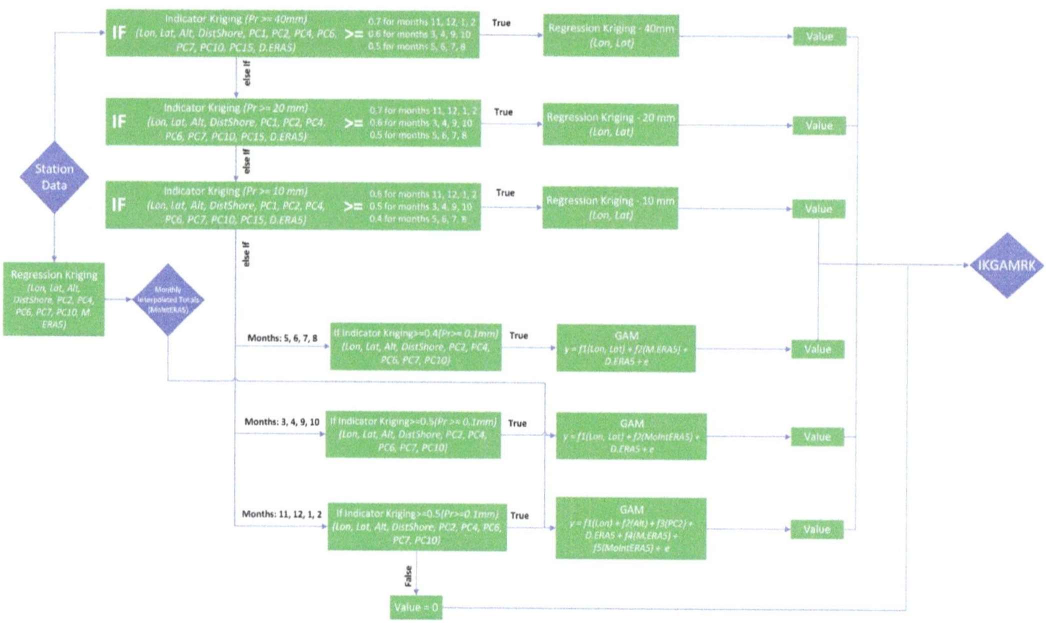

Figure 1. Flowchart of the interpolation method IKGAMRK. In the chart Lon stands for longitude, Lat for latitude, DistShore for distance from sea coastal line, Alt for altitude, M.ERA5 for monthly ERA5, D.ERA5 for daily ERA5, MoIntERA5 for Monthly Interpolated Totals with ERA5 data, and PC1-PC15 are the AUREHLY principal components.

The inherent sparsity of the gauge network in Greece presents a significant challenge in accurately capturing the intricate spatial patterns of precipitation across the diverse topographic features of the country. Recognizing this limitation, the decision to employ more complex models with a greater number of parameters and integrate ERA5 reanalysis data was driven by the need to improve the accuracy of precipitation estimates, particularly in regions with limited gauge coverage.

The IK approach is used in order to interpolate the daily occurrence of rainfall. The station data are first categorized by setting days where precipitation is less than 0.1 mm/day as zero and days where precipitation is equal to or greater than 0.1 mm/day as one. These data are interpolated using regression kriging on the grid and a threshold is set based on the interpolated values to assign a wet day to each grid point. This approach has been used before to interpolate daily precipitation and develop high-quality gridded datasets [58,59] usually using a 0.5 IK value as a threshold for assigning precipitation. In our research, we found that using different IK thresholds can be useful for the Greek area, especially in the drier months where we achieved slightly better results by utilizing a lower threshold. The thresholds used along the different models are shown in the flowcharts in Figures 1, S2 and S3.

During the training stages of the models, we found that all interpolation models could not accurately simulate extreme precipitation, in particular precipitation over 20 mm/day. To bypass this problem, we chose to add three more IK models, one for a 10 mm/day threshold, one for 20 mm/day, and one for 40 mm/day (0 or 1 depending on whether daily precipitation equals or exceeds 10, 20 or 40 mm/day). The data, similarly to assigning a wet day, are interpolated using regression kriging, and a threshold is used to assign an extreme precipitation day. To predict precipitation in the points that the IK has assigned an extreme precipitation day a regression kriging model is used that is trained on only station data that exceed the 10, 20, or 40 mm/day threshold, respectively. The model uses only longitude and latitude as independent variables for all different thresholds, since during the training stages we found that adding more variables does not result in significant improvements in performance. Given the rarity of extreme precipitation events exceeding 10, 20, and 40 mm/day in the Greek region and the limited number of rain gauges available during the model training period, there may not be enough data to effectively train the model for days identified by IK as having extreme precipitation. To circumvent this constraint, a progressive data augmentation technique was utilized. When the available data for a particular day proved inadequate for regression kriging model training, data from the preceding and subsequent days were appended to the dataset, with the condition that no data points originated from the same gauge station. The process was implemented until there were enough data to train the model. In order to blend the different models, an iterative approach is used. For each cell of the grid, the threshold for the most extreme precipitation is checked first and if the grid point exceeds the threshold the appropriate model is used. To set the thresholds for each period an iterative approach was used where all IK thresholds started from 0.5 and progressively added and subtracted 0.1 to find the most optimal value. Results from this process are displayed in Annex S2 of the Supplementary Materials section.

Building upon the previously outlined data and modeling considerations, three distinct interpolation models were formulated (Figures 1, S2 and S3) to facilitate a comparative analysis of their performance. Firstly, for all the generated models, a monthly background precipitation total dataset is generated that is used as an independent variable in the different models employed. For the IKGAM the monthly background totals are generated utilizing only station data, while for the IKGAMV2 and the IKGAM RK, the background totals also utilize ERA5 data. The parameters utilized for the background monthly totals are also presented in the model flowcharts (Figures 1, S2 and S3). Regarding the parameters used for the GAM and the IK models, IKGAM presented in Figure S2 does not incorporate any AUREHLY parameters or ERA5 data and sets a homogenous 0.5 threshold for assigning a wet day. Additionally, the final value for each cell is calculated by multiplying the IK value and the GAM value. In Figure S3, the IKGAMV2 model incorporates the AUREHLY and ERA5 data when constructing the GAM daily models. Furthermore, the months are split into three groups and each group has a separate IK threshold and a separate GAM. Finally, in Figure 1, the IKGAMRK uses three different IK models that use different thresholds for three different groups of months. Each IK is interpolated using a different precipitation threshold and for each precipitation threshold, a different model is used for interpolation. IKGAMRK's modeling framework aims at addressing the challenges in interpolating accurately the frequency of days where precipitation exceeds 10 and 20 mm/day by incorporating specific models and modular IK thresholds. The variables used for each model are given inside the frame of the flowchart.

2.3.2. Validation Process

To ensure the robustness and comparability of the model results of this study, a Leave-One-Out Cross-Validation (LOOCV) technique was selected for model evaluation. According to the methodology, each station is left out of the training process of the model, and after training, the model is used to predict the station values. Since in all of the interpolation models used in this study a monthly interpolation is performed first, therefore in order to have the most accurate results, the monthly interpolation also excludes each station that is validated during the LOOCV process. This essentially means that a monthly and a daily model must be trained for each station, which is a very time-consuming process. However, this process gives the most accurate and comparable results between the models, since all 836,194 available gauge data points are validated through the same process. The main shortcoming of the LOOCV process is that the results depend largely on the density of the station network [60]. In our research, the same station dataset is used to construct all the models therefore the results are comparable; however, caution should be taken when comparing the results of different station datasets.

2.3.3. Statistical Metrics

The metrics that will be used to compare the different interpolation models and the ERA5 datasets are the R^2 coefficient and the Root Mean Squared Error (RMSE). These statistics will be calculated for monthly and annual totals of Precipitation, wet days (where precipitation exceeds 1 mm/day), and the number of times precipitation exceeds 20 mm/day (P20). It is worth noting that some stations have missing data therefore all statics will be weighted and averaged when presented for each month or annually. To enhance readability, the presentation of statistical results for precipitation events exceeding 10 mm/day and 40 mm/day has been omitted, as these results exhibit strong similarities to those observed for the 20 mm/day threshold.

Additionally, the contingency statistics probability of detection (POD), false alarm ratio (FAR), and Critical Success Index (CSI) will be calculated for wet days and the number of times precipitation exceeded 20 mm/day. The contingency scores are calculated using the number of hits (A), false alarms (B), and misses (C) for each time precipitation hits a threshold in the predicted data compared to the station data. Specifically, POD is the ratio of observed events that are estimated correctly and it ranges from 0 (poor) to 1 (good). FAR is the ratio of false alarms, ranging from 0 (good) to 1 (poor). CSI is a combination of the hit ratio and the false alarm ratio in one score which gives a holistic metric of performance and it ranges from 0 (poor) to 1 (good). These metrics have been used by Nastos et al. [37]

to evaluate precipitation and are described in detail in their research. The formulas for all metrics and models used in this study are also presented in the Annex SI—Formulas of the Supplementary Materials sections.

To compare the distribution of the resulting datasets a two-sided non-parametric Mann–Whitney–Wilconox test was performed, and their effect size metrics were compared. The non-parametric test hypothesis assumes that two variable distributions come from the same population, and in the two-sided version of the test, the alternative hypothesis stipulates that the first group's data distribution differs from the second group. To quantify the magnitude of the observed effect, the Wilcoxon effect size (r) was calculated by dividing the Z statistic by the square root of the sample size. A smaller absolute value of r suggests a greater similarity between the groups, whereas a larger absolute value indicates a more substantial difference. Finally, all presented metrics underwent statistical significance testing using appropriate R functions or their respective specialized metric libraries. All results were found to be significant at the 1% level.

The combination of the aforementioned validation metrics and techniques will collectively enable an in-depth assessment of the models' performance. Additionally, figures comparing station data and predicted data from the models will be shown in the Section 3 in order to visualize the regional differences in performance. More specifically, figures of the annual distribution for the different Greek regions will be presented along with maps of the stations' RMSE and the mean annual totals of precipitation, wet days and P20.

3. Results

3.1. Precipitation Totals

In Table 1 the improvements in performance in the monthly and annual totals are presented. We observe that the standalone ERA5 dataset performs better than the IKGAM dataset but when blended with our interpolation technique in IKGAMV2, the results are better than the ERA5 on an annual basis. IKGAMV2's performance relative to ERA5 varied throughout the individual months, exhibiting superior performance from January through April, September, October, and December while underperforming in the remaining months. In the summer months, all interpolation methods record their worst performance, and they are the only months where ERA5 is able to outperform the interpolation models. Finally, the IKGAMRK performs better than the rest of the interpolation methods, especially on an annual basis and in the Winter and Autumn months. In Summer, the model has a very similar performance to the rest of the interpolation methods, especially when comparing the RMSE metrics of the models. Similar results are also observed for the daily metrics presented in Table S2.

In Figure S4, histograms of the distribution of annual precipitation are presented. Through the observed distribution of the models studied and the metrics presented in Tables 1 and S2, we can conclude that the ERA5 model seems to be overestimating precipitation in our station dataset on an annual basis. In Figure S4, the differences are especially pronounced in the west-central Macedonia region and western Greece. It is worth noting that most of the stations in our dataset are located in lower-elevation areas which does influence the results. In Crete, where the dataset contains a lot of stations in higher elevation, the ERA5 mean values are lower than the observed precipitation depicted in Figure S4. Previous evaluations of the ERA5 dataset also conclude with our results [61] that the ERA5 overestimates precipitation overall but underestimates precipitation in very high-elevation areas. The IKGAM dataset underestimates precipitation in most areas while the IKGAMV2, which has ERA5 data blended into the interpolation method, interpolates precipitation more accurately in most areas. Finally, the IKGAMRK interpolates precipitation more closely to the station data out of all the models, as observed through the metrics presented in Tables 1 and S2.

The maps of mean annual precipitation totals in Figure 2 further confirm the results described above. The ERA5 dataset simulates a very strong longitudinal shift in precipitation which also exists in the IKGAMV2 and IKGAMRK datasets but to a lesser extent. The IKGAM dataset records notably lower precipitation across the whole Greek region. Finally, the IKGAMV2 and IKGAMRK record precipitation much higher in the peaks of the mountains compared to the ERA5 dataset. This difference is especially pronounced in Crete where the ERA5 dataset simulates notably less precipitation in the mountain peaks, compared to the IKGAMV2 and the IKGAMRK datasets. Additionally, the IKGAM interpolates a much smoother precipitation compared to IKGAMV2 and IKGAMRK which interpolates much more refined peaks of precipitation across the Pindos mountain range. This can be attributed to the fact that IKGAM utilized very few parameters and the station dataset that we use does not have a lot of stations in the Pindos mountain range or in very high-elevation areas in general.

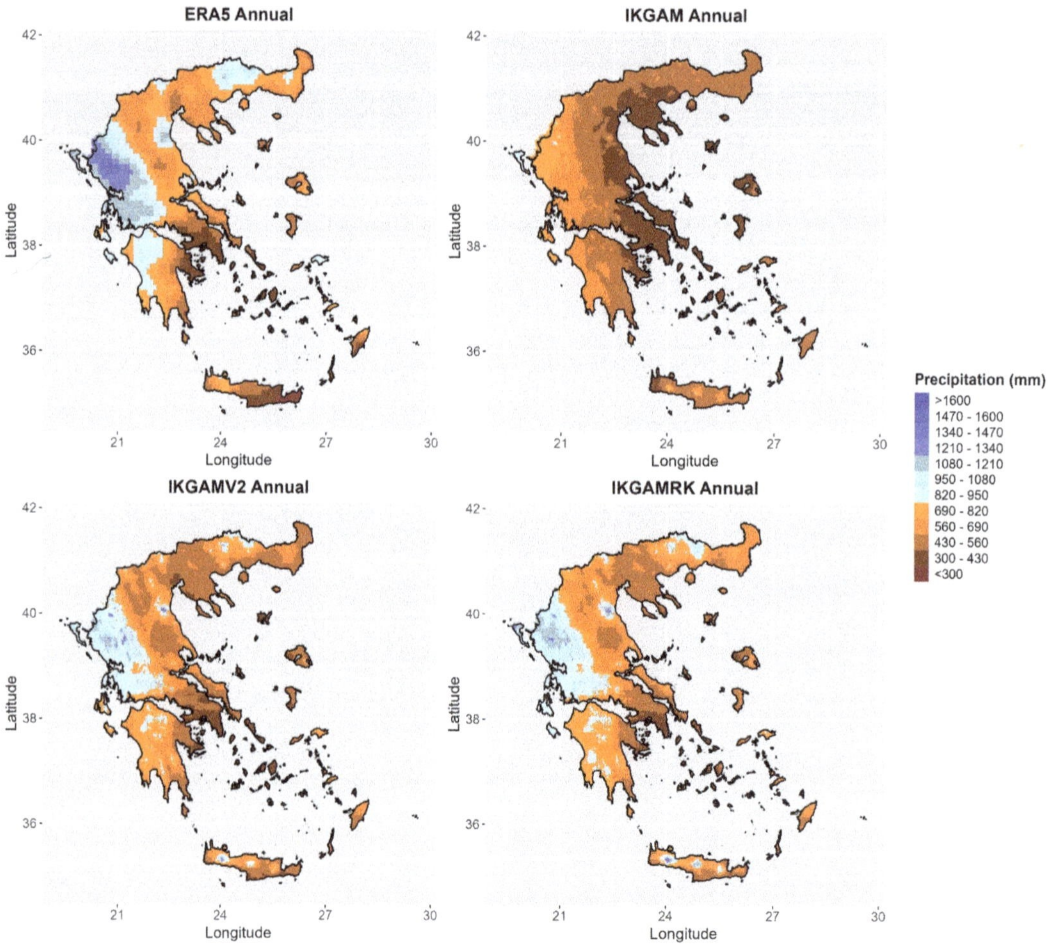

Figure 2. Maps of mean annual precipitation totals of the different interpolation methods and the ERA5 data.

When comparing the spatial distribution of the station RMSE in Figure 3, we can see that the IKGAMRK and IKGAMV2 perform notably better than the rest of the models. The differences are more pronounced in Crete, northeastern Greece, Ionian Islands, and the Evia region. The IKGAM has notable differences in the Ionian Islands and Evia region compared to the rest of the models. In the Aegean islands, the model also performs worst in the region compared to the rest of the models. The ERA5 model also has notably higher RMSE than IKGAMV2 and IKGAMRK in Crete, eastern Aegean, and the Ionian islands. Finally, it is worth noting that in the RMSE maps, we can see the influence that station density has on performance. In Athens, where the station network is denser, all the models record some of their best performance, while, in the rest of Greece, the RMSE is higher in all the models.

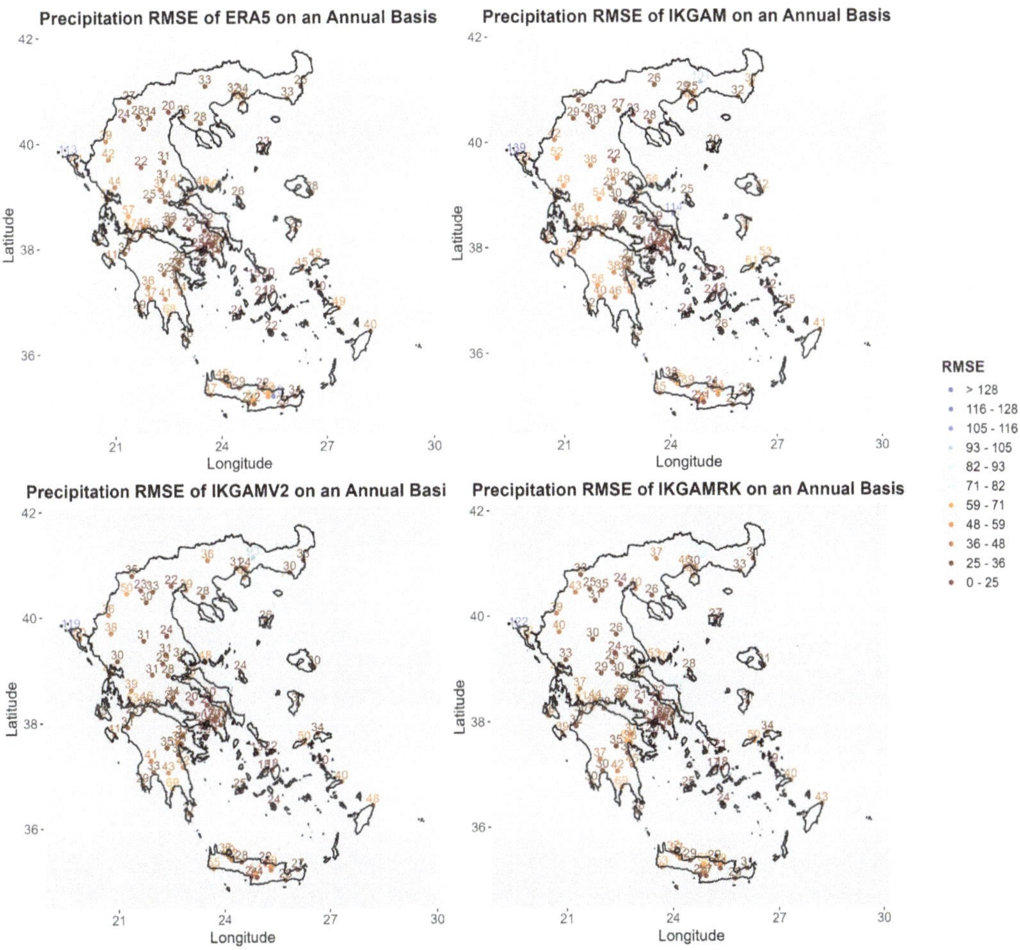

Figure 3. Annual precipitation RMSE maps of each station for the interpolation methods and the ERA5 data.

Table 1. Metrics of annual and monthly precipitation totals.

Model	ERA5		IKGAM		IKGAMV2		IKGAMRK	
Months	R^2	RMSE	R^2	RMSE	R^2	RMSE	R^2	RMSE
January	0.51	55.81	0.54	56.87	0.58	52.09	0.58	52.12
February	0.48	47.03	0.43	51.28	0.50	45.36	0.49	46.42
March	0.47	40.01	0.44	42.09	0.50	37.79	0.51	37.95
April	0.51	28.00	0.48	28.88	0.51	26.40	0.53	26.40
May	0.52	21.58	0.45	24.06	0.51	20.69	0.49	22.34
June	0.42	19.06	0.36	20.60	0.25	23.06	0.28	23.99
July	0.45	14.45	0.27	17.31	0.27	18.10	0.35	17.30
August	0.40	16.13	0.33	17.78	0.25	19.06	0.28	19.51
September	0.46	26.32	0.49	27.38	0.51	25.34	0.50	26.02
October	0.52	43.24	0.49	45.41	0.53	41.95	0.52	43.85
November	0.44	59.99	0.39	64.02	0.43	60.58	0.43	60.97
December	0.41	65.96	0.38	69.31	0.45	62.29	0.46	62.13
Annual	0.45	229.71	0.42	267.81	0.48	212.87	0.49	212.49

3.2. Wet Days

Moving forward, in Table 2, the performance in the monthly and annual number of wet days is presented. It is worth noting that when comparing the number of wet days and the number of days precipitation exceeded 20 mm the RMSE and the R^2 should be analyzed in conjunction with the contingency metrics (Table 4) since they can also give important information about the performance of the models. For wet days, the ERA5 may present a high R^2 but the RMSE is notably higher than all the interpolation methods. IKGAM has the lowest RMSE of all the models and, in Table 4, we can see that FAR is the highest in ERA5 while the lowest in IKGAM and IKGAMRK, while at the same time, CSI is the highest in IKGAM and IKGAMRK and lowest in ERA5. When comparing the metrics of IKGAM and ERA5, the R^2 values are much closer between the two models when compared to the RMSE. This difference indicates that while the ERA5 data capture the overall trend of monthly and annual wet days correctly the prediction error is higher than the rest of the models. This indicates that the ERA5 model simulates a lot more wet days and the interpolation methods are able to correct this overestimation. Moreover, the performance of the models between the different months is much more uniform, especially when compared to the models' performance in precipitation. It is worth noting that while IKGAM performed much worse than IKGAMV2 in precipitation, on wet days, IKGAM outperformed IKGAMV2. This difference can be attributed to the different thresholds used in the IK model. In IKGAM the threshold is 0.5 while in IKGAMV2 the threshold is 0.4 and 0.3 (Figures S2 and S3). This means that while a higher threshold can be used to accurately predict the number of wet days it impairs the performance of the total precipitation. This does not affect the IKGAMRK performance because while it does use a higher threshold for wet days it also utilizes extreme precipitation models which add more precipitation to the totals.

In Figure S5, the distribution of the annual wet day totals is presented for all the regions in Greece. The ERA5 model displays a higher average number of wet days across all regions of Greece. This provides a more nuanced interpretation of the contingency metrics with respect to the ERA5 data, which exhibit a high probability of detection (POD) but also the highest false alarm rate (FAR) among all evaluated models. It is worth noting that the IKGAMV2 also records a higher number of wet days on an annual basis in Figure S5. The region that has the highest deviations is the northern Aegean region similar to the precipitation.

Moving on to the maps of the different models in Figure 4, where the ERA5 model simulates a large number of wet days compared to the rest of the models. This difference is especially pronounced across the Greek mainland. The IKGAMV2 and IKGAMRK models interpolate a lot more wet days in the mountain peaks of the Pindos mountain range and the mountains of Crete compared to IKGAM. In Figure 5, the bias of each station is presented for all the models. When cross-examining Figures 4 and 5, we can see that the large number of wet days simulated by the ERA5 in the Greek Mainland is increasing the RMSE in the region. At the same time, the performance of the rest of the interpolation models is very similar. The best model overall seems to be IKGAM recording the lowest RMSE with IKGAMRK having a slightly higher RMSE in the Greek mainland.

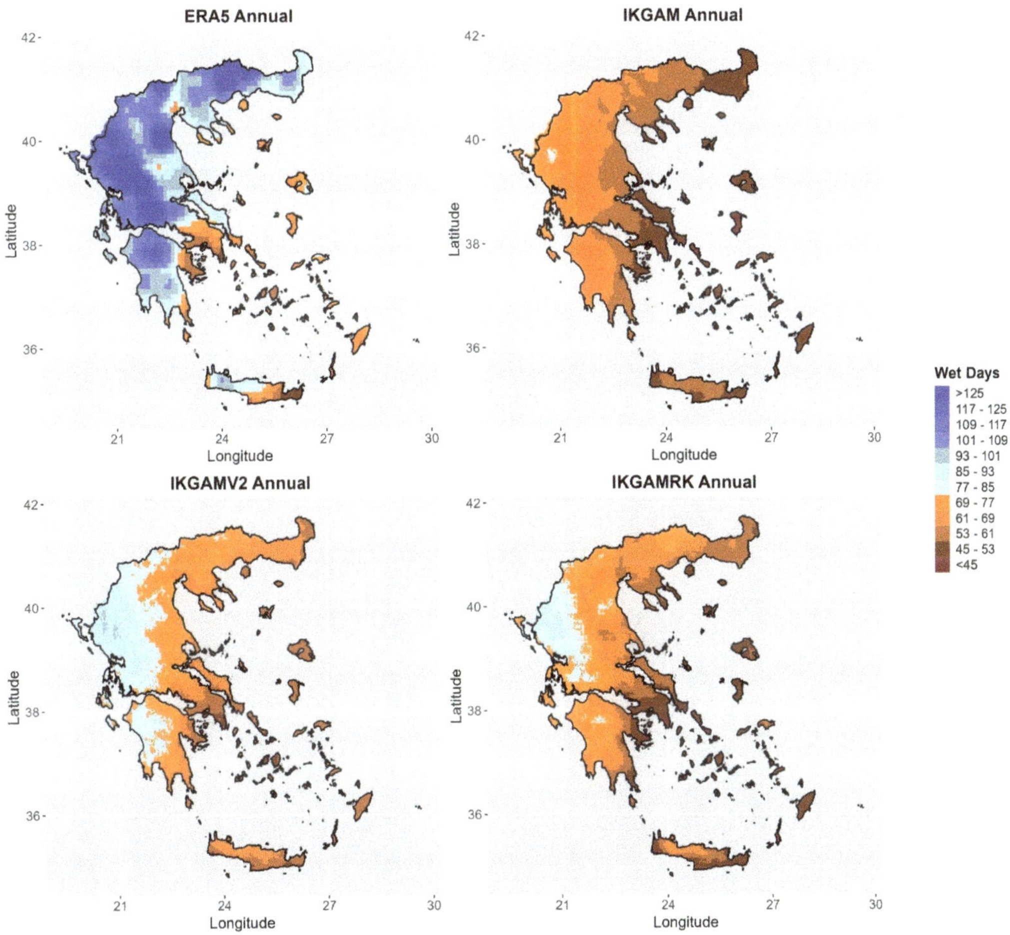

Figure 4. Maps of mean annual wet days totals of the different interpolation methods and the ERA5 data.

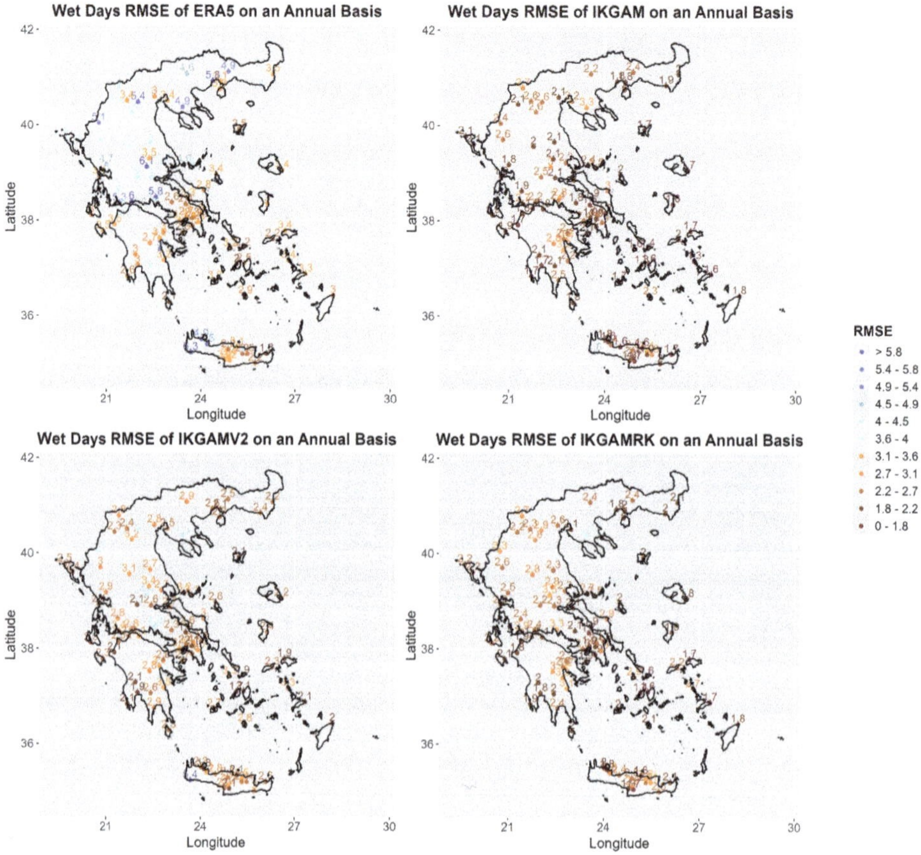

Figure 5. Annual wet days RMSE maps of each station for the interpolation methods and the ERA5 data.

Table 2. Metrics of annual and monthly wet days totals.

Model	ERA5		IKGAM		IKGAMV2		IKGAMRK	
Months	R^2	RMSE	R^2	RMSE	R^2	RMSE	R^2	RMSE
January	0.65	4.54	0.69	2.71	0.58	3.49	0.61	3.12
February	0.58	4.67	0.58	2.80	0.51	3.30	0.54	3.00
March	0.57	4.57	0.57	2.44	0.51	3.28	0.54	2.63
April	0.58	3.94	0.57	2.14	0.52	2.82	0.55	2.33
May	0.60	3.40	0.57	1.92	0.54	2.25	0.57	2.04
June	0.59	2.34	0.55	1.44	0.52	1.63	0.55	1.48
July	0.54	2.11	0.48	1.34	0.56	1.23	0.60	1.15
August	0.57	2.07	0.41	1.36	0.50	1.30	0.52	1.21
September	0.63	2.50	0.69	1.49	0.61	1.97	0.64	1.64
October	0.55	3.38	0.62	1.98	0.41	2.99	0.47	2.49
November	0.50	3.84	0.56	2.43	0.47	3.00	0.51	2.65
December	0.54	4.40	0.55	3.04	0.42	3.96	0.49	3.44
Annual	0.58	30.59	0.63	11.67	0.47	16.58	0.53	13.36

3.3. Number of Days Where Precipitation Exceeds 20 mm

Prior research [36,62,63] has established that daily precipitation exceeding 20 mm is a relatively rare occurrence in the Greek region, particularly during the summer and spring months. Consequently, the scale of the metrics presented for these extreme events will exhibit a marked contrast compared to the totals for overall precipitation and the number of wet days. In Table 3, metrics of performance for the different models are presented. Firstly, the differences in performance are much smaller compared to the rest of the parameters presented above. This occurs due to the rarity of P20 events in the Greek region. Moreover, there is an obvious drop in performance in the summer and spring months across all models. The model that has the best performance overall is the IKGAMRK, which records the best R^2 and the lowest RMSE out of all the models. Moreover, the POD and CSI metrics (Table 4) are notably higher for the IKGAMRK model compared to the rest of the models. At the same time, the FAR metric is slightly higher for ERA5 and IKGAMRK, but the differences are not as large as the POD and CSI metrics. The worst performance is recorded by the IKGAM which has the lowest CSI and POD metrics.

Table 3. Metrics of annual and monthly number of days where precipitation exceeds 20 mm totals.

Model	ERA5		IKGAM		IKGAMV2		IKGAMRK	
Months	R^2	RMSE	R^2	RMSE	R^2	RMSE	R^2	RMSE
January	0.23	1.47	0.27	1.46	0.33	1.36	0.36	1.33
February	0.21	1.27	0.12	1.37	0.21	1.26	0.24	1.24
March	0.21	1.08	0.14	1.16	0.22	1.09	0.27	1.08
April	0.17	0.78	0.10	0.80	0.19	0.74	0.22	0.74
May	0.12	0.58	0.09	0.59	0.12	0.58	0.15	0.64
June	0.08	0.45	0.02	0.45	0.06	0.45	0.10	0.52
July	0.07	0.39	0.06	0.39	0.09	0.39	0.10	0.44
August	0.09	0.41	0.12	0.40	0.10	0.42	0.15	0.45
September	0.19	0.64	0.10	0.66	0.16	0.67	0.20	0.69
October	0.32	1.08	0.24	1.08	0.26	1.12	0.30	1.15
November	0.32	1.43	0.23	1.50	0.30	1.43	0.29	1.46
December	0.23	1.68	0.19	1.69	0.27	1.57	0.30	1.53
Annual	0.23	5.65	0.23	6.17	0.27	5.45	0.28	5.08

Figure S6 illustrates the distribution of annual P20 totals, revealing that the ERA5, IKGAM, and IKGAMV2 models have lower mean totals of P20 events across the entire region. This underestimation appears to be more pronounced for the IKGAM and ERA5 models, while the IKGAMRK model's P20 distribution and mean values correspond closer to observed station data. Similar to the rest of the parameters studied, the largest deviations are recorded in western Greece and the Northern Aegean. These results along with the metrics calculated indicate that all the models, besides IKGAMRK, underestimate the number of P20 events across the Greek region. Furthermore, we can also conclude that the extreme models used in IKGAMRK help increase the predictive accuracy of the interpolation approaches in the region.

In Figure 6, maps of the mean annual P20 totals are presented. The IKGAM interpolates notably fewer P20 events across the whole region. The ERA5 simulates most P20 events across northwestern Greece and smaller amounts in northern Greece. Similar patterns are also recorded in the IKGAMV2 model. As stated before, the station dataset used is quite sparse and it does not have a lot of data in higher elevation areas which record most of the P20 events occurring, therefore it is very difficult for the IKGAM to interpolate the P20 events accurately. The differences between IKGAM and IKGAMV2 show the influence that the ERA5 data have when they are blended in the interpolation process. Finally, the IKGAMRK is able to interpolate the P20 events more accurately across the whole region. In the Pindos mountain range, the model interpolates more P20 events in the mountain peaks. In northeastern Greece, the IKGAMRK also records more P20 events than the rest of the models. In Crete the IKGAMV2 and IKGAMRK record more events in the mountain peaks compared to IKGAM and the ERA5.

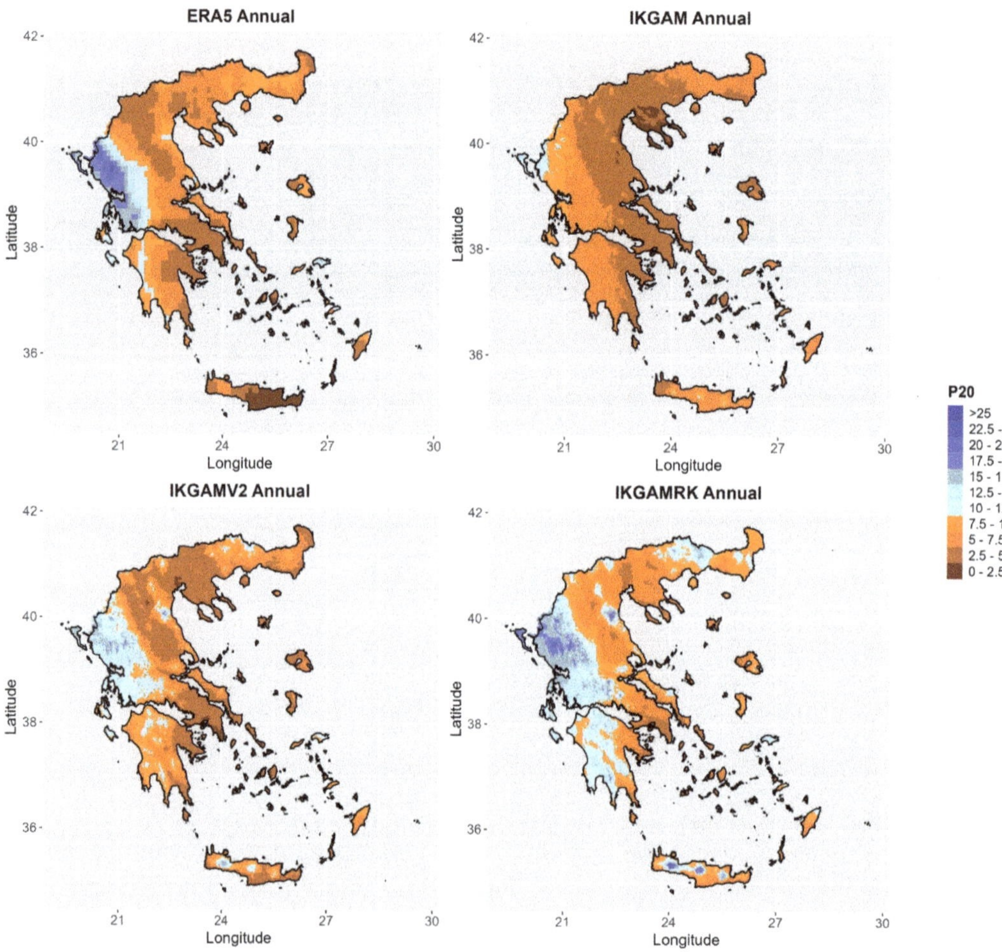

Figure 6. Maps of mean annual P20 totals of the different interpolation methods and the ERA5 data.

In Figure 7, we observe that the IKGAM has higher RMSE across northeastern Greece and the northwestern mountainous regions. In those regions, the IKGAMRK outperforms the rest of the models by a wide margin. These mountainous regions record some of the highest numbers of P20 in Greece; therefore, the models' performance is crucial in those regions. In Crete, the ERA5 data have a higher bias than the rest of the models, most notably in the eastern mountainous regions. Finally, while the IKGAMV2 has a lower bias than IKGAM and ERA5, the best model overall is IKGAMRK because it has the lowest bias in the whole Greek region and records the best metrics by a wide margin.

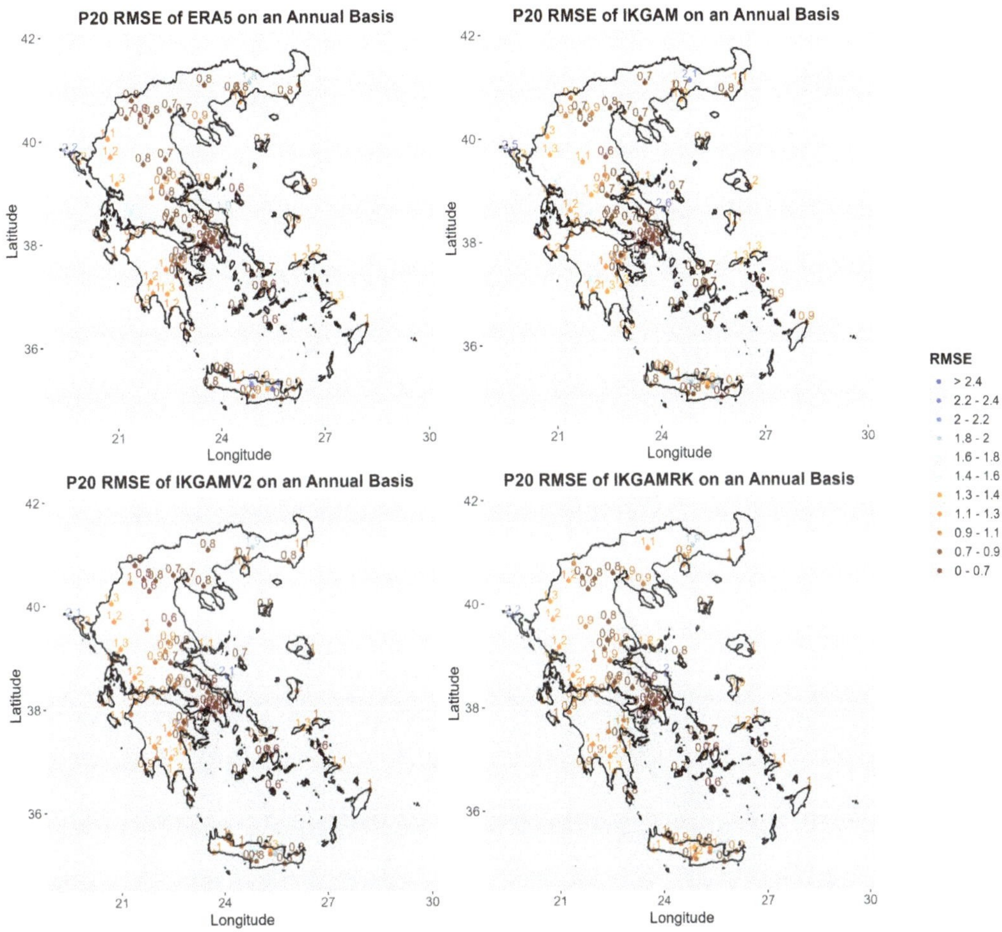

Figure 7. Annual P20 RMSE maps of each station for the interpolation methods and the ERA5 data.

4. Discussion

According to the metrics presented in the Section 3 and the distributions shown in Figure 8, the ERA5 model demonstrates biases in its precipitation estimates across the Greek region, characterized by an overestimation of total precipitation and the number of wet days, coupled with an underestimation of P20 events. This pattern is further corroborated by the higher RMSE values observed for wet days and P20 (Tables 2 and 3) in the ERA5 model relative to the interpolation methods. Furthermore, the ERA5 model exhibits a notably lower CSI score (Table 4), indicating inaccuracies in simulating wet days and P20 events.

The interpolation method IKGAM displays increased accuracy in simulating the number of wet days while underestimating P20 events as seen in the POD, FAR, and CSI metrics in Table 4. This may be due to the fact that IKGAM utilizes a higher IK threshold for wet days which may be accurate for the number of wet days but it hinders the performance of Precipitation and the number of P20 in the region as seen in Figures 2 and 6, where there are differences between IKGAM and the rest of the datasets. In Tables 1–3, we can also observe that the IKGAM has very high RMSE in the precipitation totals and in the number of P20 events but has the lowest bias in the number of wet days. When analyzing the maps of the interpolated variables (Figures 2, 4 and 6) in conjunction with the mapped RMSE values (Figures 3, 5 and 7), we can conclude that an important limitation of all the interpolation methods is the data quality, spatial distribution and density of the station dataset that is used. In areas with denser precipitation networks, such as Cities in our station dataset, the interpolation methods perform their best.

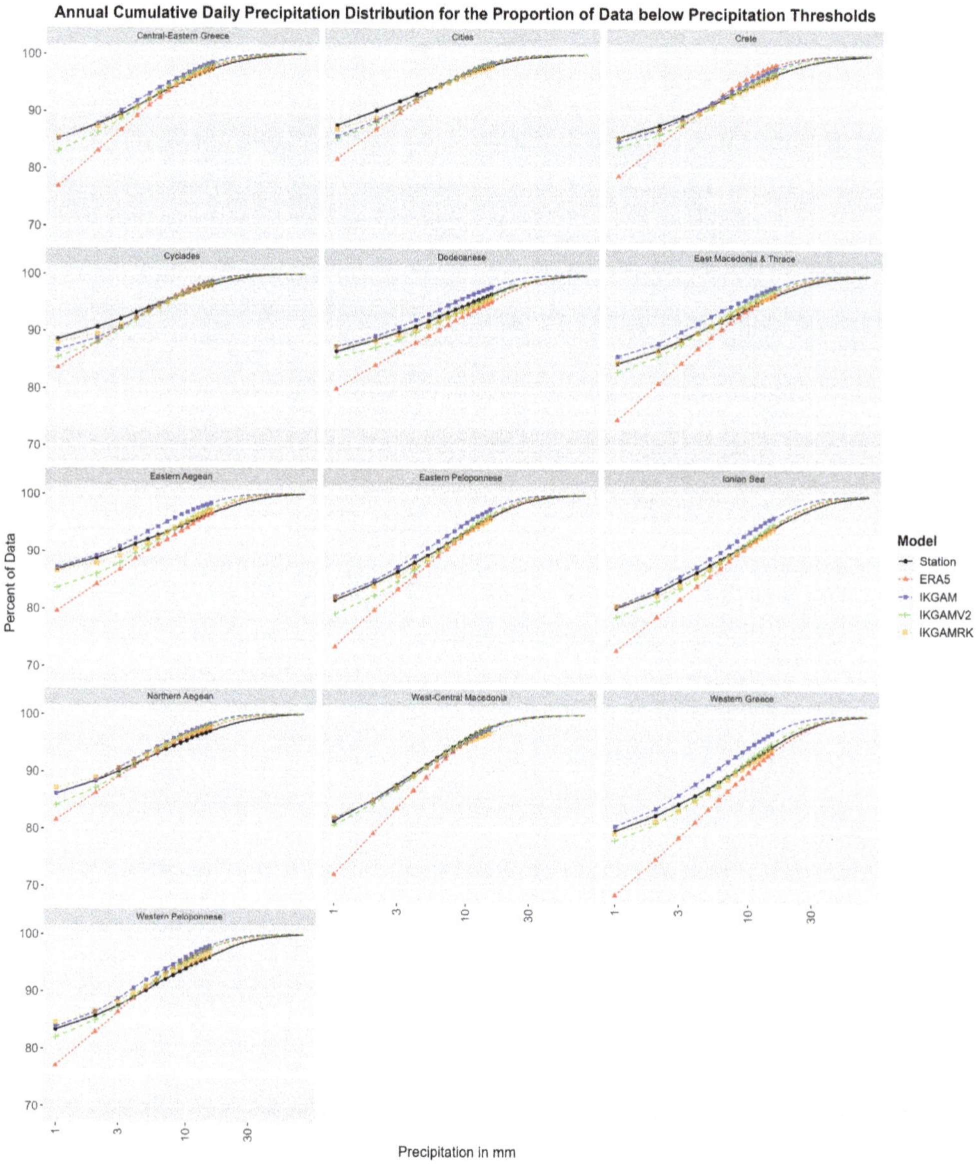

Figure 8. Comparison of cumulative precipitation distribution between the gauge dataset, the ERA5 reanalysis, and the interpolation methods for the number of values below precipitation thresholds.

Table 4. Contingency metrics of the ERA5 reanalysis and the models studied.

Models	Wet Days			P20		
	POD	FAR	CSI	POD	FAR	CSI
ERA5	0.82	0.45	0.49	0.35	0.54	0.25
IKGAM	0.72	0.28	0.56	0.25	0.48	0.20
IKGAMV2	0.74	0.34	0.53	0.36	0.49	0.27
IKGAMRK	0.70	0.29	0.55	0.45	0.55	0.29

The IKGAMV2 model had ERA5 blended in the interpolation process in order to add information to the sparse station dataset and utilized more parameters in all the models used. Due to the sparse density of the gauge network that exists in Greece, the ERA5 data that were used for the IKGAMV2 methodology improved the results of the model (Tables 1, 3, 4 and S2) compared to IKGAM. Moreover, the spatial variability improved as we can see in the maps of the parameters studied. However, it is worth noting that when adding reanalysis data in the process, inevitably, the resulting dataset will be influenced by the dataset used. Additionally, in order to increase the precipitation totals, the threshold of the IK was decreased and, as a result, the performance of wet days was hindered. Moreover, in Figure 6 we can see that IKGAMV2 still underestimates the number of P20 events. The only model that was able to interpolate extreme precipitation accurately was IKGAMRK. This is due to the fact that the model utilizes specialized models for extreme precipitation. At the same time, the model can interpolate wet days very accurately because it also has a high IK threshold for wet days.

In Figure 8, the cumulative distribution of the aggregate station data is compared to the distribution of the interpolated predicted data using the LOOCV method. This graph was derived by aggregating all station and predicted data, then for each threshold value displayed on the X-axis, the Y-axis depicts the proportion of data with rainfall values falling below that threshold. In the Supplementary Section, a different version of the cumulative graph is provided for context, where instead of the percent of data points, the percent of total precipitation is calculated.

The cumulative distribution graph (Figure 8) underscores the critical role of wet days in shaping the overall precipitation distribution, findings that are consistent with the results presented in Tables 1–4. The ERA5 model, in particular, demonstrates a notable discrepancy from observed data, characterized by an underestimation of days with sub-1 mm precipitation and an overestimation of days with 1–10 mm rainfall. This observation is further substantiated by Figure S7, which reveals a disproportionately high contribution of the 1–10 mm precipitation range to the total precipitation in the ERA5 and IKGAM datasets. To further solidify the observed differences in the cumulative plots, in Table S3 the results of the Wilcoxon–ManS2–Whitney test are presented. The results show that the ERA5 distribution has the largest effect size when compared to the interpolation methods, with the differences being more pronounced on a daily basis. All results of the test are statistically significant within the 1% level, with the IKGAMRK model recording a slightly higher p-value than the rest of the models on the distribution of monthly precipitation totals. The differences between the daily and monthly results can be attributed to the overestimation of wet days in the region. While the IKGAMV2 model represents an improvement over both IKGAM and ERA5, the IKGAMRK model exhibits the most accurate representation of the observed precipitation distribution across all precipitation thresholds, with the exception of extreme events exceeding 50 mm. The reason that we did not incorporate even more extreme precipitation models coupled with more IK models in the IKGAMRK approach is that due to their rarity, we were not able to incorporate enough data to train the models. To further optimize the methodology's performance, future research should explore the utilization of different combinations of data, models, and thresholds for extreme precipitation.

In conclusion, the best-performing model overall was IKGAMRK because it can interpolate the distribution of precipitation more accurately across the whole region. The rest of the models seem to excel at predicting certain aspects of precipitation but lack the ability to accurately predict the distribution of precipitation. The differences between the approaches are also observed when comparing their results on a daily basis (Table S2) and their distributions (Figures S7 and 8, and Table S2). This is especially pronounced in the number of P20 events (Table 4) where none of the other models can accurately predict the parameter, highlighting the outperformance of the novel modeling framework in accurately interpolating extreme precipitation events. Moreover, the main strength of IKGAMRK lies in its utilization of multiple IK models for different precipitation thresholds. The adjustable thresholds within the multi-model approach enhance the accuracy of precipitation prediction across the variable distribution.

5. Conclusions

The goal of this study is to create and compare daily gridded precipitation databases for the Greek region, in order to facilitate a comparative analysis of the different methodologies and their respective strengths and limitations. Three different interpolation models were created, IKGAM utilizes only the station network data and limited geographical variables (Figure S2) to create the precipitation dataset using a combination of IK and GAM. IKGAMV2 model has ERA5 precipitation data blended into the interpolation process and uses a variety of different geographical variables (Figure S3). Finally, IKGAMRK incorporates ERA5 precipitation data and different IK models combined with GAM and specific models for extreme precipitation (Figure 1). From the metrics and figures that were presented, the IKGAMV2 and the IKGAMRK interpolation approaches outperform the ERA5 and IKGAM when estimating annual precipitation totals (Table 1). This improvement is also evident in their respective daily metrics on an annual basis (Table S2). Regarding wet days, the ERA5 dataset simulates a larger annual total of wet days (Figure 4), which seems to hinder its performance compared to the rest of the models (Table 2). Out of the three modeling approaches, the IKGAM is the most accurate in predicting the number of wet days (Tables 2 and 4), while its performance lags the rest of the methods in the precipitation totals (Table 1) and P20 totals (Tables 3 and 4). The IKGAMV2 approach is more accurate in precipitation totals; however, it is still less accurate than IKGAMRK in predicting P20 and the number of wet days. The best-performing model is the IKGAMRK since it outperforms the rest of the models when comparing their annual precipitation totals (Table 1), the daily values on an annual basis (Table S2), and the contingency metrics regarding wet days and P20 (Table 4). The outperformance of IKGAMRK is most apparent in precipitation events over 20 mm where there is a notable difference with the rest of the models in the metrics calculated (Tables 3 and 4). This outperformance is also evident in the precipitation distribution graphs (Figures 8 and S7) and through the metrics of the Wilcoxon–Mann–Whitney test (Table S3).

Finally, although the database and the modeling process are centered around Greece, important conclusions can be made for interpolating precipitation in general. More specifically, we can conclude that when using a sparse precipitation dataset blending reanalysis data in the interpolation process can increase the accuracy of the results. However, caution should be taken, because the reanalysis data will influence the results, while due to the sparsity of the gauge networks, the influence of the reanalysis can be hard to validate. Additionally, the final dataset is also dependent on the availability of the reanalysis dataset.

Furthermore, utilizing multiple IK models across different precipitation thresholds can aid in interpolating extreme precipitation more accurately. This approach seems to be more appropriate when the station data are quite sparse as the models did have more similar performances in the Cities category where the station network was denser. An important limitation in all interpolation approaches is additionally the data quality, geographical distribution, and density of the station network since this influences the results achieved as well as the ability to validate the results. Our findings demonstrate the efficacy of multi-model approaches compared to traditional interpolation techniques, suggesting that such multi-model strategies could improve regional precipitation database accuracy in other regions exhibiting similar data limitations and spatial variability to those observed in Greece. Improved precipitation predictions can lead to better preparedness for extreme weather events, informed water resource management, and a deeper understanding of climate dynamics, ultimately aiding in effective disaster response strategies. In future research, it would be beneficial to investigate the incorporation of alternative modeling approaches to improve the prediction of extreme precipitation events. Moreover, additional data sources, such as satellite data, could be considered in order to increase the accuracy of prediction, especially for extreme precipitation events. Finally, for regions characterized by significant heterogeneity in precipitation patterns, the application of this methodology on a sub-regional basis by incorporating different model threshold values into the IK models could potentially enhance the overall accuracy of the predictions.

Supplementary Materials: The following supporting information can be downloaded at https://www.mdpi.com/article/10.3390/hydrology12020031/s1, Figure S1: Elevation of the region studied along with the location of the stations used and the subregions used in the study; Figure S2: Flowchart of the interpolation method IKGAM. In the chart, Lon stands for longitude, Lat for latitude, Alt for altitude, MoInt for Monthly Interpolated Totals; Figure S3: Flowchart of the interpolation method IKGAMV2. In the chart, Lon stands for longitude, Lat for latitude, DistShore for distance from sea coastal line, Alt for Altitude, M.ERA5 for monthly ERA5, D.ERA5 for daily ERA5, MoIntERA5 for Monthly Interpolated Totals with ERA5 data and PC1-PC15 are the AUREHLY principal components; Figure S4: Comparison of distributions between predicted and station annual precipitation totals in the Greek region; Figure S5: Comparison of distributions between predicted and station annual wet day totals in the Greek region; Figure S6: Comparison of distributions between predicted and station annual P20 totals in the Greek region. Figure S7: Comparison of cumulative precipitation distribution between the gauge dataset, the ERA5 reanalysis, and the interpolation methods for the total precipitation below certain thresholds; Figure S8: Scatterplots of observed and modeled precipitation values for the ERA5 reanalysis and the interpolation methods; Table S1: Station density in the areas studied; Table S2: Daily Metrics of the ERA5 reanalysis and the interpolation methods; Table S3: Results of Wilcoxon–Mann–Whitney test for daily and monthly distributions of the ERA5 reanalysis and the interpolation methods; Table S4: Results from the wet days threshold selection process of IKGAMRK.

Author Contributions: Conceptualization, G.N., P.N. and J.K.; methodology, G.N., P.N. and J.K.; data analysis, G.N. and J.K.; validation, G.N., J.K. and K.D.; investigation, G.N., J.K. and K.D.; data curation, G.N., J.K. and K.D.; writing—original draft preparation, G.N. and J.K.; writing—review and editing, K.D. and P.N.; visualization, G.N. and K.D.; supervision, P.N. and J.K.; project administration, P.N. and J.K. All authors have read and agreed to the published version of the manuscript.

Funding: This research received no external funding.

Data Availability Statement: The gauge data used in this study were provided by the Hellenic National Meteorological Service and can only be acquired by contacting HNMS, the reanalysis data by the ECMWF, the Digital Elevation Model by TerraSAR-X satellite mission and the sea coastal line, lake and river datasets by the https://geodata.gov.gr/ (accessed on 1 June 2024) platform.

The datasets generated during the current study are available from the corresponding author upon reasonable request.

Acknowledgments: We would like to thank Hellenic National Meteorological Service for providing us with precipitation gauge data to use in our study.

Conflicts of Interest: The authors declare no conflicts of interest.

References

1. Demuzere, M.; Kittner, J.; Martilli, A.; Mills, G.; Moede, C.; Stewart, I.D.; Van Vliet, J.; Bechtel, B. A global map of Local Climate Zones to support earth system modelling and urban scale environmental science. *Earth Syst. Sci. Data* **2022**, *14*, 3835–3873. [CrossRef]
2. Jacob, D.; Teichmann, C.; Sobolowski, S.; Katragkou, E.; Anders, I.; Belda, M.; Benestad, R.; Boberg, F.; Buonomo, E.; Cardoso, R.M.; et al. Regional climate downscaling over Europe: Perspectives from the EURO-CORDEX community. *Reg. Environ. Change* **2020**, *20*, 51. [CrossRef]
3. Georgoulias, A.K.; Akritidis, D.; Kalisoras, A.; Kapsomenakis, J.; Melas, D.; Zerefos, C.S.; Zanis, P. Climate change projections for Greece in the 21st century from high-resolution EURO-CORDEX RCM simulations. *Atmos. Res.* **2022**, *271*, 106049. [CrossRef]
4. Carvalho, D.; Pereira, S.C.; Silva, R.; Rocha, A. Aridity and desertification in the Mediterranean under EURO-CORDEX future climate change scenarios. *Clim. Change* **2022**, *174*, 28. [CrossRef]
5. Hosseinzadehtalaei, P.; Tabari, H.; Willems, P. Precipitation intensity–duration–frequency curves for central Belgium with an ensemble of EURO-CORDEX simulations, and associated uncertainties. *Atmos. Res.* **2018**, *200*, 1–12. [CrossRef]
6. Lauwaet, D.; Hooyberghs, H.; Maiheu, B.; Lefebvre, W.; Driesen, G.; Van Looy, S.; De Ridder, K. Detailed urban heat island projections for cities worldwide: Dynamical downscaling CMIP5 global climate models. *Climate* **2015**, *3*, 391–415. [CrossRef]
7. Lauwaet, D.; De Ridder, K.; Saeed, S.; Brisson, E.; Chatterjee, F.; van Lipzig, N.P.; Maiheu, B.; Hooyberghs, H. Assessing the current and future urban heat island of Brussels. *Urban Clim.* **2016**, *15*, 1–15. [CrossRef]
8. Smid, M.; Costa, A.C. Climate projections and downscaling techniques: A discussion for impact studies in urban systems. *Int. J. Urban Sci.* **2018**, *22*, 277–307. [CrossRef]
9. Stoner, A.M.; Hayhoe, K.; Yang, X.; Wuebbles, D.J. An asynchronous regional regression model for statistical downscaling of daily climate variables. *Int. J. Climatol.* **2013**, *33*, 2473–2494. [CrossRef]
10. Hamadalnel, M.; Zhu, Z.; Gaber, A.; Iyakaremye, V.; Ayugi, B. Possible changes in Sudan's future precipitation under the high and medium emission scenarios based on bias adjusted GCMs. *Atmos. Res.* **2022**, *269*, 106036. [CrossRef]
11. Ngai, S.T.; Juneng, L.; Tangang, F.; Chung, J.X.; Salimun, E.; Tan, M.L.; Amalia, S. Future projections of Malaysia daily precipitation characteristics using bias correction technique. *Atmos. Res.* **2020**, *240*, 104926. [CrossRef]
12. Manzanas, R.; Fiwa, L.; Vanya, C.; Kanamaru, H.; Gutiérrez, J.M. Statistical downscaling or bias adjustment? A case study involving implausible climate change projections of precipitation in Malawi. *Clim. Change* **2020**, *162*, 1437–1453. [CrossRef]
13. Tapiador, F.J.; Moreno, R.; Navarro, A.; Sánchez, J.L.; García-Ortega, E. Climate classifications from regional and global climate models: Performances for present climate estimates and expected changes in the future at high spatial resolution. *Atmos. Res.* **2019**, *228*, 107–121. [CrossRef]
14. Tapiador, F.J.; Navarro, A.; Moreno, R.; Sánchez, J.L.; García-Ortega, E. Regional climate models: 30 years of dynamical downscaling. *Atmos. Res.* **2020**, *235*, 104785. [CrossRef]
15. Cornes, R.C.; van der Schrier, G.; van den Besselaar, E.J.; Jones, P.D. An ensemble version of the E-OBS temperature and precipitation data sets. *J. Geophys. Res.-Atmos.* **2018**, *123*, 9391–9409. [CrossRef]
16. Lussana, C.; Saloranta, T.; Skaugen, T.; Magnusson, J.; Tveito, O.E.; Andersen, J. seNorge2 daily precipitation, an observational gridded dataset over Norway from 1957 to the present day. *Earth Syst. Sci. Data* **2018**, *10*, 235–249. [CrossRef]
17. MeteoSwiss. Documentation of MeteoSwiss Grid-Data Products: Daily Mean, Minimum and Maximum Temperature: TabsD, TminD, TmaxD. Available online: https://www.meteoswiss.admin.ch/dam/jcr:818a4d17-cb0c-4e8b-92c6-1a1bdf5348b7/ProdDoc_TabsD.pdf (accessed on 3 February 2025).
18. MeteoSwiss. Documentation of MeteoSwiss Grid-Data Products: Daily Precipitation (Final Analysis): RhiresD. Available online: https://www.meteoswiss.admin.ch/dam/jcr:4f51f0f1-0fe3-48b5-9de0-15666327e63c/ProdDoc_RhiresD.pdf (accessed on 3 February 2025).
19. Sekulić, A.; Kilibarda, M.; Protić, D.; Bajat, B. A high-resolution daily gridded meteorological dataset for Serbia made by Random Forest Spatial Interpolation. *Sci. Data* **2021**, *8*, 123. [CrossRef] [PubMed]
20. Camera, C.; Bruggeman, A.; Hadjinicolaou, P.; Pashiardis, S.; Lange, M.A. Evaluation of interpolation techniques for the creation of gridded daily precipitation (1×1 km^2); Cyprus, 1980–2010. *J. Geophys. Res.-Atmos.* **2014**, *119*, 693–712. [CrossRef]

21. Varotsos, K.V.; Dandou, A.; Papangelis, G.; Roukounakis, N.; Kitsara, G.; Tombrou, M.; Giannakopoulos, C. Using a new local high resolution daily gridded dataset for Attica to statistically downscale climate projections. *Clim. Dyn.* **2023**, *60*, 2931–2956. [CrossRef]
22. Varotsos, K.V.; Karali, A.; Kitsara, G.; Lemesios, G.; Patlakas, P.; Hatzaki, M.; Tenentes, V.; Katavoutas, G.; Sarantopoulos, A.; Koutroulis, A.G.; et al. High resolution observational daily gridded dataset for Greece: The CLIMADAT-hub project. In Proceedings of the EMS2024, Barcelona, Spain, 2–6 September 2024; Abstract Number EMS2024-643.
23. Akinsanola, A.A.; Ongoma, V.; Kooperman, G.J. Evaluation of CMIP6 models in simulating the statistics of extreme precipitation over Eastern Africa. *Atmos. Res.* **2021**, *254*, 105509. [CrossRef]
24. Merino, A.; García-Ortega, E.; Navarro, A.; Sánchez, J.L.; Tapiador, F.J. WRF hourly evaluation for extreme precipitation events. *Atmos. Res.* **2022**, *274*, 106215. [CrossRef]
25. Kazamias, A.P.; Sapountzis, M.; Lagouvardos, K. Evaluation of GPM-IMERG rainfall estimates at multiple temporal and spatial scales over Greece. *Atmos. Res.* **2022**, *269*, 106014. [CrossRef]
26. Li, Z.; Shi, Y.; Argiriou, A.A.; Ioannidis, P.; Mamara, A.; Yan, Z. A comparative analysis of changes in temperature and precipitation extremes since 1960 between China and Greece. *Atmosphere* **2022**, *13*, 1824. [CrossRef]
27. Nastos, P.T.; Kapsomenakis, J.; Douvis, K.C. Analysis of precipitation extremes based on satellite and high-resolution gridded data set over Mediterranean basin. *Atmos. Res.* **2013**, *131*, 46–59. [CrossRef]
28. Philandras, C.M.; Nastos, P.T.; Kapsomenakis, J.; Douvis, K.C.; Tselioudis, G.; Zerefos, C.S. Long term precipitation trends and variability within the Mediterranean region. *Nat. Hazards Earth Syst. Sci.* **2011**, *11*, 3235–3250. [CrossRef]
29. Seager, R.; Liu, H.; Kushnir, Y.; Osborn, T.J.; Simpson, I.R.; Kelley, C.R.; Nakamura, J. Mechanisms of winter precipitation variability in the European–Mediterranean region associated with the North Atlantic Oscillation. *J. Clim.* **2020**, *33*, 7179–7196. [CrossRef]
30. Deitch, M.J.; Sapundjieff, M.J.; Feirer, S.T. Characterizing precipitation variability and trends in the world's Mediterranean-climate areas. *Water* **2017**, *9*, 259. [CrossRef]
31. Lee, M.H.; Im, E.S.; Bae, D.H. Impact of the spatial variability of daily precipitation on hydrological projections: A comparison of GCM-and RCM-driven cases in the Han River basin, Korea. *Hydrol. Process.* **2019**, *33*, 2240–2257. [CrossRef]
32. Nastos, P.T.; Politi, N.; Kapsomenakis, J. Spatial and temporal variability of the Aridity Index in Greece. *Atmos. Res.* **2013**, *119*, 140–152. [CrossRef]
33. Markonis, Y.; Batelis, S.C.; Dimakos, Y.; Moschou, E.; Koutsoyiannis, D. Temporal and spatial variability of rainfall over Greece. *Theor. Appl. Climatol.* **2017**, *130*, 217–232. [CrossRef]
34. Politi, N.; Vlachogiannis, D.; Sfetsos, A.; Nastos, P.T.; Dalezios, N.R. High resolution future projections of drought characteristics in Greece based on SPI and SPEI indices. *Atmosphere* **2022**, *13*, 1468. [CrossRef]
35. Tzanis, C.G.; Pak, A.N.; Koutsogiannis, I.; Philippopoulos, K. Climatology of Extreme Precipitation from Observational Records in Greece. *Environ. Sci. Proc.* **2022**, *19*, 51. [CrossRef]
36. Ntagkounakis, G.; Nastos, P.T.; Kapsomenakis, Y. Creating High-Resolution Precipitation and Extreme Precipitation Indices Datasets by Downscaling and Improving on the ERA5 Reanalysis Data over Greece. *Eng* **2024**, *5*, 1885–1904. [CrossRef]
37. Nastos, P.T.; Kapsomenakis, J.; Philandras, K.M. Evaluation of the TRMM 3B43 gridded precipitation estimates over Greece. *Atmos. Res.* **2016**, *169*, 497–514. [CrossRef]
38. Li, L.; Zha, Y.; Wang, R. Relationship of surface urban heat island with air temperature and precipitation in global large cities. *Ecol. Indic.* **2020**, *117*, 106683. [CrossRef]
39. Steensen, B.M.; Marelle, L.; Hodnebrog, Ø.; Myhre, G. Future urban heat island influence on precipitation. *Clim. Dyn.* **2022**, *58*, 3393–3403. [CrossRef]
40. European Data. Available online: https://data.europa.eu/data/datasets/5eecdf4c-de57-4624-99e9-60086b032aea?locale=en (accessed on 1 June 2024).
41. GEODATA.gov.gr. Available online: https://geodata.gov.gr/ (accessed on 1 June 2024).
42. Muñoz-Sabater, J.; Dutra, E.; Agustí-Panareda, A.; Albergel, C.; Arduini, G.; Balsamo, G.; Boussetta, S.; Choulga, M.; Harrigan, S.; Hersbach, H.; et al. ERA5-Land: A state-of-the-art global reanalysis dataset for land applications. *Earth Syst. Sci. Data* **2021**, *13*, 4349–4383. [CrossRef]
43. ECMWF. Available online: https://www.ecmwf.int/en/forecasts/dataset/ecmwf-reanalysis-v5 (accessed on 1 June 2024).
44. Wu, G.; Qin, S.; Mao, Y.; Ma, Z.; Shi, C. Validation of precipitation events in ERA5 to gauge observations during warm seasons over eastern China. *J. Hydrometeorol.* **2022**, *23*, 807–822. [CrossRef]
45. Jiang, Y.; Yang, K.; Shao, C.; Zhou, X.; Zhao, L.; Chen, Y.; Wu, H. A downscaling approach for constructing high-resolution precipitation dataset over the Tibetan Plateau from ERA5 reanalysis. *Atmos. Res.* **2021**, *256*, 105574. [CrossRef]
46. Nacar, S.; Kankal, M.; Okkan, U. Evaluation of the suitability of NCEP/NCAR, ERA-Interim and, ERA5 reanalysis data sets for statistical downscaling in the Eastern Black Sea Basin, Turkey. *Meteorol. Atmos. Phys.* **2022**, *134*, 39. [CrossRef]

47. Zhang, Y.; Li, J.; Liu, D. Spatial Downscaling of ERA5 Reanalysis Air Temperature Data Based on Stacking Ensemble Learning. *Sustainability* **2024**, *16*, 1934. [CrossRef]
48. Rivoire, P.; Martius, O.; Naveau, P. A comparison of moderate and extreme ERA-5 daily precipitation with two observational data sets. *Earth Space Sci.* **2021**, *8*, e2020EA001633. [CrossRef]
49. Ntagkounakis, G.E.; Nastos, P.T.; Kapsomenakis, Y. Statistical Downscaling of ERA5 Reanalysis Precipitation over the Complex Terrain of Greece. *Environ. Sci. Proc.* **2023**, *26*, 81. [CrossRef]
50. Bénichou, P.; Le Breton, O. AURELHY: Une méthode d'analyse utilisant le relief pour les besoins de l'hydrométéorologie. *Journ. Hydrol. l'ORSTOM Montp.* **1987**, *2*, 299–304.
51. Gofa, F.; Mamara, A.; Anadranistakis, M.; Flocas, H. Developing gridded climate data sets of precipitation for Greece based on homogenized time series. *Climate* **2019**, *7*, 68. [CrossRef]
52. Mamara, A.; Anadranistakis, M.; Argiriou, A.A.; Szentimrey, T.; Kovacs, T.; Bezes, A.; Bihari, Z. High resolution air temperature climatology for Greece for the period 1971–2000. *Meteorol. Appl.* **2017**, *24*, 191–205. [CrossRef]
53. The R Project for Statistical Computing. Available online: https://www.r-project.org/ (accessed on 1 June 2024).
54. mgcv: Mixed GAM Computation Vehicle with Automatic Smoothness Estimation. Available online: https://cran.r-project.org/web/packages/mgcv/index.html (accessed on 1 June 2024).
55. Hiemstra, P.; Hiemstra, M.P. Package 'automap'. *Compare* **2013**, *105*, 10.
56. Pebesma, E.; Bivand, R.S. Classes and methods for spatial data: The sp package. *R. News* **2005**, *5*, 9–13.
57. Pebesma, E.J. Multivariable geostatistics in S: The gstat package. *Comput. Geosci.* **2004**, *30*, 683–691. [CrossRef]
58. Herrera, S.; Gutiérrez, J.M.; Ancell, R.; Pons, M.R.; Frías, M.D.; Fernández, J. Development and Analysis of a 50-Year High-Resolution Daily Gridded Precipitation Dataset Over SPAIN (Spain02). *Int. J. Climatol.* **2012**, *32*, 74–85. [CrossRef]
59. Herrera, S.; Cardoso, R.M.; Soares, P.M.; Espírito-Santo, F.; Viterbo, P.; Gutiérrez, J.M. Iberia01: A new gridded dataset of daily precipitation and temperatures over Iberia. *Earth Syst. Sci. Data* **2019**, *11*, 1947–1956. [CrossRef]
60. Parajka, J.; Merz, R.; Skøien, J.O.; Viglione, A. The role of station density for predicting daily runoff by top-kriging interpolation in Austria. *J. Hydrol. Hydromech.* **2015**, *63*, 228. [CrossRef]
61. Crossett, C.C.; Betts, A.K.; Dupigny-Giroux, L.A.L.; Bomblies, A. Evaluation of daily precipitation from the ERA5 global reanalysis against GHCN observations in the northeastern United States. *Climate* **2020**, *8*, 148. [CrossRef]
62. Nastos, P.T.; Zerefos, C.S. On extreme daily precipitation totals at Athens, Greece. *Adv. Geosci.* **2007**, *10*, 59–66. [CrossRef]
63. Nastos, P.T.; Zerefos, C.S. Decadal changes in extreme daily precipitation in Greece. *Adv. Geosci.* **2008**, *16*, 55–62. [CrossRef]

hydrology

MDPI

Article

Assessing the Risk of Natural and Socioeconomic Hazards Caused by Rainfall in the Middle Yellow River Basin

Yufeng Zhao [1,†], Shun Xiao [2,†], Xinshuang Wu [3], Shuitao Guo [1] and Yingying Yao [1,*]

1 Institute of Global Environmental Change, Xi'an Jiaotong University, Xi'an 710049, China; zhaoyuf@stu.xjtu.edu.cn (Y.Z.); 4123122025@stu.xjtu.edu.cn (S.G.)
2 School of Geography and Tourism, Shaanxi Normal University, Xi'an 710119, China; sxiao@snnu.edu.cn
3 Power China Northwest Engineering Corporation Limited, Xi'an 710065, China; wuxins@nwh.cn
* Correspondence: yaoyy27@xjtu.edu.cn
† These authors contributed equally to this work.

Abstract: Extreme rainfall events directly increase flood risks and further trigger environmental geological hazards (i.e., landslides and debris flows). Meanwhile, rainfall-induced risks are determined by climate and geographical factors and spatial socioeconomic factors (e.g., population density and gross domestic product). However, the middle stream of Yellow River Basin, where geological hazards frequently occur, lacks systematic analyses of rainfall-induced risks. In this study, we propose a comprehensive quantification framework and apply it to the Loess Plateau of northern China based on 40 years of climate data, streamflow measurements, and multiple spatial and geographical attribute datasets. A deep learning algorithm of long short-term memory (LSTM) was used to predict runoff, and the analytic hierarchy index was utilized to evaluate the comprehensive spatial risk considering natural and socioeconomic factors. Despite a decrease in annual precipitation in our study area of 1.46 mm per year, the intensity of heavy rainfall has increased since the 1980s, characterized by increases in rainstorm intensity (+4.68%), rainfall intensity (+7.07%), and rainfall amount (+5.34%). A comprehensive risk assessment indicated that high-risk areas accounted for 20.30% of the total area, with rainfall, geographical factors, and socioeconomic variables accounting for 53.90%, 29.72%, and 16.38% of risk areas, respectively. Rainfall was the dominant factor that determined the risk, and geographical and socioeconomic properties characterized the vulnerability and resilience of disasters. Our study provided an evaluation framework for multi-hazard risk assessment and insights for the development of disaster prevention and reduction policies.

Keywords: rainfall; Yellow River Basin; deep learning; spatiotemporal; risk of geo-hazards

1. Introduction

Rainfall is one of the most important triggers of geo-hazards, which are harmful to human life, damage key infrastructure, and destroy cultural and natural heritage [1]. In the event of heavy or persistent rainfall, river discharge can rapidly occur and cause flooding, landslides, and debris flows [2,3]. In addition, global warming increases the spatial heterogeneity of rainfall events in terms of the total amount of rainfall, frequency of rainfall events, and maximum amount of rainfall [4–6]. Consequently, disasters are more likely to occur in arid and semi-arid regions where disaster management and control capabilities are inadequate [7]. According to the National Geological Disaster Bulletin 2009–2016, there are approximately 14,000 geological disasters in China annually. Natural factors (e.g., rainfall, earthquakes, and the freezing and thawing of snow and ice) cause

93.4% of these disasters, which result in an average annual economic loss of 4.84 billion yuan [8]. Heavy rainfall causes intense surface runoff scouring and seepage erosion, which pre-existing drainage systems originally designed for semi-arid and arid regions are ill-equipped to manage [9,10]. In addition, debris flows and landslides caused by excessive rainfall result in heavy losses of life and property for existing communities [11,12]. Accordingly, it is crucial to assess the risks associated with multiple hazards resulting from rainfall, particularly in semi-arid and arid regions.

The risk evaluations for floods and the induced geo-hazards of landslides and debris flows show differences in scale and methodology. Flood risk is typically evaluated at a regional scale, ranging from less than 100 km^2 to over 10,000 km^2, whereas landslides and debris flows are evaluated at the location or slope level; the majority of relevant studies are conducted within 10 km^2 on average for one to occur [13–15]. Both types of hazards were assessed and projected using physical-based or data-driven models in conjunction with observed records and remote sensing imagery, but at different time scales. Models for flood prediction focus on dynamic runoff simulations for months to years, characterizing the flow components and geographical attributes (i.e., precipitation, evaporation, soil, and land use or land cover type) of the hydrological cycle [16,17]. Models for landslides simulate the moving volumes and velocities, but models for debris flows are concerned with the volume of mobilized flow, with resolutions ranging from seconds to years, requiring extensive computations [18]. These predictions are based on rainfall frequency and intensity [19]. Although the current research on both types of hazards is beneficial in theory and practice, they are typically viewed as independent. They are rarely combined to provide a comprehensive assessment of both risks.

The early detection of rainfall-induced hazards can be acquired by analyzing and simulating floods, landslides, and debris flows caused by natural drivers. In contrast, risk assessment and management should consider spatial socioeconomic factors. Risk vulnerability evaluations at the regional level apply to urban areas, population density, economic status (i.e., gross domestic product, GDP), and other infrastructure and cultural heritage [20]. These socioeconomic factors are highly heterogeneous. For flood risk, socioeconomic vulnerability assessments have been developed and adopted in previous studies, whereas the vulnerability of landslides and debris flows is primarily focused on mortality and damage to structures at the site of occurrence because of the research scale [21,22]. There are substantial scientific and practical challenges associated with developing a comprehensive framework that integrates hazards on different scales with socioeconomic vulnerability to identify spatial risk patterns.

To evaluate the regional spatial risks of multiple rainfall-caused hazards, we selected the Guanzhong Basin in the Middle Yellow River Basin of northern China as a representative case for the following reasons. The selection of the Guanzhong Basin as a case study is underpinned by several key considerations: it is a typical semi-arid region characterized by a dense population, significant agricultural activity, and an exceptionally rich cultural and natural heritage of international importance, making it highly sensitive to environmental changes and disaster impacts. Furthermore, the basin frequently experiences heavy summer rainfall which markedly increases the risk of concurrent flooding, landslides, and debris flows, particularly in its mountainous southern and western areas [23–26]. As this region has experienced more frequent heavy rainfall events in recent years, the existing urban drainage systems in urban areas have become inadequate [27,28]. In the meantime, the spatial risks of landslides and debris flows in this region are yet to be quantified. The existing research has made significant contributions to understanding and assessing various hazard risks. For instance, studies in the Yellow River Basin have focused on modeling rainstorm-induced floods and large-scale flooding events [24,26,28]. In terms of geological hazards, significant work has been completed on landslide susceptibility and risk in Shaanxi Province and the Loess Plateau using various GIS-based, statistical, and machine learning techniques, often complemented by field investigations [29–31], and furthermore, national-level disaster risk indexing and frameworks integrating socioeconomic factors into flood disaster risk in the Yellow River Basin [32,33].

However, the current approaches often address these components in isolation or lack a robust method for combining diverse data sources into a unified multi-hazard risk structure. This study aims to address this gap by proposing and applying an integrated multi-hazard risk assessment framework. The novelty herein lies in the unique combination of deep learning (LSTM for runoff prediction) with the analytic hierarchy process (AHP) to systematically integrate diverse factors including rainfall characteristics (based on daily rainfall records from 43 stations), geographical attributes, and critical socioeconomic variables. Furthermore, we compared our results with those from the locations of geo-hazards occurrences identified by field survey records. Major tourism destinations in high-risk areas were also noted. This comprehensive framework for assessing rainfall-induced hazards provides an effective tool for other regions, and the results of spatial patterns provide novel insights for the planning of urban infrastructure and the management of risks.

2. Materials and Methods

2.1. Description of the Study Area

The Guanzhong Basin is the core of the Loess Plateau and is situated in the hinterland of China in the middle stream of the Yellow River Basin (Figure 1). It encompasses a total area of 20,576.57 km^2 and constitutes a narrow corridor between the Qinling Mountains in the south and the Bei Mountains in the north. The Guanzhong Basin is a part of the lower drainage basin of the Wei River, which is a tributary of the Yellow River. It contains four catchments: Linjiacun (LJC), Zhangjiashan (ZJS), Zhuangtou (ZT), and Huaxian (HX). The mean annual rainfall is approximately 590.1 mm/year, and the surface runoff is 81.47 mm/year. The Loess Plateau is densely populated in the Guanzhong Basin, which contains Xi'an, the largest and most historically important city in northern China. The study area varies in elevation from 1779 m in the southern Qinling Mountains to 323 m in the center oases. The mean annual temperature varies between −3 °C in January and 27 °C in July.

Figure 1. (**a**) Study area; the historical heritage of (**b**) the mausoleum of the first Qin emperor and (**c**) the ancient city wall; and the hazards of (**d**) flooding, (**e**) debris flows, and (**f**) landslides occurring within the study area.

Because the Guanzhong Basin is an encompassing drainage area, the surge of river flow induced by heavy summer rainfall discharges into this area, increasing the risk of flooding. Furthermore, rainfall is often responsible for triggering geological hazards (e.g., landslides and debris flows) in the southern mountains. There are 764.8 km^2 of main urban land, 1636.9 km^2 of rural settlements and historical sites, and 142 tourist attractions.

2.2. Rainfall Analysis and Runoff Simulation

2.2.1. Rainfall Analysis

We collected the 1980–2019 daily rainfall data for 43 stations from the National and Shaanxi Meteorological Bureaus. The spatial pattern of rainfall variation was examined using four variables: rainfall days, which was the total number of days with daily rainfall greater than 0.1 mm; rainfall intensity, which represented the mean amount of daily rainfall; maximum daily rainfall amount; and the rainstorm amount, which was the daily rainfall amount greater than 50 mm per day [34–36].

The Mann–Kendall trend test [37] was applied to each station to determine the changing trends in rainfall days, rainfall intensity, maximum daily rainfall, and rainstorm amount.

2.2.2. Runoff Simulation

Runoff was simulated using an LSTM network, a type of recurrent neural network adept at learning temporal dependencies from time-series data [38–41]. The core LSTM architecture comprised an input layer, a single LSTM layer with 256 hidden units, and an output linear layer. Key hyperparameters for training included 1000 epochs a mini-batch size of 128, a learning rate of 0.001, and the Adam optimizer. In addition to an LSTM trained solely on local sub-basin data, six transfer learning (TL) configurations were evaluated. These TL models were pre-trained on either the full CAMELS US dataset (671 catchments; models TL_A, TL_B, TL_C) or a selected subset of 531 CAMELS catchments (models TL_A_531, TL_B_531, TL_C_531). The TL strategies differed in their fine-tuning approach on local data; TL_A involved freezing core LSTM weights and fine-tuning only input/output layers; TL_B allowed the fine-tuning of LSTM recurrent connections and output layers; and TL_C allowed full fine-tuning of all pre-trained weights. The inputs to LSTM for the simulations included climate forcing (e.g., precipitation, solar radiation, maximum temperature, and minimum temperature), as well as attribute data (e.g., soil type, land cover, and geology). The training period spanned from 1971 to 1978, the validating period from 1979 to 1986, and the predicting period between 1987 and 2019. The Nash–Sutcliffe efficiency coefficient (NSE) was used as a goodness-of-fit measure for runoff prediction by LSTM [42]. In order to evaluate the performance of the models in extreme cases, we calculated the percentage deviation between the top 2% high-flow range (FHV) and the bottom 30% low-flow range (FLV) [43]. The FHV and FLV were calculated using the following equation:

$$FHV(FLV) = \frac{\sum(R^S - R^O)}{\sum R^O} \quad (1)$$

where R^s represents the simulated runoff value and R^o denotes the observed runoff value.

2.3. Risk Evaluation

2.3.1. Risk of Hazards and Vulnerability Index

We adopted a comprehensive risk index (***R***) to assess the hazards caused by rainfall and the vulnerability to socioeconomic effects [44]. ***R*** could be expressed using the following equation:

$$R = \sum_{i=1}^{n} C_i \omega_i \quad (2)$$

$$\sum_{i=1}^{n} \omega_i = 1 \tag{3}$$

where C_i represents the indicator of hazards and vulnerability and ω_i denotes the weight identified to that indicator, $i = 1, 2, \ldots, n$. The sum of all the weights is equal to 1.

Three types of indicators were identified in this study, as shown in Figure 2 [45]. The first was climate-relevant indicators, which included climate-forcing variables. Climate forcing primarily consisted of annual rainfall, rainfall days, rainfall intensity, rainstorm intensity, and maximum rainfall amount. The second type comprised geographic attribute indicators, which comprised landscape (i.e., elevation and slope), geology (i.e., soil type), and hydrology (i.e., drainage density) indicators. The third component is the Socio-Hazard Risk Index, which, in this study, quantifies the risk arising from the exposure and concentration of socioeconomic elements (e.g., population density, GDP, urban areas, historical sites) to potential rainfall-induced natural hazards. A higher index value signifies areas where a natural hazard event would likely lead to greater societal and economic losses due to the characteristics of the elements at risk. This index therefore focuses on the societal consequences and susceptibility to damage rather than societal factors as primary hazard triggers, and is related to the broader concept of socioeconomic vulnerability [46].

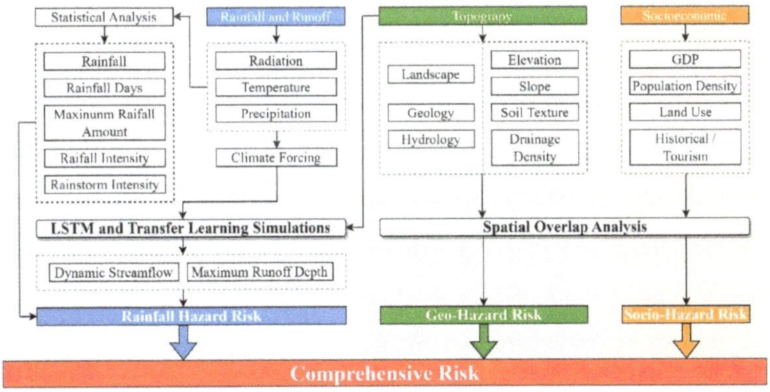

Figure 2. Framework of evaluation for rainfall-induced risk.

2.3.2. Determining Risk Weights

The AHP was employed in this study to determine the weights for each indicator in the composite index of R, as this index simultaneously considered multiple indicators of an interrelated nature and human systems. The AHP is a multi-criteria decision-making model that is capable of integrating the effects of multiple factors into a single conclusion [47,48]. It is applied in risk assessment for natural hazards [49,50]. Factors can be qualitative, quantitative, independent, or dependent on one another [51], such as the interrelated factors of rainfall amount and surface runoff.

The AHP procedure for identifying the weights is summarized in Table 1. In this study, climate-relevant attributes, geographic attributes, and socioeconomic indicators were organized into a unified hierarchy. Then, a pairwise comparison matrix was constructed based on a comparative scale [48], which included integer numbers between 1 and 9, with 1 signifying that the indicators were of equal importance and 9 indicating that one indicator was much more important than its pairwise indicator [52].

Table 1. Procedure for AHP calculation.

Step	Procedure
1	Develop a hierarchy of indicators and criteria
2	Develop a pairwise comparison matrix
3	Check for consistency
4	Obtain the relative weights

The discordances in the pairwise comparison results were then evaluated using the consistency ratio (***CR***), and the weights were accepted when the ***CR*** was less than 0.1. ***CR*** could be expressed as follows:

$$CR = \frac{CI}{RI} \tag{4}$$

where ***CI*** represents the consistency index in Equation (4) and ***RI*** denotes the random index, which serves as the average random consistency index and is obtained from Table 2.

$$CI = \frac{\lambda_{\max} - n}{n - 1} \tag{5}$$

where $\lambda_{\max}$ is the maximum eigenvalue of the judgment matrix and n denotes the order of the matrix. The method for assigning the value of n is based on Table 2.

Table 2. Mean random consistency index ***RI*** value [53].

n	1	2	3	4	5	6	7	8	9
RI	0	0	0.52	0.89	1.12	1.24	1.36	1.41	1.46

When the ***CR*** < 0.10, the consistency of the judgment matrix was regarded as within an acceptable range; otherwise, the elements in the judgment matrix should have been properly revised to ensure adequate consistency.

2.3.3. Data Integration

Data sources and descriptions are provided in Table 3. The data for climate-relevant indicators consisted of time-series datasets that included 43 rainfall stations and 20 meteorological stations. The indicators of geographic attribution were primarily spatial datasets in a grid format, as depicted in Figure 3. The described dataset was converted into numeric values for evaluation purposes. Soil texture was expressed as a percentage of the soil/sand content ratio. A digital elevation model (DEM) was used in conjunction with a grid system in a geographic information system (GIS) to extract data about river networks, which were then represented as drainage density for evaluation purposes. Statistical datasets from national and provincial statistical yearbooks, as well as spatial data with grids, were used for socioeconomic indicators. Land use types were categorized on a scale from 1 to 6 and normalized based on forest land, grassland, cultivated land, housing development, undeveloped land, and water area. Spatial distribution data of geological hazard locations from 1960 to 2019 were collected in order to validate the calculated risk area, which covered seven types of geo-hazards (i.e., collapse, debris flow, land subsidence, ground fissures, landslides, and slopes). During the data integration process for constructing the AHP framework and for LSTM model inputs, careful attention was paid to data quality and completeness. For time-series data such as daily rainfall from meteorological stations, short, isolated periods of missing values were imputed using linear interpolation or data from the nearest neighboring station with a high correlation, where appropriate.

Table 3. Overview of data sources and descriptions.

Name	Description	Format	Source	Resolution
Precipitation	Daily precipitation (from 20:00 to 20:00 the next day) data from 1980 to 2019	Text	National Meteorological Bureau and Shaanxi Meteorological Bureau	Daily
Temperature	Air temperature	Text	Meteorological Bureau	Monthly
DEM	Electronic topographic map	Grid	http://www.csdb.cn/	30 m
Slope	The slowness of surface units	Grid	Based on DEM	30 m
Land use	Cultivated land, woodland, grassland, waters, residential, and unused land	Grid	https://www.resdc.cn/	1 km
Drainage network	Ratio of total river length to basin area of main and tributaries.	Grid	Based on DEM	1 km
Sand content in soil	Use percentage to reflect the content of sand particles.	Shapefile	https://www.resdc.cn/	1 km
GDP	Gross Domestic Product	Grid	https://www.resdc.cn/	1 km
Population density	Population per square kilometer	Grid	https://www.resdc.cn/	1 km
Streamflow	Daily or monthly runoff data	Text		Monthly
Geological hazard points	Dataset of geological hazard points in Weihe River Basin	Shapefile	http://www.ncdc.ac.cn/	1 km
Tourism points	Location and description of tourist spots	Shapefile		1 km

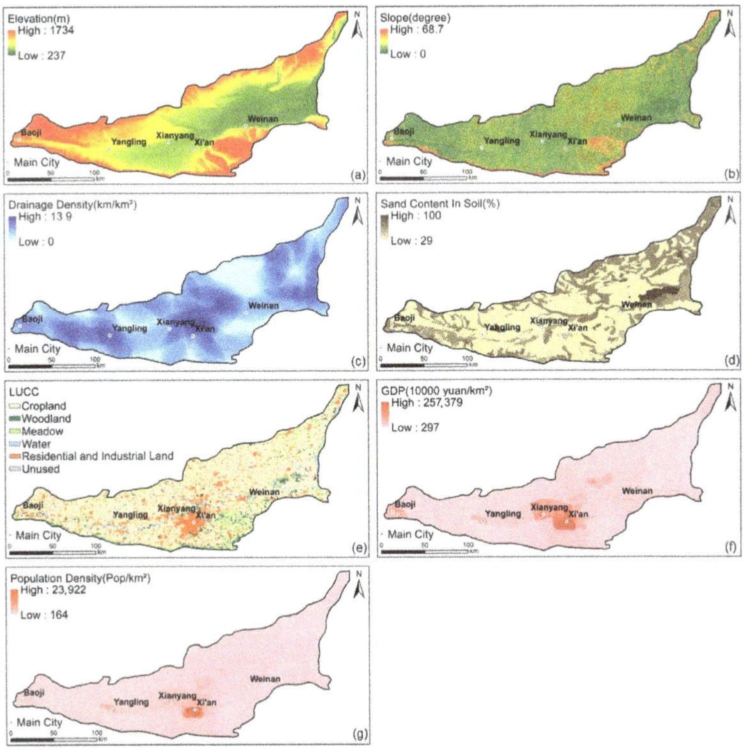

Figure 3. Spatial pattern of geographical and socioeconomic indicators. (**a**) Elevation (m), (**b**) slope (°), (**c**) drainage density (km/km^2), (**d**) sand content in soil (%), (**e**) land use, (**f**) gross domestic product (10,000 yuan/km^2), and (**g**) population density (Pop/km^2).

As the three types of indicators had distinct units, we converted each indicator into a standardized value for calculation. The standard normalization method that was used to map the data to a range of 0–1 was as follows [54]:

$$X' = \frac{X - X_{min}}{X_{max} - X_{min}} \quad (6)$$

where X represents the value of the indicator and X_{max} and X_{min} are the maximum and minimum values of X, respectively.

3. Results

3.1. Spatiotemporal Variation in Rainfall

The changing trend in rainfall from the 43 meteorological stations and the spatial patterns of the annual average are presented in Figure 4. In this study, we adopted a significance level of 95% and a critical test value of ±1.96. Annual rainfall in the Guanzhong Basin ranged from 723.5 mm in the southern mountains to 503.1 mm in the northern alluvial plain. The annual rainfall trend analysis using the Mann–Kendall trend test indicated a decreasing trend for the entire study area between 1980 and 2019, with a mean z-statistic of −0.64; nevertheless, the rainfall in the southern mountains exhibited an increasing trend, with a z-statistic ranging between 0.20 and 1.22 (Figure 4a). Annual rainfall days varied from 105 d in the southwestern region to 77 d in the northeastern region. Over the past 40 y, the annual rainfall days in the study area showed a downward trend, with an average z-value of −1.83. Since the 1980s, the average number of rainfall days for the entire study area has decreased by 9.96%. Over the past 40 y, the annual rainfall days in the entire study area have decreased, with an average z-value of −1.83 (Figure 4b).

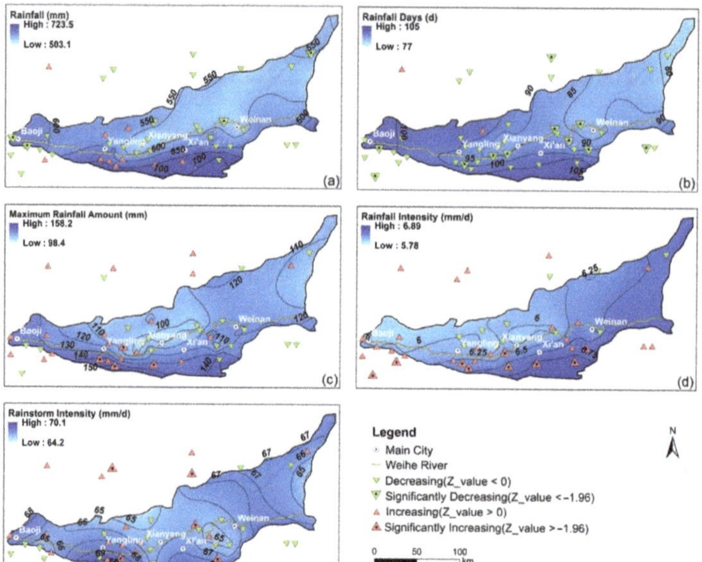

Figure 4. Spatiotemporal variation in rainfall between 1980 and 2019. (**a**) Annual rainfall (mm), (**b**) rainfall days (d), (**c**) maximum rainfall amount (mm), (**d**) rainfall intensity (mm/d), and (**e**) rainstorm intensity (mm/d). The triangle indicates the result of the Mann–Kendall trend test, with red indicating an upward trend and green indicating a downward trend. The size of the triangle reflects the statistical significance of the trend at a 95% confidence level.

The difference in the spatial distribution of the maximum rainfall amount (the day with the maximum rainfall) ranged from 98.4 mm to 158.2 mm. In contrast to the decreasing trend in annual rainfall, the maximum amount of precipitation for the entire study area showed an upward trend, with a mean z-statistic of 0.52 between 1980 and 2019. The maximum rainfall at 22 of the 43 meteorological stations showed an upward trend, particularly in the southern and western mountainous regions, as shown in Figure 4c. From the western to the eastern part of the study area, rainfall intensity (mean rainfall amount for rain days) decreased. The rainfall intensity for the entire study area showed a significant increase between 1980 and 2019, with a mean z-value of 2.37. The rainfall intensity has increased by 9.20% over the past decade (2010–2019) (Figure 4d). Similar to the maximum rainfall amount, the intensity of rainstorms across the entire study area exhibited a significant upward trend during the test period, with an average z-value of 1.94 (Figure 4e).

Figure 5 illustrates the seasonal variation in monthly rainfall. The average annual rainfall was 590 mm, with the highest rainfall occurring in July and August. A total of 52.4% of the annual rainfall occurred between July and September. In addition, the intensity of rainfall and rainstorms reached their peaks in August. A summer rainfall of 46.5% implied the characteristics of a continental climate on the Loess Plateau.

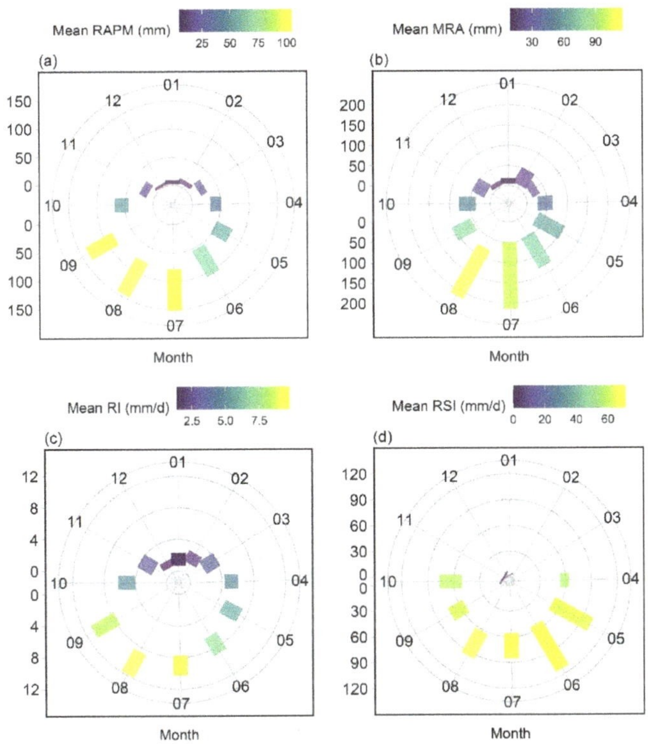

Figure 5. Seasonal variation in rainfall. (**a**) Rainfall amount per month (mm); (**b**) maximum rainfall amount (mm); (**c**) rainfall intensity (mm/d); and (**d**) rainstorm intensity (mm/d). The histogram represents the range, and the colors represent the average.

3.2. Variation in Runoff

The Weihe River Basin mainly includes four sub-basins, and the four hydrological stations are the outlets along the upstream tributaries (i.e., LJC, ZJS, and ZT) and mainstream (i.e., HX). The results of the runoff simulation and prediction for all sub-basins are depicted in Figure 6. The application of transfer learning generally led to an improved runoff simulation performance compared to LSTM, although the optimal TL strategy varied by sub-basin. At the LJC station, the highest NSE of 0.67 was achieved by the TL_C model (pre-trained on 671 catchments with full fine-tuning). For the ZJS station, the TL_A_531 model (pre-trained on 531 catchments with minimal core fine-tuning) demonstrated the best performance with an NSE of 0.67. At the ZT station, the TL_A model (pre-trained on

671 catchments with minimal core fine-tuning) provided the highest NSE of 0.63. Finally, at the HX mainstream station, the TL_A_531 model again proved most effective, achieving a robust NSE of 0.78. While these models were selected based on their optimal NSE, an examination of their corresponding FHV and FLV metrics reveals varied performance in simulating extreme flows. For instance, the TL_C model at LJC, despite its high NSE (0.67), showed an underestimation of peak flows (FHV −23.13%) but reasonable low-flow simulation (FLV 4.85%). The TL_A_531 model at ZJS (NSE 0.67) slightly overestimated peak flows (FHV 17.36%) and slightly underestimated low-flows (FLV −5.19%). At ZT, the best NSE model (TL_A, NSE 0.63) simulated peak flows accurately (FHV −1.92%) but underestimated low-flows (FLV −20.29%). Finally, the TL_A_531 model at HX (NSE 0.78) also tended to underestimate peak flows (FHV −19.43%) while significantly overestimating low-flows (FLV 67.54%). This indicates that even models performing well on overall water balance (high NSE) may exhibit different biases for high- and low-flow extremes, underscoring the importance of multi-metric assessment for specific applications. The performance of our simulations using LSTM (NSE is 0.78) was consistent with the results obtained using physical hydrological models (NSE from 0.76 to 0.82) [55–57]. The results showed that the runoff varied from 44.2 m^3/s (42 mm/year) in the upstream to 187 m^3/s (169.29 mm/year) in the downstream. During the wet season (i.e., between July and September), the upstream runoff increased to 76 m^3/s (71.2 mm/year), while the downstream runoff increased to 344 m^3/s. Despite these observed variations and specific biases in peak-flow and low-flow simulations, the ability of the best-performing models to capture the overall hydrological regime and seasonal runoff dynamics (evidenced by NSE values above 0.63 for all sub-basins and 0.78 for mainstream HX) provides a sufficiently robust and quantitative basis for their use in evaluating the relative risk of rainfall-induced flooding across the study area.

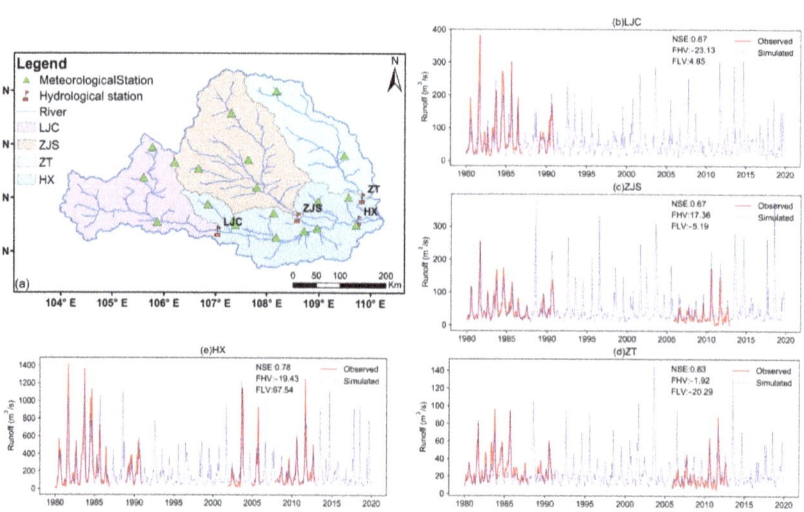

Figure 6. Streamflow simulations and predictions for sub-basins. According to the trend of the river and the distribution of hydrological stations, the (**a**) Weihe River Basin is divided into four sub-basins: Linjiacun (LJC), Zhangjiashan (ZJS), Zhuangtou (ZT), and Huaxian (HX). Runoff simulation and prediction in the (**b**) LJC, (**c**) ZJS, (**d**) ZT, and (**e**) HX watersheds. The purple lines represent the measured runoff curves, whereas the red lines denote the runoff prediction curves.

Figure 7 shows the monthly variations in runoff for four sub-basins. From 1980 to 2019, the annual average runoff depth in the upper stream of the Weihe River was 59.0 mm for LJC, 37.4 mm for ZJS, and 29.6 mm for ZT, and it was 169.3 mm for downstream HX. The runoff from July to September accounted for 42.00% of the total runoff.

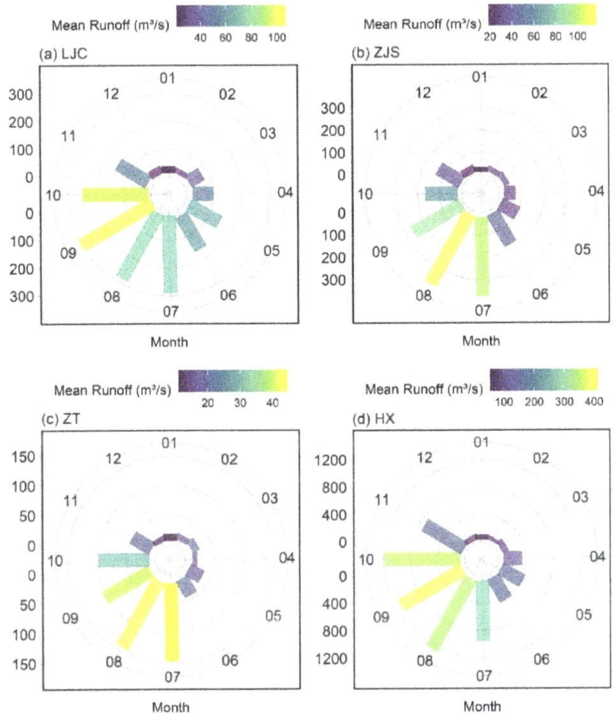

Figure 7. Monthly variations in runoff for four sub-basins: (**a**) Linjiacun (LJC); (**b**) Zhangjiashan (ZJS); (**c**) Zhuangtou (ZT); and (**d**) Huaxian (HX). The histogram represents the range, and the colors represent the average.

Over the period 1980–2019, the upper stream of the Weihe River averaged 1.81 billion m^3 of runoff for LJC, 1.62 billion m^3 for ZJS, and 0.76 billion m^3 for ZT, while it averaged 5.92 billion m^3 for downstream HX. Approximately 42.00% of the total runoff occurred between July and September. Converted to runoff depth, the annual average runoff depth in the upper stream of the Weihe River was 59.0 mm for LJC, 37.4 mm for ZJS, and 29.6 mm for ZT, while it was 169.3 mm for downstream HX.

3.3. Spatial Risk Assessment

The risk patterns for rainfall hazards, geo-hazards, and socio-hazards are shown in Figure 8. In accordance with the normalization principle and natural breakpoint grading method, the risk index was graded from 0 to 1, with 1 representing the highest level of risk, as shown in Table 4. The rainfall hazard risk map was based on rainfall and flooding, including annual rainfall, rainfall days, rainfall intensity, rainstorm intensity, maximum rainfall amount, average runoff depth, and flood season runoff depth. The map indicated a high risk of climate-related areas in the western and middle regions, which reflected the high rainfall and induced flooding in these areas. There was a 6.87% share of the total area that was deemed to have high- and very-high risk. Geo-hazard risk maps were based on several geographic attributes, including elevation, slope, soil type, and drainage density. The distribution of high-risk areas was relatively dispersed, reflecting that disasters were affected by multiple factors. The area at risk from geological attributes accounted for

29.72% of the total area. The socio-hazard risk map (Figure 8c), derived from indicators reflecting socioeconomic exposure (GDP, population density, land cover, historical sites), highlights areas where potential societal and economic impacts from natural hazards would be most significant. As expected under this definition, high socio-hazard risk areas are predominantly located in urban centers such as Xi'an, reflecting the concentration of valuable assets and population. Only 13.43% of the total area was considered high-risk.

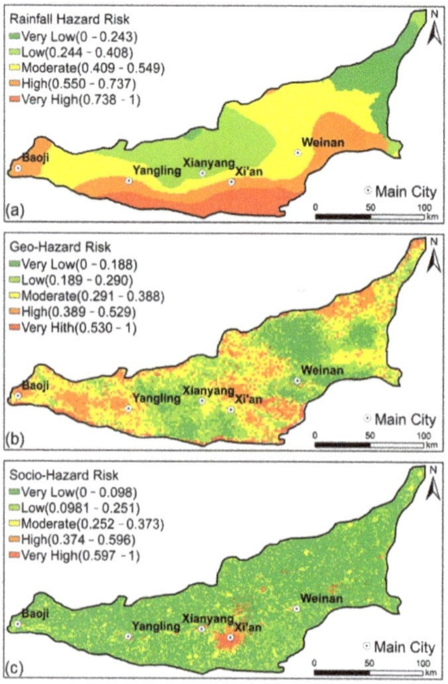

Figure 8. Risk map of (**a**) rainfall hazards, (**b**) geo-hazards, and (**c**) socio-hazards.

Table 4. Grading of various risk levels.

Level	Rainfall Hazard Risk	Geo-Hazard Risk	Socio-Hazard Risk
Very Low	0–0.243	0–0.188	0–0.098
Low	0.244–0.408	0.189–0.290	0.0981–0.251
Moderate	0.409–0.549	0.291–0.388	0.252–0.373
High	0.550–0.737	0.389–0.529	0.374–0.596
Very High	0.738–1	0.530–1	0.597–1

The weights for each type of factor and the comprehensive weight were calculated, as shown in Table 5. The sum of the comprehensive weights for each type of factor represented their contribution to the cause of the hazards. Thus, rainfall and runoff contributed 53.9%, geographical attributes contributed 29.72%, and socioeconomic variables contributed 16.38% of the hazards. Figure 9 depicts the comprehensive risk distribution and corresponding area statistics for each risk level, which were calculated by superimposing the comprehensive weight. The high-risk area encompassed 2762.56 km^2, which represented 13.43% of the total study area. The very high-risk area was 1414.4 km^2, which represented 6.87% of the total study area. High-risk and very high-risk areas were concentrated in the southern mountainous area and were primarily located over valley terrain. Generally, the northern region was subject to a low level of risk owing to its relatively flat terrain and steady rainfall.

Table 5. Calculation of the weights of factors.

Target	Variable		Weight of Criterion Layer	Index		Weight of Index Layer	Comprehensive Weight
Risk of floods, landslides, mudslides, and other disasters caused by rainfall	Rainfall and Runoff	(B_1)	0.5390	Rainfall (mm)	(C_1)	0.0582	0.0314
				Rainfall Days (d)	(C_2)	0.0397	0.0214
				Rainfall Intensity (mm/d)	(C_3)	0.1280	0.0690
				Maximum Rainfall Amount (mm)	(C_4)	0.1783	0.0961
				Rainstorm Intensity (mm/d)	(C_5)	0.2092	0.1127
				Average runoff depth (mm)	(C_6)	0.0740	0.0399
				Runoff depth in flood season (mm)	(C_7)	0.3126	0.1685
	Geographical	(B_2)	0.2972	Elevation (m)	(C_8)	0.1187	0.0353
				Slope (degree)	(C_9)	0.1671	0.0497
				Soil Texture	(C_{10})	0.2427	0.0721
				Drainage Density (km/km^2)	(C_{11})	0.4715	0.1402
	Socio-economic	(B_3)	0.1638	Population Density (pop/km^2)	(C_{12})	0.5247	0.0859
				GDP	(C_{13})	0.3338	0.0547
				Land Use	(C_{14})	0.1416	0.0232

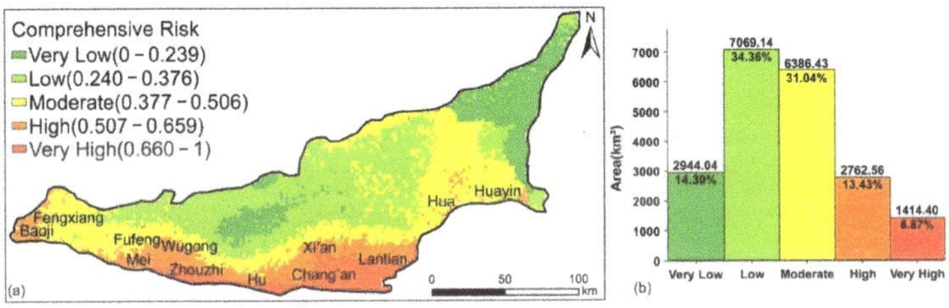

Figure 9. Comprehensive risk overlaying rainfall, geo-hazard, and socio-hazard risks based on calculated weight. (**a**) Spatial risk map, and (**b**) area statistics for each risk level.

Figure 10 illustrates a comparison of our findings with historical geo-hazard locations identified by field surveys, and Table 6 displays the number of identified geo-hazard locations within regions with varying levels of risk. Approximately 91.76% of historical geological disasters occurred in moderate, high, and very high-risk areas, whereas no disasters occurred in very low-risk areas. The 8.24% of historical geological disasters that occurred in our defined low-risk areas caused a limited degree of damage to houses and roads, which was within the acceptable range. This comparison demonstrated that our risk map, which combined climate, geographic, and social factors, was reliable for providing a spatial assessment to control and prevent rainfall-induced hazards.

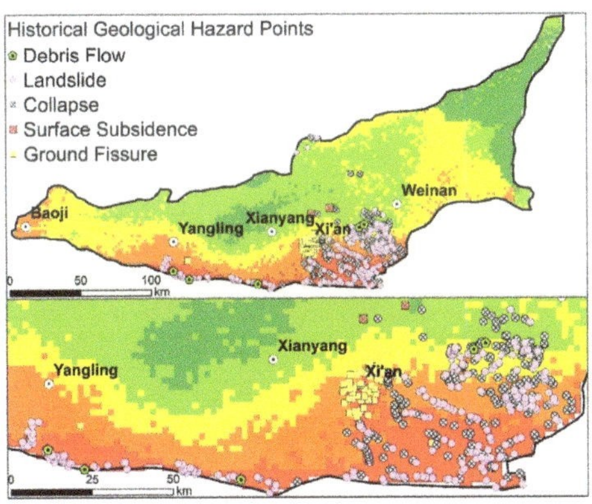

Figure 10. A comparison between mapped risk areas and identified historical geological hazard locations via field surveys. The 1960–2019 spatial distribution data of geological disaster points were compiled by the data center of resource and environmental science and data (https://www.resdc.cn/).

Table 6. Proportion of historical disaster points in each risk level.

Risk Level	Number of Historical Geological Hazard Points/Piece	Proportion/%
Very Low	0	0
Low	42	8.24
Moderate	96	18.82
High	158	30.98
Very High	214	41.96

4. Discussion and Implications

The analysis of historical rainfall trends facilitates the assessment of the impacts of spatiotemporal rainfall changes on geo-hazards as a result of climate change. In this study, our application of the Mann–Kendall trend test to rainfall data from 1980 to 2019 (Figure 4) revealed significant overall shifts in rainfall patterns across the study area. Specifically, for annual rainfall, the Mann–Kendall test yielded an average z-statistic of −0.64 across all stations, indicating a statistically significant general decreasing trend. This trend has resulted in an approximate 11.63% reduction in the characteristic magnitude of annual rainfall over the 40-year period. Similarly, a statistically significant and substantial decreasing trend was identified for annual rainfall days (average z-statistic of −1.83), which has led to an approximate 25.58% reduction in the number of rain days per year. Conversely, average rainfall intensity exhibited a statistically significant increasing trend (average z-statistic of +2.37), causing an approximate 20.93% increase. Average rainstorm intensity also showed a statistically significant increasing trend (average z-statistic of +1.94), resulting in an approximate 11.63% increase. This indicated that the frequency of rainfall in the study area decreased, and the amount of rainfall became more concentrated. The increased frequency of extreme rainfall events resulted in landslides, and debris flows as a result of the increased concentration and rainstorm intensity. Other computationally complex and data-demanding mechanistic models, as well as extensive field surveys and experiments in the study area and surrounding areas, also illuminate the impact of rainfall changes [31,58,59]. While this study did not use specific local hazard models to distinguish between hazard types or rainfall-induced hazards, rainfall analysis could serve as an initial early warning approach for geological hazards.

Flooding is a form of hydrological hazard that responds directly to changes in rainfall. The risk of flooding can be determined by simulating the river runoff and the depth of runoff from upstream to downstream. As shown in Figures 6 and 7, peak flows may occur from July to September and could exceed three to four times the monthly median flow in the upstream region (LJC, ZJS, and ZT stations), whereas the flood risk may persist into October for the downstream HX station. Although there was a high risk of overbank flooding in the study area because of the confluence of the three rivers on the three sides, the occurrence frequency of peak flows for the HX station was reduced. This was consistent with previous conclusions [58,60]. Additionally, this result could be used as spatial attribute data to overlay analyses with other types of risks associated with hazards.

Hazards, such as debris flows caused by rainfall, were heavily influenced by geological factors such as lithology, elevation, slope, and aspect. Land use was also an important risk factor, and the dense distribution of mountains and water networks in the study area provided physical prerequisites for mudflow origination. Rainfall and geo-environmental factors represented measures of hazard risk, whereas socioeconomic factors (population and land use) denoted measures of the vulnerability of the carrier itself. In the study area, the spatial distribution of precipitation and its geo-environmental and socioeconomic impacts varied widely. Rainfall-driven impacts were mainly concentrated in the southern mountainous areas, geo-environmental impacts were more widespread, and socioeconomic factors were mainly clustered in urban areas (Figure 8). Additionally, the results after overlaying the AHP weight reflected the composite spatial distribution of disaster risk. The resulting map in Figure 9 indicates that this distribution is highly correlated with rainfall. The greater the indicator weight coefficient, the more significant the influence of the indicator on the occurrence of geo-hazards. In this study, runoff depth, rainstorm intensity, and drainage density were the main factors influencing the risk assessment outcomes, based on the comprehensive indicator weight calculation.

Our socioeconomic analysis focused primarily on population and GDP, whereas the study area, as a historical city in western China, contained a large number of historical heritage sites. By superimposing a map of tourist attractions of heritage sites on the comprehensive risk results map, we determined that 39 tourist attractions were located within the high-risk and the very high-risk areas (Figure 11). Tourist attractions in high-risk areas represented 27.46% of the total number of tourist attractions included in the entire study area. According to statistics, more than 60% of global heritage sites are under threat from geological hazards [61]. Historical sites demand special attention in disaster assessment, mainly because of their irreplaceable cultural value and difficult recovery. Our risk assessment indicated the risk faced by the current spatial distribution of heritage sites in the study area and implied that this issue should be incorporated in comprehensive risk index evaluations in future hazard assessments.

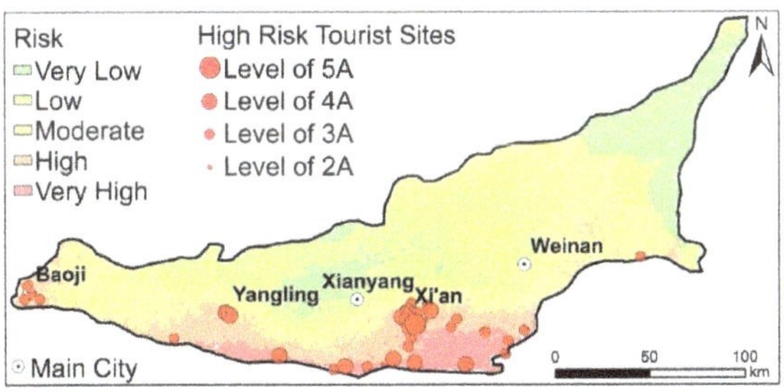

Figure 11. Distribution of high-risk tourist destinations. The size of red dots represents the level of tourist attractions.

5. Conclusions

In this study, we evaluated the risk index of rainfall-induced disasters on the Loess Plateau using integrated approaches of deep learning, AHP, and statistical analysis. The main conclusions were as follows:

(1) Over the past four decades, the annual rainfall in the Guanzhong Basin of the Loess Plateau decreased at a rate of 1.46 mm/year, rainfall days decreased by 10.89%, the maximum rainfall amount increased by 5.34%, rainfall intensity increased by 7.66%, and rainstorm intensity increased by 4.68%.

(2) The performance of our simulations using LSTM was acceptable, which indicated that runoff varied between 44.2 m^3/s (42 mm/year) in the upstream and 187 m^3/s (169.29 mm/year) in the downstream. A peak flow period was observed between July and September, which could exceed three to four times the monthly median flow in the upstream region. The downstream region could be at risk of flooding until October.

(3) The high-risk and very high-risk areas encompassed 4176.96 km^2, which covered 20.3% of the total study area and were mainly located in the southern Qinglin Mountain terrain. Rainfall and runoff dominated the risk of hazards with a weight fraction

of 53.9%, while geographical attributes contributed 29.72% and socioeconomic contributed 16.38%.

Our study provided a comprehensive evaluation framework that quantified the spatial–temporal variations in rainfall and runoff, geographical attributes, and socioeconomic factors in order to assess the multiple rainfall-induced geo-hazards. The results provided valuable insights for the development of future geo-hazard monitoring networks, as well as for the protection of historical and cultural heritage monuments on the Loess Plateau and other global historical cities.

Author Contributions: Conceptualization, Y.Y.; methodology, Y.Z., X.W. and Y.Y.; resources, S.X.; writing—original draft preparation, Y.Z., S.X. and Y.Y.; writing—review and editing, Y.Z., S.X., X.W., S.G. and Y.Y.; visualization, Y.Z. and X.W.; supervision, Y.Y.; project administration, Y.Y.; funding acquisition, S.X. and Y.Y. All authors have read and agreed to the published version of the manuscript.

Funding: This study was supported by the National Natural Science Foundation of China (Grant No. 42422701, 42171158, 41771220) and the Young Top Talents Support Program of Xi'an Jiaotong University (Grant No. HJ6J011).

Data Availability Statement: The data supporting the findings of this research are available in the manuscript, as shown in Table 3.

Conflicts of Interest: The authors declare no conflicts of interest.

References

1. Dano, U.L. An AHP-based assessment of flood triggering factors to enhance resiliency in Dammam, Saudi Arabia. *GeoJournal* **2022**, *87*, 1945–1960. [CrossRef]
2. Borga, M.; Stoffel, M.; Marchi, L.; Marra, F.; Jakob, M. Hydrogeomorphic response to extreme rainfall in headwater systems: Flash floods and debris flows. *J. Hydrol.* **2014**, *518*, 194–205. [CrossRef]
3. Tang, C.; Zhu, J.; Ding, J.; Cui, X.F.; Chen, L.; Zhang, J.S. Catastrophic debris flows triggered by a 14 August 2010 rainfall at the epicenter of the Wenchuan earthquake. *Landslides* **2011**, *8*, 485–497. [CrossRef]
4. Dominguez, F.; Rivera, E.; Lettenmaier, D.; Castro, C. Changes in winter precipitation extremes for the western United States under a warmer climate as simulated by regional climate models. *Geophys. Res. Lett.* **2012**, *39*, 5. [CrossRef]
5. Zhang, Y.; Zhou, D.; Guo, X. Regional climate response to global warming in the source region of the Yellow River and its impact on runoff. *Sci. China Earth Sci.* **2024**, *67*, 843–855. [CrossRef]
6. Lu, J.; Qin, T.; Yan, D.; Zhang, X.; Jiang, S.; Yuan, Z.; Xu, S.; Gao, H.; Liu, H. Drought and wetness events encounter and cascade effect in the Yangtze River and Yellow River Basin. *J. Hydrol.* **2024**, *639*, 131608. [CrossRef]
7. Yao, Y.; Zhang, M.; Deng, Y.; Dong, Y.; Wu, X.; Kuang, X. Evaluation of environmental engineering geology issues caused by rising groundwater levels in Xi'an, China. *Eng. Geol.* **2021**, *294*, 106350. [CrossRef]
8. Chen, Y.; Zhao, L.; Wang, Y.; Wang, C. Review on Forecast Methods of Rainfall-induced Geo-hazards. *J. Appl. Meteorol. Sci.* **2019**, *30*, 142–153. [CrossRef]
9. Jiang, W.; Tan, Y. Heavy rainfall-related excavation failures in China during 1994 to 2018: An overview. *Eng. Fail. Anal.* **2021**, *129*, 105695. [CrossRef]
10. Yao, Y.; Khan, Z.A. Predicting Pakistan's next flood. *Science* **2022**, *378*, 483. [CrossRef]
11. Chong, Y.; Chen, G.; Meng, X.; Yang, Y.; Shi, W.; Bian, S.; Zhang, Y.; Yue, D. Quantitative analysis of artificial dam failure effects on debris flows—A case study of the Zhouqu '8.8'debris flow in northwestern China. *Sci. Total Environ.* **2021**, *792*, 148439. [CrossRef] [PubMed]
12. Wang, G. Lessons learned from protective measures associated with the 2010 Zhouqu debris flow disaster in China. *Nat. Hazards* **2013**, *69*, 1835–1847. [CrossRef]
13. Herrera, G.; Mateos, R.M.; García-Davalillo, J.C.; Grandjean, G.; Poyiadji, E.; Maftei, R.; Filipciuc, T.-C.; Jemec Auflič, M.; Jež, J.; Podolszki, L. Landslide databases in the Geological Surveys of Europe. *Landslides* **2018**, *15*, 359–379. [CrossRef]
14. Miao, Q.; Yang, D.; Yang, H.; Li, Z. Establishing a rainfall threshold for flash flood warnings in China's mountainous areas based on a distributed hydrological model. *J. Hydrol.* **2016**, *541*, 371–386. [CrossRef]
15. O'Connor, J.E.; Costa, J.E. Spatial distribution of the largest rainfall-runoff floods from basins between 2.6 and 26,000 km^2 in the United States and Puerto Rico. *Water Resour. Res.* **2004**, *40*, 1. [CrossRef]

16. Liu, Z.; Zhang, H.; Liang, Q. A coupled hydrological and hydrodynamic model for flood simulation. *Hydrol. Res.* **2019**, *50*, 589–606. [CrossRef]
17. Liu, S.; Yao, Y.; Kuang, X.; Zheng, C. A preliminary investigation on the climate-discharge relationship in the upper region of the Yarlung Zangbo River basin. *J. Hydrol.* **2021**, *603*, 127066. [CrossRef]
18. Intrieri, E.; Carlà, T.; Gigli, G. Forecasting the time of failure of landslides at slope-scale: A literature review. *Earth Sci. Rev.* **2019**, *193*, 333–349. [CrossRef]
19. Wang, X.; Huang, G.; Liu, J. Projected increases in intensity and frequency of rainfall extremes through a regional climate modeling approach. *J. Geophys. Res. Atmos.* **2014**, *119*, 13271–13286. [CrossRef]
20. Mallick, J.; Alqadhi, S.; Talukdar, S.; AlSubih, M.; Ahmed, M.; Khan, R.A.; Kahla, N.B.; Abutayeh, S.M. Risk assessment of resources exposed to rainfall induced landslide with the development of GIS and RS based ensemble metaheuristic machine learning algorithms. *Sustainability* **2021**, *13*, 457. [CrossRef]
21. Papathoma-Köhle, M.; Gems, B.; Sturm, M.; Fuchs, S. Matrices, curves and indicators: A review of approaches to assess physical vulnerability to debris flows. *Earth Sci. Rev.* **2017**, *171*, 272–288. [CrossRef]
22. Pollock, W.; Wartman, J. Human vulnerability to landslides. *GeoHealth* **2020**, *4*, e2020GH000287. [CrossRef] [PubMed]
23. Riedel, J.L.; Sarrantonio, S.M. Debris flow magnitude, frequency, and precipitation threshold in the eastern North Cascades, Washington, USA. *Nat. Hazards* **2021**, *106*, 2519–2544. [CrossRef]
24. Hu, L.; Zhang, Q.; Singh, V.P.; Wang, G.; He, C.; Zhao, J. Escalating rainstorm-induced flood risks in the Yellow River Basin, China. *Environ. Res. Lett.* **2024**, *19*, 064006. [CrossRef]
25. Lan, H.; Zhao, Z.; Li, L.; Li, J.; Fu, B.; Tian, N.; Lai, R.; Zhou, S.; Zhu, Y.; Zhang, F. Climate change drives flooding risk increases in the Yellow River Basin. *Geogr. Sustain.* **2024**, *5*, 193–199. [CrossRef]
26. Sun, Y.; Zhang, Q.; Singh, V.P. Flooding in the Yellow River Basin, China: Spatiotemporal patterns, drivers and future tendency. *J. Hydrol. Reg. Stud.* **2024**, *52*, 101706. [CrossRef]
27. Myhre, G.; Alterskjær, K.; Stjern, C.W.; Hodnebrog, Ø.; Marelle, L.; Samset, B.H.; Sillmann, J.; Schaller, N.; Fischer, E.; Schulz, M. Frequency of extreme precipitation increases extensively with event rareness under global warming. *Sci. Rep.* **2019**, *9*, 16063. [CrossRef]
28. Zhao, Z.; Huo, A.; Liu, Q.; Yang, L.; Luo, C.; Ahmed, A.; Elbeltagi, A. Assessment of urban inundation and prediction of combined flood disaster in the middle reaches of Yellow river basin under extreme precipitation. *J. Hydrol.* **2024**, *640*, 131707. [CrossRef]
29. Xu, S.; Zhang, M.; Ma, Y.; Liu, J.; Wang, Y.; Ma, X.; Chen, J. Multiclassification method of landslide risk assessment in consideration of disaster levels: A case study of Xianyang City, Shaanxi Province. *ISPRS Int. J. Geo-Inf.* **2021**, *10*, 646. [CrossRef]
30. Liang, S.; Chen, D.; Li, D.; Qi, Y.; Zhao, Z. Spatial and temporal distribution of geologic hazards in Shaanxi Province. *Remote Sens.* **2021**, *13*, 4259. [CrossRef]
31. Zhuang, J.; Peng, J.; Wang, G.; Javed, I.; Wang, Y.; Li, W. Distribution and characteristics of landslide in Loess Plateau: A case study in Shaanxi province. *Eng. Geol.* **2018**, *236*, 89–96. [CrossRef]
32. Zhou, Y.; Liu, Y.; Wu, W.; Li, N. Integrated risk assessment of multi-hazards in China. *Nat. Hazards* **2015**, *78*, 257–280. [CrossRef]
33. Hu, L.; Zhang, Q.; Wang, G.; Singh, V.P.; Wu, W.; Fan, K.; Shen, Z. Flood disaster risk and socioeconomy in the Yellow River Basin, China. *J. Hydrol. Reg. Stud.* **2022**, *44*, 101272. [CrossRef]
34. Liu, Z.; Xu, Z.; Charles, S.P.; Fu, G.; Liu, L. Evaluation of two statistical downscaling models for daily precipitation over an arid basin in China. *Int. J. Climatol.* **2011**, *31*, 2006–2020. [CrossRef]
35. Ma, S.; Zhou, T.; Dai, A.; Han, Z. Observed changes in the distributions of daily precipitation frequency and amount over China from 1960 to 2013. *J. Clim.* **2015**, *28*, 6960–6978. [CrossRef]
36. Penalba, O.C.; Robledo, F.A. Spatial and temporal variability of the frequency of extreme daily rainfall regime in the La Plata Basin during the 20th century. *Clim. Chang.* **2010**, *98*, 531–550. [CrossRef]
37. Neeti, N.; Eastman, J.R. A contextual mann-kendall approach for the assessment of trend significance in image time series. *Trans. GIS* **2011**, *15*, 599–611. [CrossRef]
38. Feng, D.; Fang, K.; Shen, C. Enhancing streamflow forecast and extracting insights using long-short term memory networks with data integration at continental scales. *Water Resour. Res.* **2020**, *56*, e2019WR026793. [CrossRef]
39. Gauch, M.; Mai, J.; Lin, J. The proper care and feeding of CAMELS: How limited training data affects streamflow prediction. *Environ. Model. Softw.* **2021**, *135*, 104926. [CrossRef]
40. Kratzert, F.; Klotz, D.; Shalev, G.; Klambauer, G.; Hochreiter, S.; Nearing, G. Benchmarking a catchment-aware long short-term memory network (LSTM) for large-scale hydrological modeling. *Hydrol. Earth Syst. Sci. Discuss* **2019**, *2019*, 5089–5110. [CrossRef]
41. Yao, Y.; Zhao, Y.; Li, X.; Feng, D.; Shen, C.; Liu, C.; Kuang, X.; Zheng, C. Can transfer learning improve hydrological predictions in the alpine regions? *J. Hydrol.* **2023**, *625*, 130038. [CrossRef]
42. Man, Y.; Yang, Q.; Shao, J.; Wang, G.; Bai, L.; Xue, Y. Enhanced LSTM Model for Daily Runoff Prediction in the Upper Huai River Basin, China. *Engineering* **2022**, *24*, 229–238. [CrossRef]

43. Yilmaz, K.K.; Gupta, H.V.; Wagener, T. A process-based diagnostic approach to model evaluation: Application to the NWS distributed hydrologic model. *Water Resour. Res.* **2008**, *44*, 9. [CrossRef]
44. Aydin, M.C.; Sevgi Birincioğlu, E. Flood risk analysis using gis-based analytical hierarchy process: A case study of Bitlis Province. *Appl. Water Sci.* **2022**, *12*, 122. [CrossRef]
45. He, Y.; Wang, Q.; Xu, Y.; Li, Z.; Yuan, J.; Lu, M.; Lin, Z. Climate change increased the compound extreme precipitation-flood events in a representative watershed of the Yangtze River Delta, China. *Stoch. Environ. Res. Risk Assess.* **2022**, *36*, 3803–3813. [CrossRef]
46. Zhang, Y.; You, W. Social vulnerability to floods: A case study of Huaihe River Basin. *Nat. Hazards* **2014**, *71*, 2113–2125. [CrossRef]
47. Saaty, T.L. What is the analytic hierarchy process. In *Mathematical Models for Decision Support*; Springer: Berlin/Heidelberg, Germany, 1988; pp. 109–121.
48. Saaty, T.L. How to make a decision: The analytic hierarchy process. *Eur. J. Oper. Res.* **1990**, *48*, 9–26. [CrossRef]
49. Lyu, H.; Sun, W.; Shen, S.; Arulrajah, A. Flood risk assessment in metro systems of mega-cities using a GIS-based modeling approach. *Sci. Total Environ.* **2018**, *626*, 1012–1025. [CrossRef]
50. Dou, J.; Yunus, A.P.; Bui, D.T.; Merghadi, A.; Sahana, M.; Zhu, Z.; Chen, C.; Khosravi, K.; Yang, Y.; Pham, B.T. Assessment of advanced random forest and decision tree algorithms for modeling rainfall-induced landslide susceptibility in the Izu-Oshima Volcanic Island, Japan. *Sci. Total Environ.* **2019**, *662*, 332–346. [CrossRef]
51. Stefanidis, S.; Stathis, D. Assessment of flood hazard based on natural and anthropogenic factors using analytic hierarchy process (AHP). *Nat. Hazards* **2013**, *68*, 569–585. [CrossRef]
52. Salo, A.A.; Hämäläinen, R.P. On the measurement of preferences in the analytic hierarchy process. *J. Multi-Crit. Decis. Anal.* **1997**, *6*, 309–319. [CrossRef]
53. Tang, J. Prioritizing User Requirements based on Analytic Hierarchy Process. *Proc. J. Phys. Conf. Ser.* **2021**, *1861*, 012021. [CrossRef]
54. Pollesch, N.L.; Dale, V.H. Normalization in sustainability assessment: Methods and implications. *Ecol. Econ.* **2016**, *130*, 195–208. [CrossRef]
55. Chang, J.; Wang, Y.; Istanbulluoglu, E.; Bai, T.; Huang, Q.; Yang, D.; Huang, S. Impact of climate change and human activities on runoff in the Weihe River Basin, China. *Quat. Int.* **2015**, *380*, 169–179. [CrossRef]
56. Li, Y.; Chang, J.; Jin, W.; Guo, A.; Wang, D. Runoff simulation in subzones of the Wei River Basin based on the SWAT model. *J. Northwest A F Univ. (Nat. Sci. Ed.)* **2017**, *45*, 204–212.
57. Yang, S.; Kang, T.; Bu, J.; Chen, J.; Wang, Z.; Gao, Y. Detection and Attribution of Runoff Reduction of Weihe River over Different Periods during 1961–2016. *Water* **2020**, *12*, 1416. [CrossRef]
58. Qi, Q.; Ju, Y.W. Study on Rainfall-induced Loess Geological Hazards Characteristics and Preventive Measures in Shanxi Province. *Proc. Appl. Mech. Mater.* **2015**, *744–746*, 1741–1744. [CrossRef]
59. Zhuang, J.; Iqbal, J.; Peng, J.; Liu, T. Probability prediction model for landslide occurrences in Xi'an, Shaanxi Province, China. *J. Mt. Sci.* **2014**, *11*, 345–359. [CrossRef]
60. Tian, S.; Luo, W.; Jia, Z.; Butalia, R.S. Temporal variation of over-bank flooding of Wei River and its impact on a riparian wetland in Xi'an, China. *Hydrol. Process.* **2010**, *24*, 1296–1307. [CrossRef]
61. Pavlova, I.; Makarigakis, A.; Depret, T.; Jomelli, V. Global overview of the geological hazard exposure and disaster risk awareness at world heritage sites. *J. Cult. Herit.* **2017**, *28*, 151–157. [CrossRef]

hydrology

MDPI

Article

Regional Frequency Analysis Using L-Moments for Determining Daily Rainfall Probability Distribution Function and Estimating the Annual Wastewater Discharges

Pau Estrany-Planas [1], Pablo Blanco-Gómez [2,*], Juan I. Ortiz-Vallespí [3], Javier Orihuela-Martínez [1] and Víctor Vilarrasa [4]

1 Empresa Municipal de Agua y Alcantarillado (EMAYA), Son Pacs, Camí dels Reis, 400, 07010 Palma, Spain
2 Vielca Ingenieros, S.A., Avda. Aragón 17, 46010 Valencia, Spain
3 Vielca Medio Ambiente, S.L., Avda. Aragón 17, 46010 Valencia, Spain
4 Global Change Research Group, IMEDEA, CSIC-UIB, C/ Miquel Marqués, 21, 07190 Esporles, Spain
* Correspondence: pablo.b@vielca.com

Abstract: The spatial distribution of precipitation is one of the major unknowns in hydrological modeling since meteorological stations do not adequately cover the territory, and their records are often short. In addition, regulations are increasingly restricting the amount of wastewater that can be discharged each year. Therefore, understanding the annual behavior of rainfall events is becoming increasingly important. This paper presents Rainfall Frequency Analysis (RainFA), a software package that applies a methodology for data curation and frequency analysis of precipitation series based on the evaluation of the L-moments for regionalization and cluster classification. This methodology is tested in the city of Palma (Spain), identifying a single homogeneous cluster integrated by 7 (out of 11) stations, with homogeneity values less than 0.6 for precipitation values greater than or equal to 0.4 mm. In the evaluation of the prediction capacity, the selected cluster of 7 stations performed in the first quartile of the 120 possible combinations of 7 stations, both for the detection of the occurrence of rainfall—in terms of Probability of Detection (POD), False Alarm Ratio (FAR), Critical Success Index (CSI) and Bias Score (BS) statistics—and for the accuracy of rainfall—according to Root Mean Square Error (RMSE), Nash–Sutcliffe Efficiency coefficient (NSE) and Percent Bias (PBIAS). The cluster was also excellent for predicting different rainfall ranges, resulting in the best combination for both light—i.e., [1, 5) mm—and moderate—i.e., [5, 20) mm—rainfall prediction. The Generalized Pareto gave the best probability distribution function for the selected region, and it was used to simulate daily rainfall and system discharges over annual periods using Monte Carlo techniques. The derived discharge values were consistent with observations for 2023, with an average discharge of about 700,000 m^3 of wastewater. RainFA is an easy-to-use and open-source software programmed using Python that can be applied anywhere in the world.

Keywords: regional frequency analysis; cluster analysis; L-moments; spatio-temporal distribution; Python software

1. Introduction

The Mediterranean climate is characterized by complex spatial and seasonal variability, with significant interannual fluctuations in precipitation levels [1]. Furthermore, in the context of anthropogenic climate change, there is a consensus that the frequency and intensity of heavy precipitation events have increased since the 1950s and will continue to increase [2]. In urban contexts, the occurrence of precipitation events, coupled with the intensification of urbanization activities, underscores the necessity to enhance the capacity of urban drainage systems. It is, therefore, evident that the statistical characterization of precipitation constitutes an indispensable component of the implementation of reliable hydrological and hydraulic models [3].

Frequency analysis of rainfall from weather station records depends on both the quality of the data and the length of the historical series. Despite the importance of developing in situ frequency analysis [4,5] both for extreme events [6,7] or drought periods [8], the lack of sufficiently long records limits the robustness of the required statistical data treatment [9]. Neighboring climate observatories are often used for quality control and gap filling, requiring long data series with few missing records and significant time overlap with the data series to be completed [10].

A notable improvement in accuracy can be achieved through regional frequency analysis [4,6,11], using data from the neighboring observatories to complete the information. In terms of statistical analysis, the L-moments method is particularly well-suited to the task, offering a number of advantages over alternative techniques [6,9,11–14]. These advantages include [4]: (1) their robustness in the presence of outliers, (2) a more effective discrimination of probability density functions (PDF), (3) a less biased estimation method, and (4) successful implementation experiences worldwide. The implementation of such techniques for hydrological applications has been widely distributed as an R package under an open-source license [3,4,10,12] and, to the best of our knowledge, it is not common practice to find such analysis using Python libraries.

This paper presents a new software package coded in Python for Rainfall Frequency Analysis (RainFA) [15] and applies its procedures to the case study of the city of Palma (Spain), where the insufficient capacity of its sewerage system determines the number of discharges into the Mediterranean Sea, affecting the water quality and increasing the possibility of beach closures, which ultimately affects the tourism sector [16,17]. Starting with subdaily rainfall data from 11 stations for the period 2015–2021, RainFA is used to homogenize to daily and curate the registers and, after a statistical analysis and clustering process, to determine a homogeneous group of 7 stations that can be considered to be a region. Next, a PDF is fitted for both the daily rainfall and the number of rainy days per year. The resulting runoff generates pollution vectors that have a significant impact on fresh and estuary waters; consequently, efforts to minimize their impact are required, in accordance with European regulations [18,19] and their respective state transpositions [20]. In response to these challenges, RainFA implements a methodology to quantify the annual discharges into the Mediterranean Sea, which will improve the capabilities of the future digital twin of the sewerage system that EMAYA is currently developing in the area [16].

2. Materials and Methods

2.1. RainFA

RainFA develops a methodology for data curation and frequency analysis of rainfall series, programmed in Python, that allows (see Figure 1): (1) preliminary data analysis, by means of (a) homogenizing the time-step; (b) performing data quality control—i.e., detecting anomalous values based on user-defined thresholds as well as through statistical analysis to identify outliers, performing homogeneity tests, and using double-mass curves and HDR boxplots between tuples of data series; (2) clustering, supported by the Silhouette width and Mantel statistics, as well as the Ward's dendrogram and evaluating the Discordancy and Homogeneity measures using the L-moments method; (3) IDW interpolation, for geospatial and temporal distribution of precipitation within the homogeneous regions; (4) frequency analysis, to adjust the best-fitting PDF function for the daily precipitation determination; and (5) wastewater discharge estimation, according to a simplified methodology based on the Monte Carlo simulations of *N* years of daily precipitation events as described in Section 2.7. This methodology has been applied to the time series (2015–2021) of 11 weather stations and the urban drainage system of the metropolitan area of Palma (Spain).

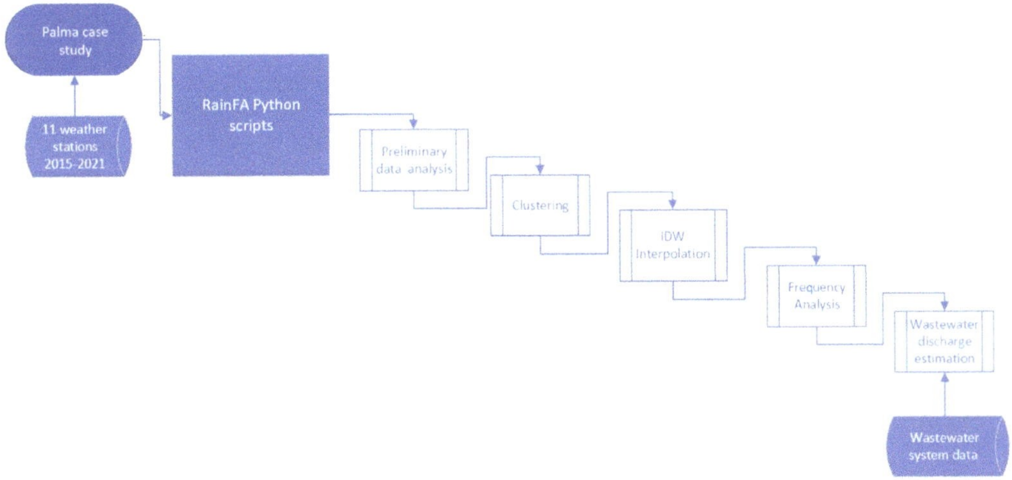

Figure 1. Flowchart of the adopted methodology.

2.2. Preliminary Data Analysis

Data continuity and stationarity are two of the most important concerns associated with historical series analysis [4,12]. In terms of continuity and data quality, a set of control tests are planned, including user-defined homogenization of the temporal discretization and rainfall thresholds, descriptive statistics—e.g., boxplots and outliers—and double-mass graphs. In addition, the non-parametric Mann–Kendall [21,22] trend test is applied for stationarity.

2.2.1. Temporal Homogenization

First of all, RainFA adjusts the sample size of the data using the resample function of the Pandas library of Python [23]. To do this, the time-step is determined with the least common multiple of the different series, and the sum is used to complete the series in cases where the original frequency is smaller.

2.2.2. Daily Quality Control

After the time homogenization, a series of different tests and procedures for quality control are proposed to ease the detection of unusual values that should be discarded from the data series. A daily timestamp has been selected for the data curating procedure in accordance with [3,10,12,24] and the different methods are described in the following paragraphs.

User-Defined Thresholds

A user-defined threshold option has been implemented within the RainFA software package (v1.0.1) to allow a manual series cleaning procedure to discard some values that may influence the following statistical analysis. Default values are set in *Not a Number* (*NaN*) for negative numbers and will require a completeness procedure in further steps of the method. Additionally, an equivalent to 40 mm for 10 min cumulative rainfall has been adopted for the definition of the upper threshold.

Outlier Detection and Bivariate HDR Boxplots

Rainfall time series are cataloged as functional data to provide measurements in a temporal continuum that can be plotted on curves or surfaces for their analysis [25]. According to [26], a side benefit of some of the functional data analysis techniques is the identification of outliers that may not be obvious from a standard statistical analysis or a graphical representation of the original data. An early approach to this representation was John Tukey's box and whiskers plot (or boxplot) that represents, in a compact form, the distribution summary statistics [27]. RainFA introduces boxplots of each data series to identify the potential outliers of the dataset, after the definition of a lower threshold that would be used to trigger rainfall. The default considered value is 0.2 mm, as it corresponds to a common error in a telemetry rain gauge device.

In addition, ref. [28] introduced the concept of 2D dimensions to the boxplot analysis, relying on a density estimate of the selected variable to compare between two different data series. To implement this comparison, ref. [29] presented a method of calculating and plotting Highest Density Regions (HDR), which are often considered the most appropriate subset to use to summarize a probability distribution. According to [26], the Bivariate HDR boxplot is constructed using a bivariate kernel density estimate $\hat{f}(z)$, which is defined as:

$$\hat{f}(z) = \frac{1}{n}\sum_{i=1}^{n} K_{h_i}(z - Z_i), \tag{1}$$

where z correspond with the data points of the time series, Z_i represents a set of bivariate points, K_{h_i} is the kernel function, and h_i is the bandwidth for dimension i.

Next, from the kernel density estimates, an HDR is defined as:

$$R_\alpha = \left\{ z : \hat{f}(z) \geq f_\alpha \right\}, \tag{2}$$

where f_α is such that $\int_{R_\alpha} \hat{f}(z)dz = 1 - \alpha$, which corresponds to a region with probability coverage $1 - \alpha$, meaning that all points within the region have a higher density estimate than any point outside its boundaries. RainFA considers an outer 90% highest density region for the outer contour, which defines the outlier consideration.

Double-Mass Curves

Double-mass curves were presented as a tool for evaluating the consistency of hydrological data—e.g., streamflow, rainfall, sediment series, etc.—from a single station by comparing it with a pattern composed of several nearby stations [30]. The principle of the double-mass curve is based on the fact that the cumulative graph of one variable against the accumulation of another variable over the same period is represented as a straight line as long as the two quantities are proportional, with the slope of this line being the constant of proportionality between them.

A change in the slope of the curve could mean a variation in the constant of proportionality or perhaps that there is no relationship. For this reason, it is common practice in hydrology to plot the double-mass curves of each individual time series against the average registers of all stations.

RainFA considers both the Bivariate HDR boxplots—as mentioned in the previous sub-section—and the double-mass curves to study the rainfall patterns and the existing proportionality between the different stations, expecting to detect potential anomalies and the cluster relationships between stations.

Trend Analysis

Non-stationarity of hydroclimatic series is one of the main concerns in the analysis of historical data and has been particularly considered in the regional frequency analysis of rainfall and flood series [4,12]. As noted in [31–33], the Mann–Kendall test is a robust procedure to the influence of extremes and suitable for use with skewed data because it does not require that the data be normally distributed, only that they be independent [31,34].

The null hypothesis,H_0, assumes that the data (x_1 ,..., x_n) are samples of n independent and equally distributed random variables [4,31]. The alternative hypothesis, H_1, of a two-sided test assumes that the distributions of x_j and x_k are not identical for all $j, k \leq n$ with $j \neq k$ [31]. First, the statistic S, which has mean zero and is asymptotically normal [31,33], is computed with the following equations:

$$S = \sum_{k=1}^{n-1} \sum_{j=k+1}^{n} sgn(x_j - x_k), \tag{3}$$

$$sgn(x_j - x_k) = \begin{cases} +1 & \text{if } sgn(x_j - x_k) > 0 \\ 0 & \text{if } sgn(x_j - x_k) = 0 \\ -1 & \text{if } sgn(x_j - x_k) < 0 \end{cases} \tag{4}$$

The variance of S is determined with the following equation [32], which takes into account that ties might be present:

$$Var(S) = \frac{1}{18}\left[n(n-1)(2n+5) - \sum_{p=1}^{q} t_p(t_p - 1)(2t_p + 5)\right], \tag{5}$$

where q corresponds with the number of tied groups and t_p is the number of data values in the pth group.

And finally, statistic S and its variance—i.e., $Var(S)$—are used to determine the test statistic Z as:

$$Z = \begin{cases} \frac{S-1}{\sqrt{Var(S)}} & \text{if} \quad S > 0 \\ 0 & \text{if} \quad S = 0 \\ \frac{S+1}{\sqrt{Var(S)}} & \text{if} \quad S < 0 \end{cases} \tag{6}$$

The Z value evaluates the presence of a statistically significant trend, where a positive trend corresponds to a positive Z value and a negative trend corresponds to a negative Z value. Since Z has a normal distribution, the goodness of the H_0 hypothesis associated with a α-confidence level is determined by the expression $|Z| \leq Z_{1-\frac{\alpha}{2}}$, corresponding the $Z_{1-\frac{\alpha}{2}}$ with 1.96 for a confidence level of 95%. RainFA uses the existing *pyMannKendall* software package to apply the Mann–Kendall trend test [35].

2.3. *Identification of Candidate Homogeneous Regions*

2.3.1. Cluster Overview

Cluster analysis is used to identify the candidate homogeneous regions by comparing different site characteristics, either related to spatial or meteorological conditions, which are analyzed using multivariate statistics and classified according to their similarity [11,12,36]. RainFA considers the spatial characteristics, which are UTM coordinates for latitude and longitude, together with the elevation in meters. It also includes three rainfall characteristics from the AEMET gauging stations: the maximum rainfall in 24 h per year, the average annual rainfall, and the average number of days with daily rainfall above 1 mm [37]. If rainfall characteristics are not available for the site, they can be extracted from the rain gauge data series (site statistics).

As recommended in [12], RainFA rescales the features using the Min-Max normalization methodology, which involves determining the minimum and maximum values of the different parameters and the features are ranged between 0—used for the minimum value—and 1—set for the maximum value—according to the aforementioned methodology. Then, the optimal number of clusters is identified using the Silhouette width and Mantel curve comparison methodologies [38], while the homogeneous regions are determined by applying Ward's dendrogram [39]. Finally, the homogeneity of the selected regions is evaluated using the statistics derived from the L-moments [11]: (1) the Discordancy measure (D_i), which is a measure of the dissimilarity of a given site from the rest of the sites within a group; and (2) the Heterogeneity measure (H), or *H-statistic*, which compares the dispersion of the L-moments from what would be expected for a homogeneous region.

2.3.2. Cluster Optimization

The optimal number of clusters is determined by measuring the degree to which an object—in this case, a gauging station—belongs to its cluster compared to its membership in the next closest cluster [38]. According to [40], the Silhouette width, $s(i)$, is defined as:

$$s(i) = \frac{b(i) - a(i)}{\max\{a(i), b(i)\}}, \tag{7}$$

where $a(i)$ corresponds to the average dissimilarity of object i to all other objects of cluster A and $b(i)$ is determined as

$$b(i) = \min_{C \neq A}(d(i, C)), \tag{8}$$

where C represents a cluster different from A and $d(i, C)$ corresponds to the average dissimilarity of object i to all other objects in cluster C. Thus, $b(i)$ corresponds to the closest cluster to A and is denoted as cluster B. $s(i)$ varies between -1 and 1, which means that i is "well clustered" when $s(i)$ is close to 1 and, on the contrary, when $s(i)$ is close to -1, the object i is clearly "misclassified", and it is on average much closer to cluster B than to cluster A. An intermediate situation is reached when $s(i)$ is close to 0, and it is not clear whether the object i should be assigned to cluster A or B [40].

The Silhouette width extends this cluster classification to all the objects in the dataset and determines the average of the clustering procedure, allowing the optimal number of groups to be determined. The Silhouette width is recognized as one of the strongest indices for data clustering [41].

In addition, RainFA implements Mantel's correlation coefficient test, which examines two distance matrices from the same samples. In this case, the initial distance matrix and the distribution into clusters matrix are compared, and, as a result, a correlation matrix is determined using the Pearson r coefficient for calculations [38]. The higher the correlation between the two matrices, the better the Mantel correlation.

2.3.3. Ward's Dendrogram

The minimum variance hierarchical clustering algorithm, or Ward's dendrogram [39], is recommended for the identification of homogeneous regions in a clustering process such as that applied by the RainFA package for the evaluation of rainfall data in a specific geographical context [11]. Ward's dendrogram algorithm uses the W function to minimize the sum of the squares of the deviations of the object vectors from the centroid of their respective clusters [42]:

$$W = \sum_{k=1}^{K} \sum_{j=1}^{m} \sum_{i=1}^{N_k} \left(y_{ij}^k - y_{\cdot j}^k \right)^2, \tag{9}$$

where K is the number of clusters, m corresponds with the dimension of the attribute vector, N_k represents the number of feature vectors in cluster k, y_{ij}^k denotes the rescaled value of the attribute j in the feature vector i assigned to cluster k, while $y_{\cdot j}^k$ is the mean value of the feature j for cluster k.

The algorithm starts with single station groups where the centroids coincide with the feature vectors and W is zero. At each step of the algorithm, a group fusion is performed, and the smallest increment in the function W is the fusion criterion. Ward's dendrogram results are very useful for identifying homogeneous regions in regionalization processes [11,39]. However, it does not allow for the reallocation of stations that may have been poorly classified in the early stages of the analysis [42].

2.3.4. L-Moments Homogeneity Measure

RainFA evaluates the homogeneity of the regions using two statistics based on L-moments and described in [6,11,43]: (1) the Discordancy (D_i) of a station i with respect to the rest of the stations in a group; and (2) the Heterogeneity (H_i) measure, which checks whether the variation in L-moments between the sites of a region is similar to what would be expected for a homogeneous region [4,8].

The Discordancy (D_i) is defined according to:

$$D_i = \frac{1}{3}\left[(u_i - \bar{u})^T (u_i - \bar{u}) S^{-1} \right], \tag{10}$$

where u_i is the vector of L-moment ratios corresponding to the site i—and includes the L-variation ($L\text{-}CV$), the L-skewness (τ_3) and the L-kurtosis τ_4—, $\bar{u}$ is the vector of average L-moments ratios in the selected cluster and S is the sample covariance matrix of the L-moments of all sites in the cluster. In general terms, sites with $D_i \geqslant 3$ are considered discordant [11,12].

The second statistic, the Heterogeneity (H_i), requires the Monte Carlo simulations of regions with equal L-moments and length registers, which may allow the determination of homogeneous regions to compare with [11,44]. This is carried out by first fitting a four-parameter *kappa* distribution function that is adjusted to the previously mentioned L-moments ratios, i.e., L-variation ($L\text{-}CV$), L-skewness (τ_3) and L-kurtosis (τ_4); and then the computed function is used to carry out the Monte Carlo simulations [4,11].

The Heterogeneity (based on the L-variation ratio [6,43]) is then calculated using:

$$H = \frac{V - \mu_v}{\sigma_v}, \tag{11}$$

where μ_v and σ_v are the mean and standard deviation of the simulated values, and the *V-statistic* is determined from the observed regional data as:

$$V = \left\{ \frac{\sum_{i=1}^{N} n_i [t^{(i)} - t^R]^2}{\sum_{i=1}^{N} n_i} \right\}^{\frac{1}{2}}, \tag{12}$$

where i refers to a station with a registration length of n_i, t^i corresponds to the L-variation ratio, N is the total number of stations of the homogeneous region and t^R represents the average L-variation ratio of the region, which is determined as:

$$t^R = \frac{\sum_{i=1}^{N} n_i \cdot t^{(i)}}{\sum_{i=1}^{N} n_i}, \quad (13)$$

According to the original criterion [11], the region can be considered "acceptably homogeneous" if $H < 1$, "possibly heterogeneous" if $1 \leq H < 2$, and "definitely heterogeneous" if $H \geqslant 2$.

2.4. Generalized IDW Interpolation for Homogeneous Regions

The Inverse Distance Weighted (IDW) is a widely used interpolation method for both image and spatial data contexts [45]. The main characteristic of this method is that all the points on the Earth's surface are considered to be interdependent, based on a relationship that is inversely proportional to the distance to each point [46]. By generalizing this relationship in the selected cluster regions, it is possible to determine an influence weight parameter for each station and thus to compound the variable to be simulated, which in this case is rainfall. Since rainfall varies over time, it is advisable to first determine the weights and then apply them to the time series, using the following procedure in a Geographical Information System (GIS) program: (1) represent the stations according to their coordinates; (2) create a new field called *Weights* in the station point layer; (3) assign the value 1 to station *i* and 0 to the rest of the stations; (4) apply the IDW interpolation method to calculate the weighted raster map of station *i*; (5) repeat the steps (3) and (4) for all the stations; (6) combine the calculated weighted raster maps into a multiband raster map, integrating the weights of all the stations into a single file.

2.5. Probability Distribution Functions Selection for Homogeneous Regions

The regional probability distribution function is selected according to the goodness-of-fit statistic Z^{Dist}, described in [11] and applied in [4,12,47], as an alternative of different statistical tests, such as Anderson-Darling (AD), Kolmogorov–Smirnov (KS) or Chi-Squared (CS), as described and applied in [9,44,48]. The different candidate distributions are fitted to the regional average L-moments, and its L-kurtosis (τ_4) is determined. As indicated in [4] and its sources, and in accordance with the determination of the Heterogeneity measure, the simulated length records of each site correspond to the observed data. The Z^{Dist} is then calculated as:

$$Z^{Dist} = \frac{\tau_4^{Dist} - t_4^R + B_4}{\sigma_4}, \quad (14)$$

where τ_4^{Dist} is the L-kurtosis of the fitted distribution and t_4^R the regional average L-kurtosis, while σ_4 and B_4 are the standard deviation and bias of t_4^R, which can be obtained by repeated simulation of a homogeneous region whose sites have the same fitted frequency distribution and record lengths as those in the observed data. [11] declares the fit to be adequate if Z^{Dist} is close to zero, which is a reasonable criterion: $|Z^{Dist}| \leq 1.64$.

2.6. Evaluation Criteria

2.6.1. Detection of Occurrence of Rainfall

The ability of each precipitation product to detect the occurrence of rainfall events is assessed using four categorical indices (see Table 1): (1) probability of detection (POD); (2) false alarm ratio (FAR); (3) critical success index (CSI); and (4) bias score (BS). POD indicates the ability of precipitation products to accurately capture the actual occurrence of precipitation. FAR assesses the proportion of false alarms detected by rainfall products. The ability of the data to comprehensively detect true precipitation events is measured by

CSI. CSI is an accurate and balanced detection metric and is based on POD and FAR. BS is the ratio of the estimated rainfall to observed rainfall.

In Table 1, N_{11} denotes the observed rainfall that is correctly detected, N_{01} represents the observed rainfall that has not been detected, and N_{10} is the precipitation that is detected but not observed.

Table 1. Precipitation occurrence performance metrics.

Statistic	Equation	Optimal Value
POD	$\frac{N_{11}}{N_{11}+N_{01}}$	1
FAR	$\frac{N_{10}}{N_{11}+N_{10}}$	0
CSI	$\frac{N_{11}}{N_{11}+N_{01}+N_{10}}$	1
BS	$\frac{N_{11}+N_{10}}{N_{11}+N_{01}}$	1

2.6.2. Accuracy of the Simulations

Three quantitative performance metrics are employed to measure the accuracy of the precipitation datasets in terms of daily rainfall (see Table 2): (1) Root Mean Square Error (RMSE); (2) Nash–Sutcliffe Efficiency coefficient (NSE); and (3) Percent Bias (PBIAS). The RMSE explains the deviation between two data sets, i.e., the degree of dispersion between observed and estimated precipitation data. The NSE indicates how well the correlation of observed and simulated data fits the 1:1 line when plotted together. In contrast, PBIAS indicates the degree to which the observed value is overestimated or underestimated as a percentage.

Table 2. Precipitation accuracy performance metrics.

Statistic	Equation	Value Range
RMSE	$\sqrt{\frac{\sum_{i=1}^{n}(O_i-S_i)^2}{n}}$	$[0,+\infty)$
NSE	$1-\frac{\sum_{i=1}^{n}(O_i-S_i)^2}{\sum_{i=1}^{n}(O_i-\bar{O})^2}$	$(-\infty,1]$
PBIAS	$\frac{\sum_{i=1}^{n}(O_i-S_i)\cdot 100}{\sum_{i=1}^{n}O_i}$	$(-\infty,+\infty)$

In Table 2, O_i and S_i are the observed and simulated data, respectively, i is the data index, and n is the total number of measurements, while $\bar{O}$ is the mean of the observed data.

In addition, the ability of the simulated rainfall to predict different daily rainfall intensities is evaluated according to the thresholds [49,50] shown in Table 3.

Table 3. Classification of rainfall events based on its daily rainfall intensity. Modified for daily values from the World Meteorological Organization (WMO).

Type of Event	Daily Rainfall Intensity (mm/day)
No rain	<1
Light rain	[1, 5)
Moderate rain	[5, 20)
Heavy rain	[20, 40)
Violent rain	≥40

2.7. Annual Sewage Discharge Estimation

The Royal Decree 665/2023 [20] and the proposal of adaptation of the Urban Wastewater Treatment Directive [18] establish criteria for the design of sewerage infrastructure, limiting the system's outflows and direct discharges into the sea. The calculations of the former are based on daily rainfall, while the latter considers annual flows. To evaluate them, RainFA adjusts the daily rainfall probability function for the calculated region and determines both (1) the expected number of annual sewer system discharges and (2) the total discharge volume. This approach is based on the simulation of all rainfall events occurring during the N years simulated using the Monte Carlo method.

The probability distribution function for the annual number of precipitation events can be fitted as a normal distribution, denoted as $N(\mu_V, \sigma_V^2)$, where V represents an array with the total number of significant precipitation events (greater than 0.4 mm) in one year in the selected homogeneous region:

$$\mu_V = \frac{\sum_{i=1}^{n} V_i}{n}, \tag{15}$$

$$\sigma_V^2 = \frac{\sum_{i=1}^{n} (V_i - \mu_V)^2}{n}. \tag{16}$$

n precipitation events (n = 1000) can be simulated for both probability distribution functions. For each year i, n different daily precipitation events P_{ij} are simulated with the selected probability distribution function (PDF) of the homogeneous region. The goodness of fit of the distribution is checked with the Shapiro–Wilk test [51].

The other variables that are needed to estimate the annual average discharge are:

- **Effective Area (EA)**: non-permeable area in the watershed.
- **Rainfall Inflow (RI_{ij})**: total rainfall entering the system for the event j of the year i. $RI_{ij} = EA \times P_{ij}$.
- **Attenuation and Treatment Capacity (C_{A+T})**: defined as the sum of the volume of the sewer system and attenuation tanks (C_A), and the capacity volume of the treatment plant (C_T). The parameter C_T depends on the duration of the rainfall, as shown in Figure 2.
- **Dry Volume (V_{dry})**: average daily volume of wastewater during dry days.
- **Discharge event (D_{ij})**: discharge for the event P_{ij} when the attenuation and treatment capacity is exceeded.

$$D_{ij} = \begin{cases} RI_{ij} + V_{dry} - C_{A+T} & \text{if} \quad RI_{ij} + V_{dry} - C_{A+T} > 0 \\ 0 & \text{if} \quad RI_{ij} + V_{dry} - C_{A+T} \leq 0' \end{cases} \tag{17}$$

Figure 3 shows the complete procedure for estimating the annual sewage discharge. It should be noted that the methodology does not take into account the duration of the rainfall, so it is recommended to use the time of concentration in the catchment as the duration of all rainfall events to obtain a conservative estimate of the discharge to the environment.

Finally, the defined methodology allows the estimation of the Annual Average Discharge ($\overline{AAD}$), which can then be evaluated for compliance with future European discharge limits as [18]:

$$\overline{AAD} = \sum_{i=1}^{N} \sum_{j=1}^{n} (D_{ij} > 0), \tag{18}$$

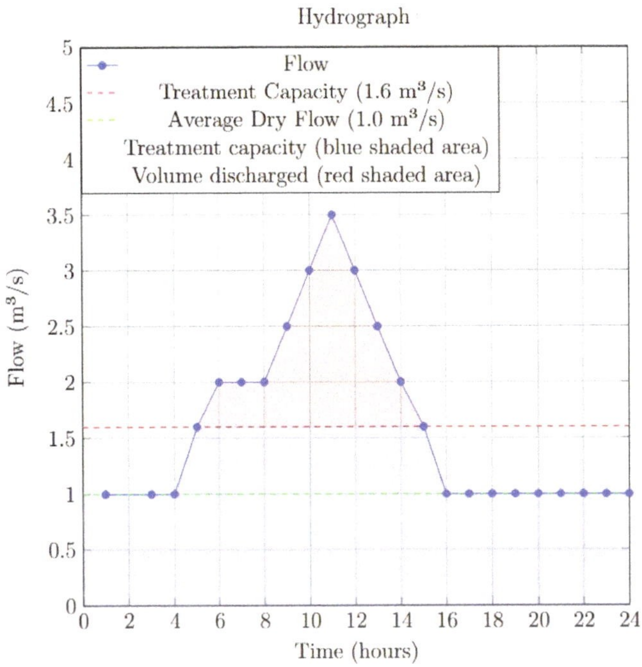

Figure 2. Example of hydrogram showing the treated water in the wastewater treatment plant and the discharged volume without being treated

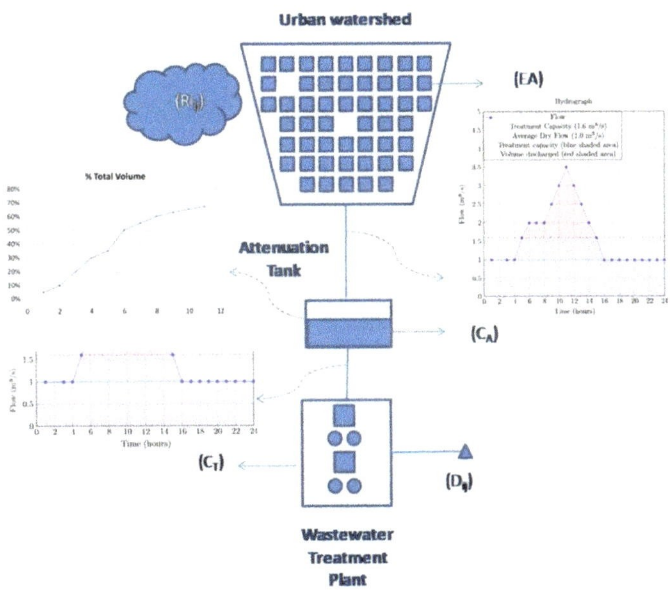

Figure 3. Example of a hydrogram showing the treated water in the wastewater treatment plant and the discharged volume without being treated.

3. Description of the Study Area and Data Sources

3.1. Case Study

The city of Palma is located in the southwest of the island of Mallorca, Spain, with an area of 208.63 km^2 and a population of 420,000 inhabitants. The metropolitan area, which discharges its wastewater into the municipal sewerage system, includes three other municipalities—Esporles, Bunyola, and Marratxí—so that the total area associated with the study area is 382.8 km^2 with 471,000 inhabitants.

Specifically, the study area is in the lower depression and estuary of the *Sa Riera*, *na Bàrbara*, and *Gros* gullies. Geomorphologically, the basin is bounded by the *Serra de Tramuntana* to the north and the *Serra de na Burguesa* to the west. It is also bordered by the lower hills near *Pòrtol*, known as *Serra de Son Mayol* to the northeast and *Costes des Xorrigo* slopes to the east [52] (Figure 4).

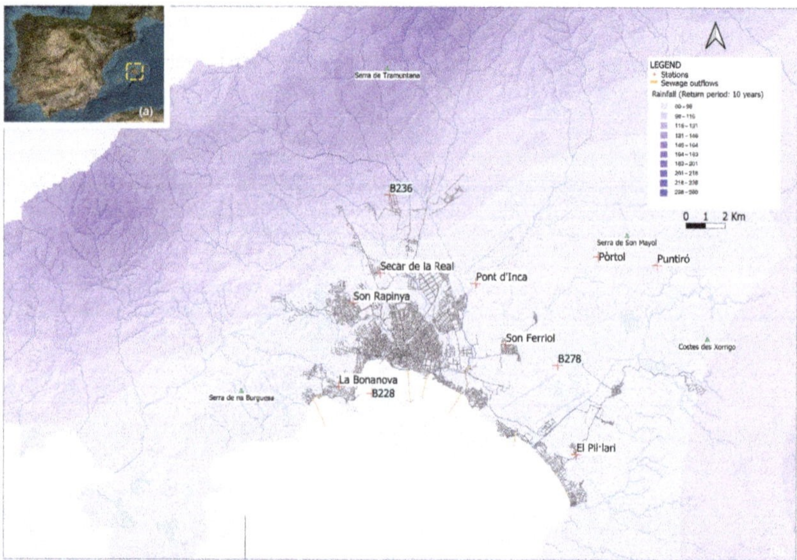

Figure 4. Location map: (**a**) Location of the city of Palma on the island of Mallorca (Spain); (**b**) Hydrometeorological stations (from AEMET, dashed in blue; and from Balearsmeteo), sewer overflows, surrounding mountains and 10-year return period daily rainfall (mm). Source: IDEIB maps.

Annual rainfall in the area has averaged 449 mm over the period 1980–2010, with a rainy season in autumn—i.e., from September to November—accounting for 52.2% of annual rainfall [53]. Furthermore, the interannual variability of rainfall is considerable, with pluviometric irregularity indices reaching 3.49 at the Portopí station (B228) for the 1980–2010 data series [54].

In addition, the maximum daily rainfall series shows a high spatial variability. The barrier effect of the Tramuntana mountain range is transformed into a reduction of the maximum daily rainfall in the area of Palma. In particular, for a 10-year return period, precipitation levels can vary considerably, from 130 mm/day in the municipality of Bunyola—in the northern part of the study area—to 70 mm/day in the coastal area of the city of Palma [55]. Specifically, the area surrounding Sant Jordi is characterized by the lowest maximum daily rainfall records due to the sheltering effect of the *Serra de Son Mayol* to the north and the *Costes des Xorrigo* to the east, which prevents the penetration of the north and east winds, respectively [56].

3.2. Data Sources

This study uses data from 11 rain gauges collected between 2015 and 2021. Three of these rain gauges are managed by AEMET [37]: (1) Palma Portopí (B228); (2) Palma Universitat Illes Balears (B236); and (3) Palma Aeropuerto (B278). The remaining 8 stations are operated by the non-profit organization BalearsMeteo [57], with which EMAYA has an agreement for access to the data recorded by their stations in the municipalities where they both operate. Table 4 shows the date-time range for each station.

In the present case study, the different meteorological stations have recorded every 1, 5, and 10 min, so the 10 min time-step was chosen for homogenization, with the sum of 2 or 10 values being used for resampling, as appropriate. The daily rainfall was used for clustering and homogeneity evaluation.

Table 4. Availability of information per rain gauge station.

Station Name	Series Start Date-Time	Series End Date-Time
El Pil·lari	31/12/2015 18:50	18/05/2022 23:00
La Bonanova	12/02/2015 11:30	18/05/2022 10:30
Pòrtol	08/10/2016 10:20	11/05/2022 00:00
Son Rapinya	21/11/2018 14:50	11/05/2022 00:00
Pont d'Inca	23/01/2009 20:00	19/05/2022 12:00
Puntiró	14/01/2017 16:40	26/06/2022 22:30
Secar de la Real	05/02/2021 21:10	23/06/2022 19:50
Son Ferriol	03/05/2021 23:40	22/06/2022 15:10
Portopí (B228)	01/01/2015 00:10	01/07/2022 00:00
UIB (B236)	01/01/2015 00:10	01/07/2022 00:00
Aeropuerto (B278)	01/01/2015 00:10	01/07/2022 00:00

Finally, the period chosen to assess the accuracy of the simulated rainfall datasets was from 2021 to 2022. All rain gauges were operational during this period.

4. Results

4.1. Data Curation and Stationarity

After the temporal homogenization to a 10-min time-step, four different data quality control tests were performed: (1) user-defined thresholds; (2) boxplots, both the traditional Box and Whisker and the Bivariate HDR boxplot; (3) double-mass curves; and (4) the Mann–Kendall trend test.

No values exceeded the 40 mm threshold defined for the 10-min time-step. Next, individual boxplots were constructed for all the stations (see Figure 5), both for precipitation values greater than 0.2 mm and values exceeding 2 mm, showing consistent outlier densities and similar ranges across the different stations, except for the *Son Ferriol* station, which had no daily records greater than 2 mm. Furthermore, both the double-mass curves (see Figure 6) and the Bivariate HDR boxplots (see Appendix A) show that the *Son Ferriol* station has no correspondence with the mean values of all the stations or with the rest of the stations in tuples. In addition, as shown in Table 4, there are only about 13 months of records. As a result of both findings, the time series of *Son Ferriol* was excluded from the rest of the study.

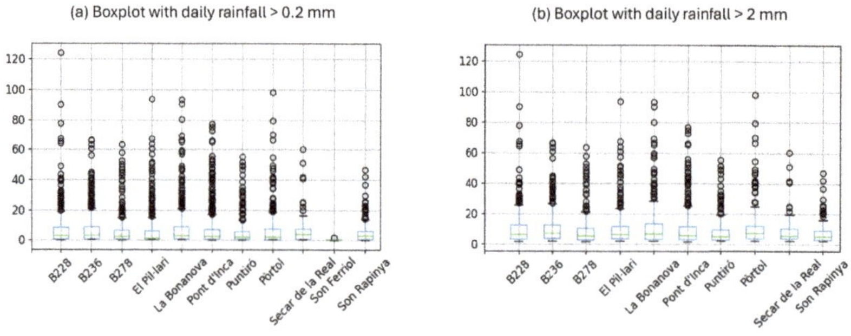

Figure 5. Boxplots with minimum precipitation thresholds of (**a**) daily rainfall > 0.2 mm; and (**b**) daily rainfall > 2 mm.

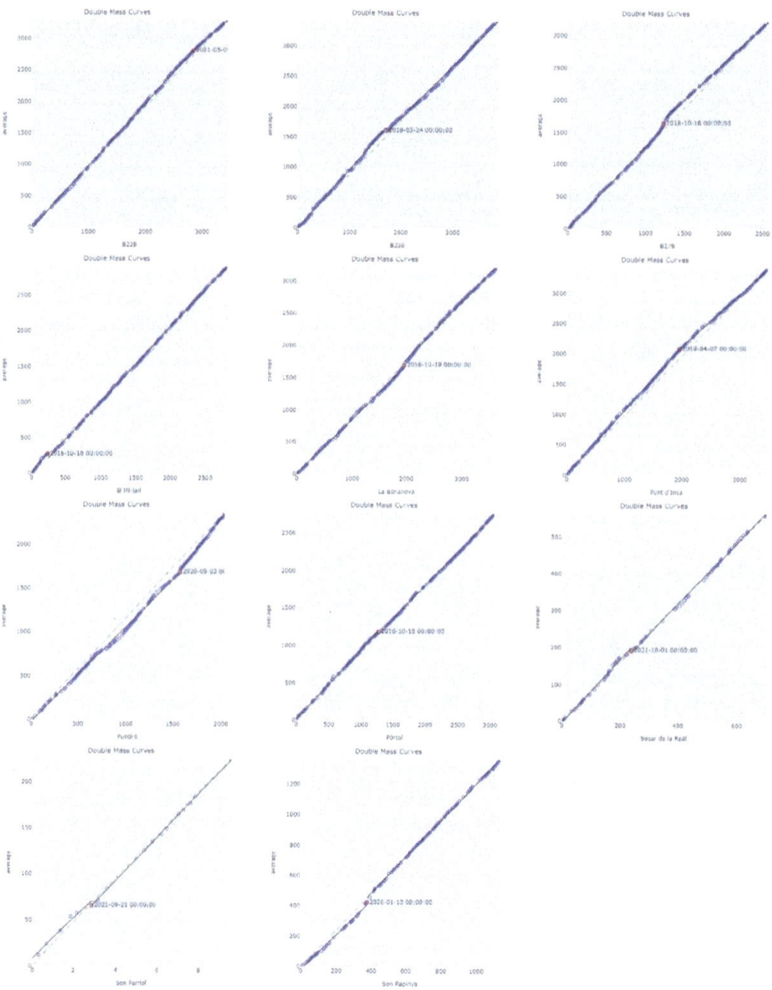

Figure 6. Double-mass curves of the different rain gauges.

From the analysis of the double-mass curves (Figure 6), a fairly uniform behavior is observed among all stations, albeit with certain peculiarities: *B228*, *B236*, *El Pil·lari*, and *Pont d'Inca* align with the overall station average; meanwhile, *La Bonanova*, *Secar de la Real*, and *Pòrtol* exhibit rainfall levels that exceed the average; finally, *B278*, *Puntiró*, and *Son Rapinya* present rainfall levels that are below the average. Furthermore, when looking at the results of the Bivariate HDR boxplot (see Appendix A), additional insights emerge both from the shape and the average of the individual density plots of each station, and especially from the common density regions, where the higher the correspondence between the values of the axes and the shape of their 50% and 90% HDR regions, i.e., with a smaller deviation from the 1:1 line, the greater the similarity between the stations.

The stationarity of the data is tested using the non-parametric Mann–Kendall test [21,22], the results of which are summarized in Table 5. Four out of ten stations show a certain trend, i.e., *Pont d'Inca* and *Puntiró*, and two other stations can be considered in the threshold area so that the region as a whole could be considered to be stationary.

Table 5. Results of trends in daily rainfall using the Mann–Kendall Test.

Station	s	Var(S)	Z	Trend
B228	−962	2.72×10^6	−0.58	No
B236	−3942	4.01×10^6	−1.97	Decreasing
B278	−1413	1.66×10^6	−1.09	No
El Pil·lari	−2213	1.38×10^6	−1.88	No
La Bonanova	−3161	2.56×10^6	−1.98	Decreasing
Pont d'Inca	7737	1.07×10^7	2.36	Increasing
Pòrtol	−2297	1.64×10^6	−0.90	No
Puntiró	−1156	1.02×10^6	−2.27	Decreasing
Secar de la Real	91	3.56×10^4	0.48	No
Son Rapinya	−835	2.47×10^5	−1.68	No

4.2. Clustering and Homogeneity

According to the Mantel Score and the Silhouette Width graphs, between two and three cluster regions could be defined in the selected stations (Figure 7). This result is consistent with the results of Ward's dendrogram (Figure 8) findings, as there is a group of six stations that could be associated with two different station tuples. However, an alternative interpretation of Ward's dendrogram result is that there is a single cluster with four possible exclusions.

According to the above results, the only region to be studied would include six rainfall stations located in the floodplain of the *Sa Riera*, *na Bàrbara* and *Gros* torrents. In addition, the station *B278* shows rainfall patterns very similar to those of this group, so it could be included, also in line with the Heterogeneity results that will be presented subsequently in this section. The selected region has a maximum distance from the sea of just over five kilometers, an average altitude of around 40 m, and an average annual rainfall of 400 (±50) mm/year. Furthermore, the region geomorphologically corresponds to a semicircular depression open to the sea in the Palma watershed [52].

Next, the L-moments homogeneity measure [11] statistics, i.e., Discordancy and Heterogeneity, are applied. First, for the Discordancy measure, all the stations within the selected region agree with the statistical data of the region. Table 6 shows the results of applying the Discordancy test to the data.

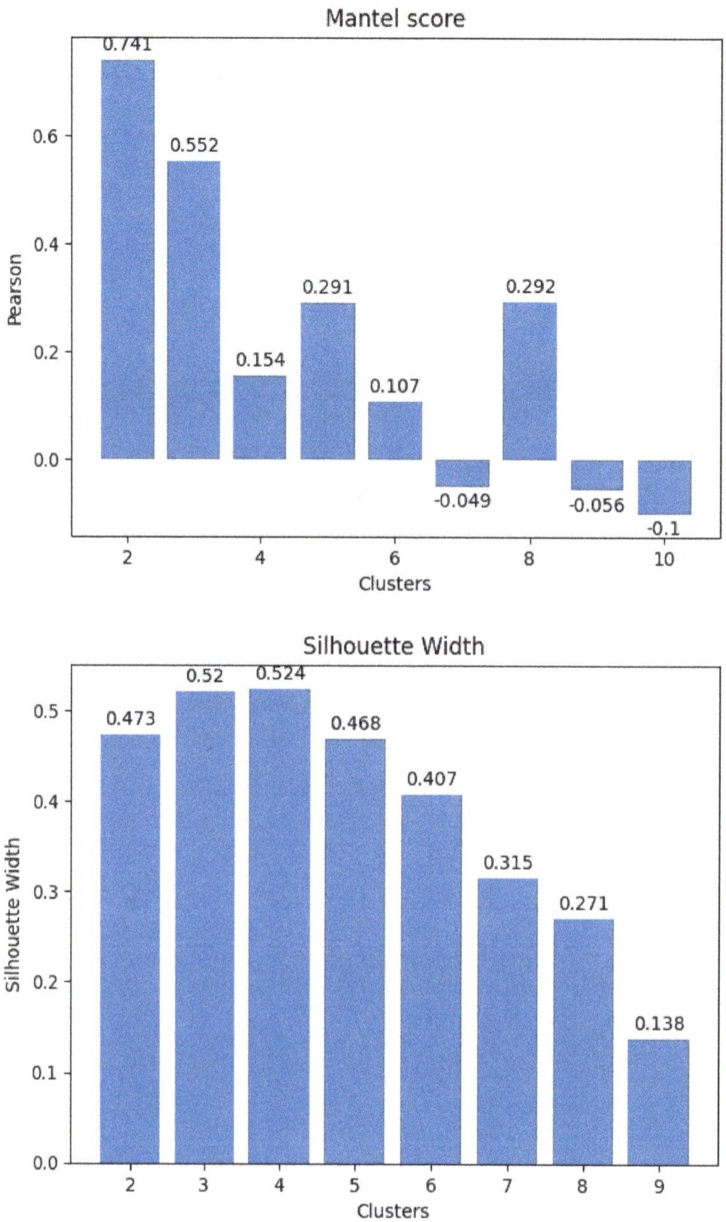

Figure 7. Mantel Score and Silhouette Width graphs.

The results of the Heterogeneity test (see Table 7) show that the extended cluster (W. group*), including the *B278* station, has an acceptable measure of homogeneity (H < 1) for daily rainfall greater than 0.4 mm. The lack of homogeneity in daily rainfall data, including the 'No rain' consideration (i.e., <1 mm), can be attributed to wind effects, dew, installation errors, or other types of error [58]. Consequently, precipitation inference calculations and PDF selection are performed for the acceptably homogeneous region ($P_d \geq 0.4$ mm).

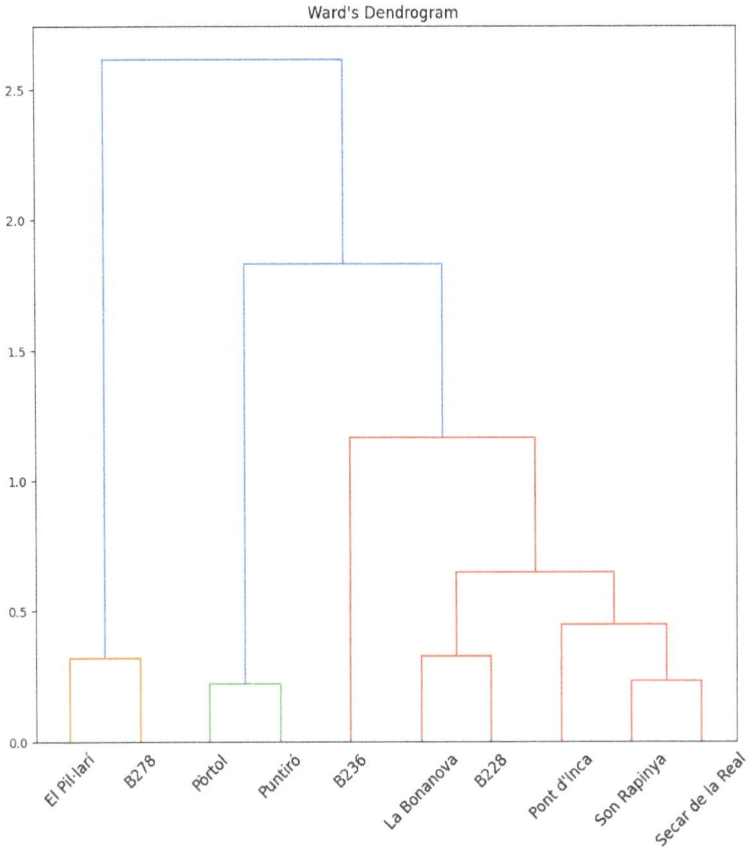

Figure 8. Dendrogram of rain gauges according to Ward's method.

Table 6. Results of the Discordancy Test.

No.	Name	Di	Critical Value	Classification
2	La Bonanova	1.08	1.92	Non-discordant
4	Son Rapinya	0.60	1.92	Non-discordant
5	Pont d'Inca	0.86	1.92	Non-discordant
7	Secar de la Real	1.55	1.92	Non-discordant
9	B228	0.76	1.92	Non-discordant
10	B236	1.20	1.92	Non-discordant
11	B278	0.95	1.92	Non-discordant

Moreover, for rainfall events greater than 1 mm, all the studied combinations could be considered to be homogeneous regions. Further investigation of these results, including the study of different combinations of stations, the addition of new stations with completely different characteristics within the island of Mallorca, and the clustering capacity of the Heterogeneity measure, would be beneficial.

Table 7. Heterogeneity results.

No.	Name	All	Ward Group	W. Group*
1	El Pil·lari	x	-	-
2	La Bonanova	x	x	x
3	Pòrtol	x	-	-
4	Son Rapinya	x	x	x
5	Pont d'Inca	x	x	x
6	Puntiró	x	-	-
7	Secar de la Real	x	x	x
9	B228	x	x	x
10	B236	x	x	x
11	B278	x	-	x
Heterogeneity for different minimum daily precipitation				
Pd ≥ 0.2 mm		9.74	3.61	3.69
Pd ≥ 0.4 mm		2.72	0.79	0.56
Pd ≥ 1 mm		−0.04	0.97	0.59
Pd ≥ 2 mm		−0.66	0.37	0.14

4.3. *Evaluation of the Precipitation Inference Model Using the IDW Method in the Selected Homogeneous Region*

All the indices referred to in Tables 1–3 (see Sections 2.6.1 and 2.6.2) are calculated not only for the selected homogeneous region but also for the 120 possible combinations of seven out of the ten available stations. This allows for the determination of whether the selected homogeneous region performs well in comparison to the other possible combinations.

4.3.1. Detection of Occurrence of Rainfall

The values of our categorical indices for the selected homogeneous region are found to be highly satisfactory, both in absolute terms and in relation to the other possible combinations of seven stations (see Table 8 and Appendix B for the complete set of results). The POD, FAR, and CSI obtained values are significantly higher than the thresholds typically considered acceptable for this type of meteorological inference. Furthermore, it is noteworthy that the combination of the homogeneous regions presents the sixth-best performance in terms of rainfall detection and false alarm avoidance, as quantified by the CSI index. The BS index suggests that the model may have a slight tendency to overestimate the probability of rainfall.

Table 8. Precipitation occurrence performance results.

Statistic	Value	Performance Rank
POD	0.95	25°
FAR	0.15	13°
CSI	0.81	12°
BS	1.11	19°

4.3.2. Accuracy of Rainfall

The NSE index shows favorable values in both absolute and relative terms, as summarized in Table 9—the complete results are presented in Appendix D. Furthermore, the

PBIAS values show favorable results, indicating a minimal degree of overestimation of the proposed inference model. Unfortunately, the RMSE for this combination is about 70% of the mean observed rainfall. Nevertheless, this suboptimal performance in absolute, though not in relative terms, could be improved by incorporating more robust inference methods, such as the Inverse Distance Weighting (IDW) method or Krigging [36,59].

Table 9. Precipitation accuracy performance results.

Statistic	Value	Performance Rank
RMSE	2.86 mm	25°
NSE	0.86	14°
PBIAS	−2.49%	58°

Figure 9 shows that the model performs suboptimally in absolute terms, especially for rainfall below 10 mm. However, the results of the homogeneous regions combination are comparable to the best of the combinations. It is also noteworthy that the model shows a slight tendency to underestimate, except for moderate daily rainfall (Pd < 10 mm), where it clearly overestimates. It is, therefore, essential to carry out a cluster analysis before implementing any inference method to ensure that the most appropriate approach is taken for each type of rainfall. This is particularly important in regions such as the case study presented here, where summer convective rainfall can be highly localized.

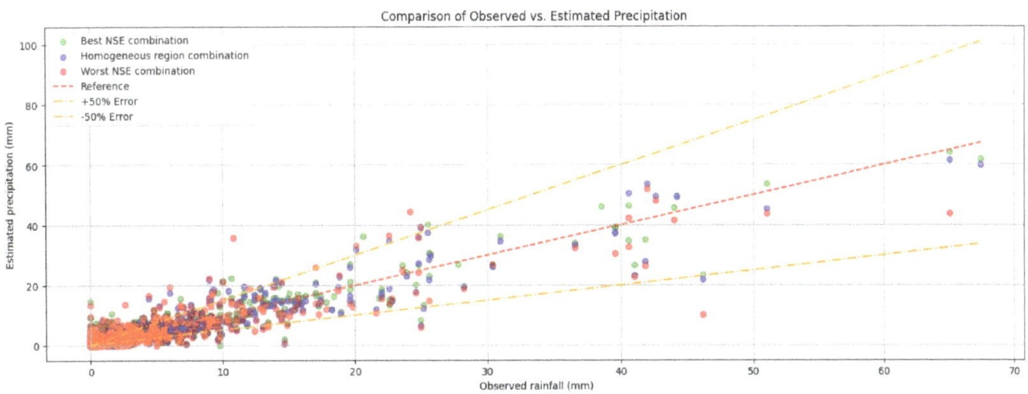

Figure 9. Comparison between simulated and observed rainfall.

The ability of the simulated precipitation to predict different ranges of daily rainfall according to WMO criteria is a remarkable achievement, both in absolute and relative terms, as shown in Table 10. The complete results are presented in Appendix C.

Table 10. Critical Success Index (CSI) results and ranking for different types of daily rainfall events.

Type of Event	Daily Rainfall Intensity (mm/day)	CSI Value	Rank
No rain	<1 mm	0.84	5°
Light rain	[1, 5) mm	0.51	1°
Moderate rain	[5, 20) mm	0.62	1°
Heavy rain	[20, 40) mm	0.44	26°
Violent rain	≥40 mm	0.7	15°

4.4. PDF Selection

A number of candidate distributions are tested, including lognormal (GLO), Generalized Extreme Value (GEV), Generalized Pareto (GPA), and the Pearson type III (PE3), with the aim of identifying the optimal PDF for the case study homogeneous region. The goodness-of-fit statistic Z^{Dist} is used to evaluate the quality of fit for the different probability functions. The results, which are summarized in Tables 11 and 12, indicate that the GPA PDF provides the best fit (Figure 10).

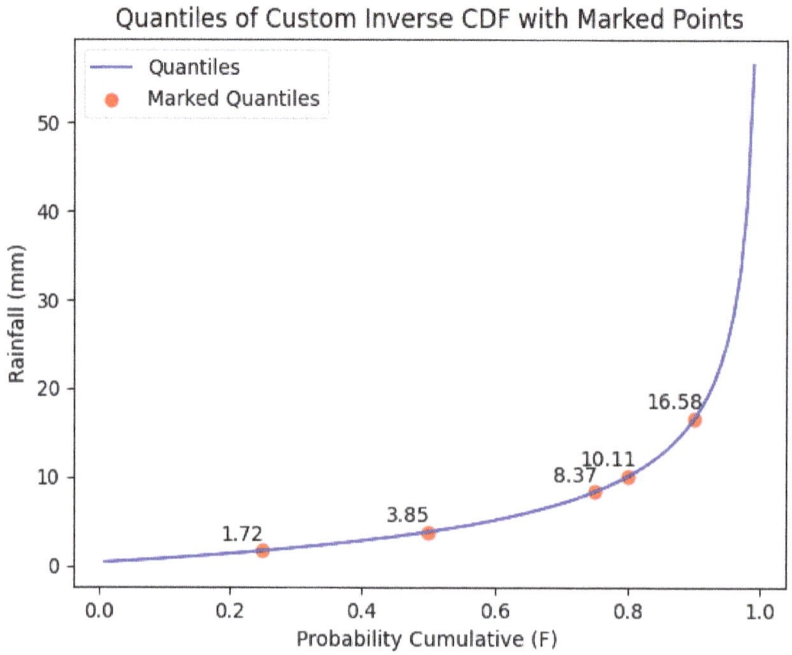

Figure 10. GPA inverse CDF.

Table 11. Goodness-of-fit results.

PDF	$Z^{Dist} < 1.64$
GLO	1.54
GEV	2.97
GPA	1.06
PE3	3.62

Furthermore, the PDF selection was applied at the different rainfall station locations, showing that the GPA PDF also provides the best local fit, according to the Z^{Dist} statistic (see Table 13). The graphical results, represented in Appendix E, range from 13.85 mm to 19.23 mm of cumulative rainfall for the 10-year return period or 0.9 marked quantile, indicating clear variability in volume among the stations despite them being part of a homogeneous group.

Table 12. Parameters of the different PDFs evaluated.

PDF	Parameters
GLO	k (shape): −1.103, α (scale): 4.708, ξ (location): 3.944
GEV	k (shape): −0.462, α (scale): 3.821, ξ (location): 2.894
GPA	k (shape): −0.392, α (scale): 4.326, ξ (location): 0.4
PE3	γ (shape): 0.321, σ (scale): 11.91, μ (location): 0.120

Table 13. Goodness-of-fit results at local scale.

Name	Z^{Dist}			
	GEV	GLO	GPA	PE3
La Bonanova	0.95	0.61	0.46	1.38
Son Rapinya	0.69	0.24	0.01	1.22
Pont d'Inca	1.32	0.84	0.57	1.69
Secar de la Real	0.39	0.04	0.06	1.21
B228	0.92	0.52	0.38	1.68
B236	1.59	1.22	0.64	0.85
B278	1.01	0.57	0.42	1.44

4.5. Sewage Discharge Estimation

The sewerage system of the city of Palma is evaluated using the discharge estimation methodology presented in Section 2.7. The effective 'impervious' area (EA) is estimated at 605 ha, which is considered a high figure for an almost completed separate sewerage system, but includes both (1) roofs that are incorrectly connected to the sewerage network and (2) an old mixed network in the city center that is challenging to remove. The latter accounts for 25% of the total contribution.

The urban wastewater treatment plant (known as EDAR II) receives a daily inflow of 78,000 m^3 (V_{dry}) of wastewater from its catchment area. In addition, the treatment capacity of this plant, combined with the support of the other urban treatment plant (known as EDAR I), is 90,000 m^3 (C_T). This estimate takes into account that the treatment plants can operate at a maximum capacity of 6000 m^3/h during a 2-h rainfall event (corresponding to the time of concentration of the tributary area). Furthermore, in terms of the attenuation capacity, a stormwater tank with a capacity of 40,000 m^3 has recently been added; this, together with the capacity of the sewerage network, including both the collection system and the treatment plants, calculated at 30,000 m^3, gives a total for attenuation and treatment capacity (C_{A+T}) of 160,000 m^3.

The wastewater discharged into the environment has an approximate mean value of 715,000 m^3 $\pm 33{,}000$ with a 95% confidence interval (see Figure 11), which is in line with the annual discharge estimate of 700,000 m^3 calculated by EMAYA using its level and flow sensors at the discharge points for the year 2023.

As can be seen in Figure 11, the volume distribution of annual discharges is considerably scattered due to the high dispersion obtained previously for the annual rainfall with the n = 1000 years of simulations using the Monte Carlo method. These results are due to the fact that the proposed methodology does not take into account the interdependence of the two statistical variables—i.e., the number of rainy days, with a mean of 56.35 ± 0.88 with a 95% confidence interval; and the daily rainfall selected in Section 4.4 (generalized Pareto). The data show that years with a high number of episodes have a lower rainfall per episode, suggesting that the expected dispersion should not be as high as observed.

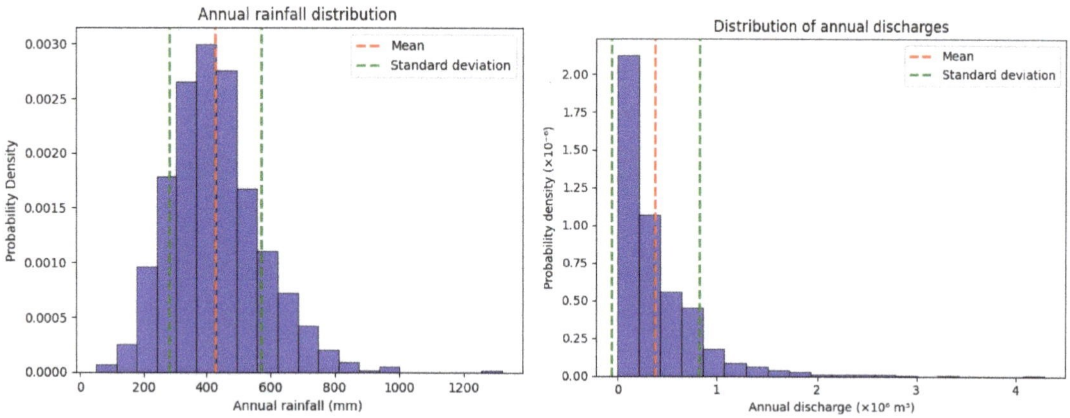

Figure 11. Distribution of rainfall and annual discharges using Monte Carlo simulations.

5. Discussion

As can be seen in the results, after the first stage of the data curation process, the time series from the *Son Ferriol* station were discarded from the evaluation both for the short length of their records and for their inconsistency with the magnitude of the precipitation records of the other stations. Next, from the analysis of the double-mass curves, and especially from the Bivariate HDR boxplots (see Appendix A, it is possible to detect relationships between stations that are consistent with the results of the clustering process. In this sense, for example, there is a clear alignment with the 1:1 line and density distribution in the Bivariate HDR boxplots of *La Bonanova* and *B228* or *Son Rapinya* and *Secar de la Real*—and even with these last two with *Pont d'Inca*—as shown in the clustering hierarchy of Ward's dendrogram (represented in Figure 8).

The two scores used to determine the number of clusters suggested the creation of either two or three distinct groups. According to Ward's dendrogram mentioned above, this results in two groups of two stations and an additional group of six stations. Geographically, *B278* and *El Pil·lari* stations are located on the periphery of the study area, in a drained marshland close to the Mediterranean Sea. On the other hand, *Pòrtol* and *Puntiró* stations are in a hilly area called *Serra de Son Mayol*, which could explain the differences in rainfall patterns with the rest of the stations (see Figure 4). Finally, the remaining stations are in the floodplain of three ephemeral streams that flow into the Bay of Palma. Notwithstanding this final conclusion, the revision of the annual rainfall, the average altitude, and the proximity to the main cluster group suggested the inclusion of *B278* in the main cluster, separating it from *El Pil·lari* station.

Regarding the PDF selection method, different statistical methods, such as KS and AD, were introduced in Section 2.5. However, the RainFA package implements the Z^{Dist} statistic as described in [11] to select the most appropriate PDF. According to this, the GPA function was selected as the best fit. It is important to note that the present case has been applied to study the sewerage capacity, which is usually designed for a return period of between 10 and 25 years, which is in line with the quantity of input data. If the objective were different and required a higher return period, additional goodness-of-fit methods would need to be applied and compared to determine the best PDF selection.

From the evaluation of the 120 possible combinations of seven out of ten available stations, the selected cluster performed outstandingly, demonstrating the ability of the present methodology to carry out regional analysis of the hydrology of regions with poor meteorological data availability.

Finally, the results and methodology for calculating the expected mean discharge values are considered acceptable and provide a reliable order of magnitude of the runoff produced in the sewer system. In any case, future studies will investigate the relationship between the rainfall intensity per event and the average annual rainfall, together with the duration of the rainfall, which could refine the results of the discharges to the environment.

The proposed methodology makes it possible to estimate the resources required for attenuation and treatment within the sewerage system, as well as for the reduction of impervious surfaces with inadequate connection to the sewerage network. The aim is to ensure compliance with existing and future European regulations [18,19].

Figure 12 illustrates possible scenarios for achieving the requirement of discharges of less than 1% of the total volume of water treated by the system, as previously required by the proposed European Directive (it should be noted that this percentage has since been increased to 2%). [19]. It can be seen that this would require a total of up to 260,000 m^3 of attenuation and treatment capacity if the current old mixed network remains connected. Conversely, a reduction of only 100 hectares would require only 230,000 m^3, which would mean increasing the volume of stormwater tanks by 70,000 m^3, while maintaining the current treatment capacity of the wastewater treatment plants.

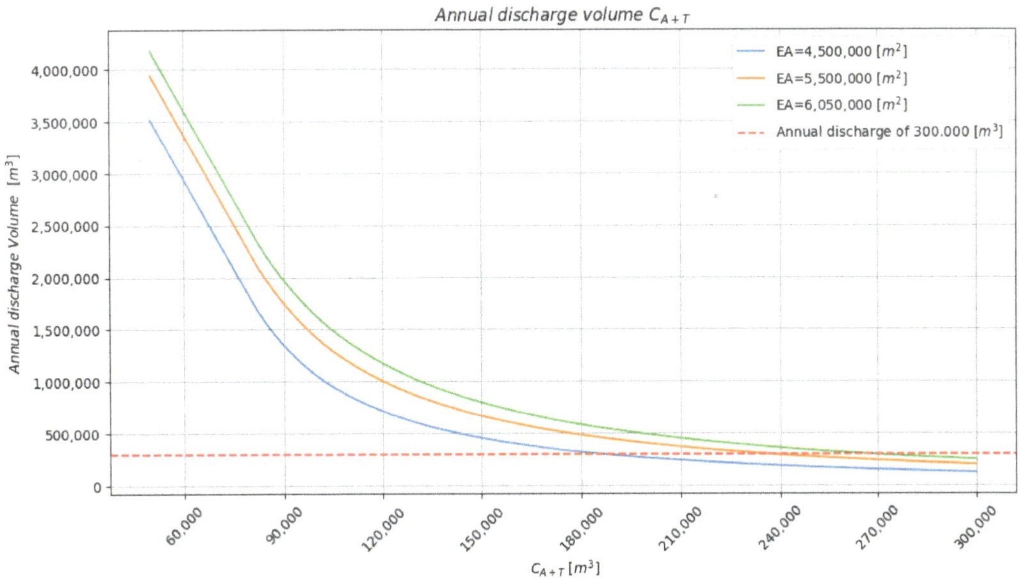

Figure 12. The annual discharge volume determined by the system's attenuation and treatment capacity.

In the present study, RainFA has been applied to a data curation process for 10-min time-step series to determine a homogeneous region in terms of daily rainfall. However, the software is prepared to and the method has been widely applied to different types of data (either historic or future projections under climate change scenarios), such as monthly, daily, or annual maximum rainfall in 24 h, streamflows or temperature [1,4,9,11–14,36,48,60–62], among others.

6. Conclusions

The spatial and temporal distribution of rainfall data is an important factor in the simulation of hydrological processes. The present study develops a comprehensive methodology for Rainfall Frequency Analysis using a Python software package called RainFA. It incorporates a number of key techniques, such as time scale homogenization, data curation, cluster analysis, L-moment homogenization measure, spatial distribution for series completion, and PDF fitting, for a complete Rainfall Frequency Analysis that could be applied worldwide. In addition, the package includes a tool for estimating sewer discharge from Monte Carlo rainfall simulations generated from the selected PDF distribution and the characteristics of the existing sewer network.

The RainFA software has been applied to the city of Palma (Spain) and has successfully identified a region with acceptable homogeneity for daily rainfall above 0.4 mm, with stations covering almost the entire catchment area of the EMAYA drainage system. Furthermore, the results show a satisfactory performance in terms of the detection of rainfall events, both in absolute and relative terms. However, the rainfall accuracy shows suboptimal performance in detecting the total amount of daily rainfall. Nevertheless, the results for our homogeneous region could be improved using more robust inference techniques. In addition, a methodology for simulating annual rainfall events and estimating annual wastewater discharges is proposed and tested to meet future European requirements [19]. This new methodology and the complete application procedure implemented in RainFA and applied to the context of the metropolitan area of Palma complements and develops the calculation procedure regulated in Spain by [20], providing a toolbox for the determination of the discharge capacity of urban drainage systems. Finally, in the case of Palma, it has allowed EMAYA to analyze its attenuation and treatment needs and to evaluate the possible measures that need to be implemented in order to comply with the objectives of the EU regulation.

Author Contributions: P.E.-P.: Conceptualization, Methodology, Investigation, Resources, Data curation, Writing—original draft, Visualization, Supervision. P.B.-G.: Conceptualization, Methodology, Software, Validation, Formal analysis, Investigation, Resources, Data curation, Writing—original draft, Writing—review and editing, Project administration. J.I.O.-V.: Software, Resources, Data curation. J.O.-M.: Conceptualization, Software, Data curation, Writing—review and editing. V.V.: Investigation, Resources, Data curation, Writing—original draft, Writing—review and editing. All authors have read and agreed to the published version of the manuscript.

Funding: This research has received funding by the Instituto Mediterráneo de Estudios Avanzados (IMEDEA). IMEDEA is an accredited "Maria de Maeztu Excellence Unit" (Grant CEX2021-001198, funded by MICIU/AEI/10.13039/501100011033).

Institutional Review Board Statement: Not applicable.

Informed Consent Statement: Not applicable.

Data Availability Statement: The original contributions presented in this study are included in the article. Further inquiries, including the rainfall data series, can be directed to the corresponding author.

Acknowledgments: The authors would like to thank EMAYA for providing the data it holds through an agreement with BalearsMeteo, as well as the data acquired through AEMET. The authors would also like to thank Juan Ortiz-Vallespí for his support in developing the RainFA code, as well as Vicente M. Candela Canales for his assistance to the R&D department of Vielca Ingenieros, S.A.

Conflicts of Interest: Author Pau Estrany-Planas and Javier Orihuela-Martínez were employed by the company "Empresa Municipal de Agua y Alcantarillado (EMAYA)", Author Pablo Blanco-Gómez was employed by the company "Vielca Ingenieros, S.A." and Author Juan I. Ortiz-Vallespí was employed by the company "Vielca Medio Ambiente, S.L.". The remaining authors declare that the research was conducted in the absence of any commercial or financial relationships that could be construed as a potential conflict of interest.

Appendix A. Bivariate HDR Boxplot Results

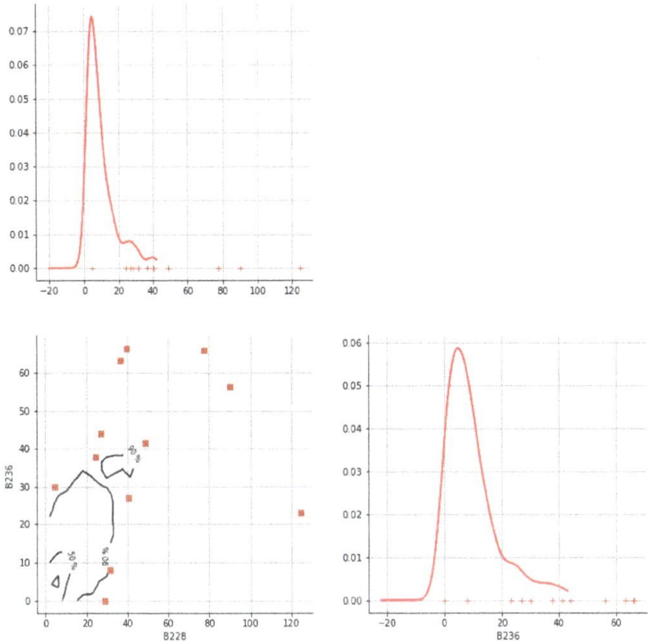

Figure A1. Bivariate HDR boxplots of B228 and B236 rain gauge stations.

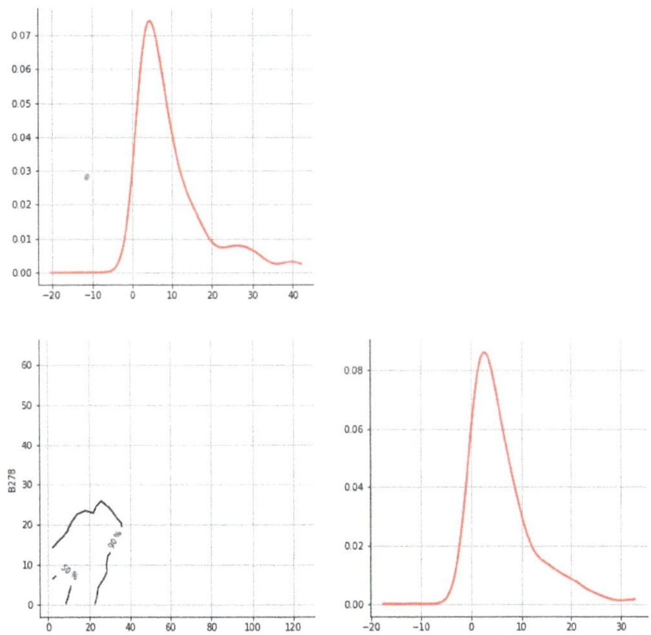

Figure A2. Bivariate HDR boxplots of B228 and B278 rain gauge stations.

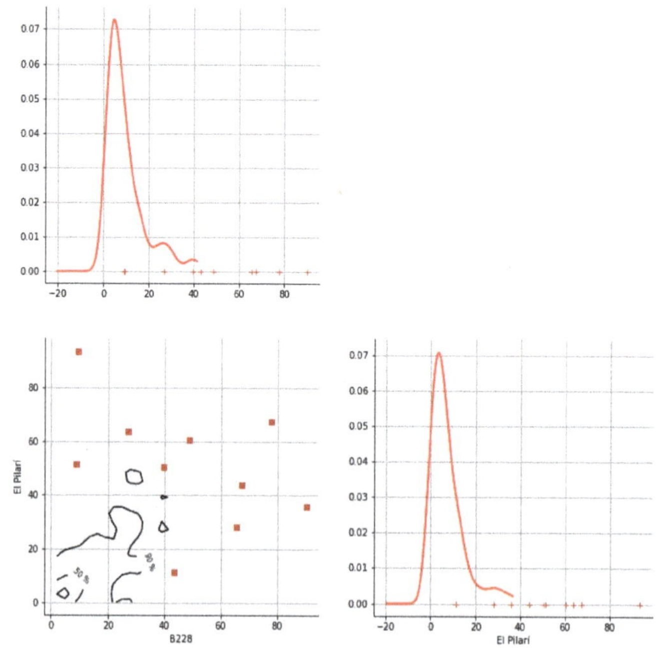

Figure A3. Bivariate HDR boxplots of B228 and El Pil·lari rain gauge stations.

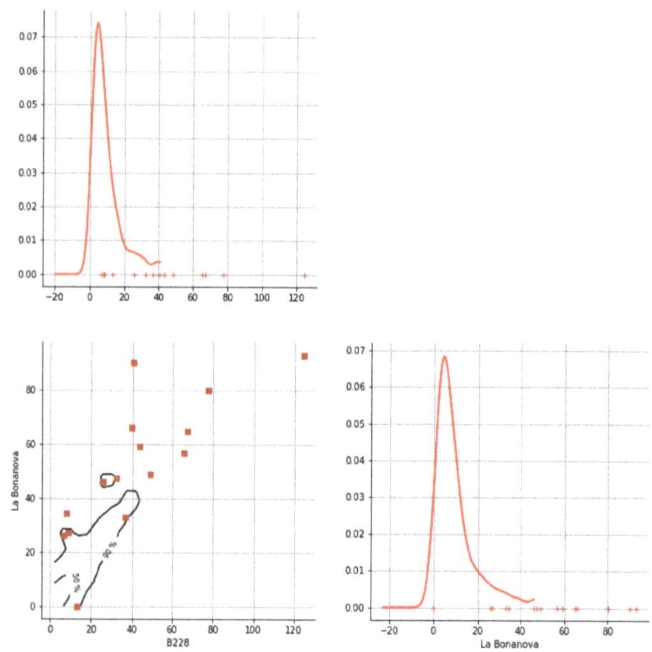

Figure A4. Bivariate HDR boxplots of B228 and La Bonanova rain gauge stations.

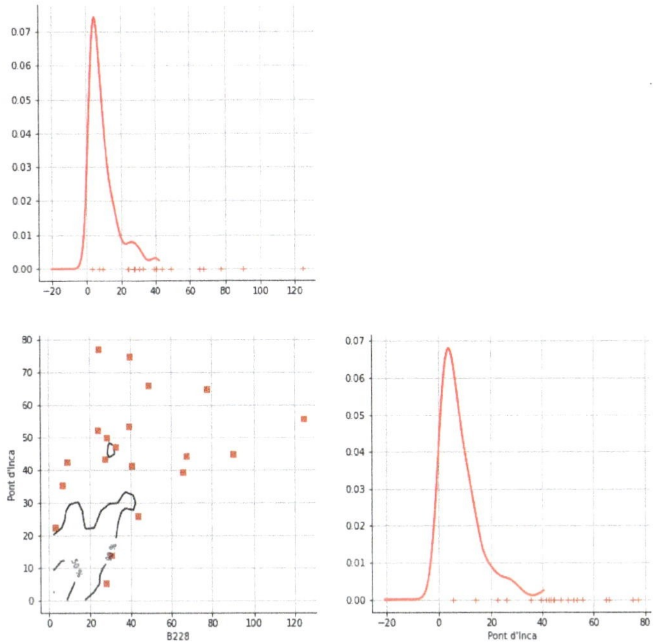

Figure A5. Bivariate HDR boxplots of B228 and Pont d'Inca rain gauge stations.

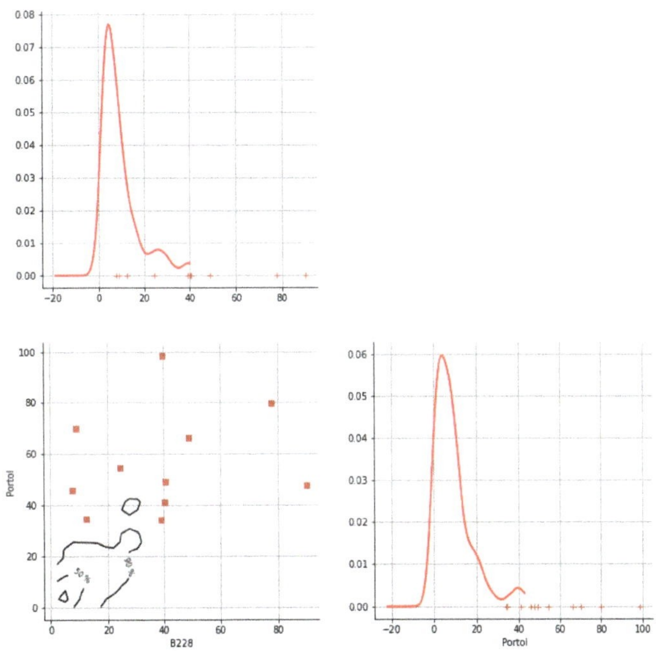

Figure A6. Bivariate HDR boxplots of B228 and Pòrtol rain gauge stations.

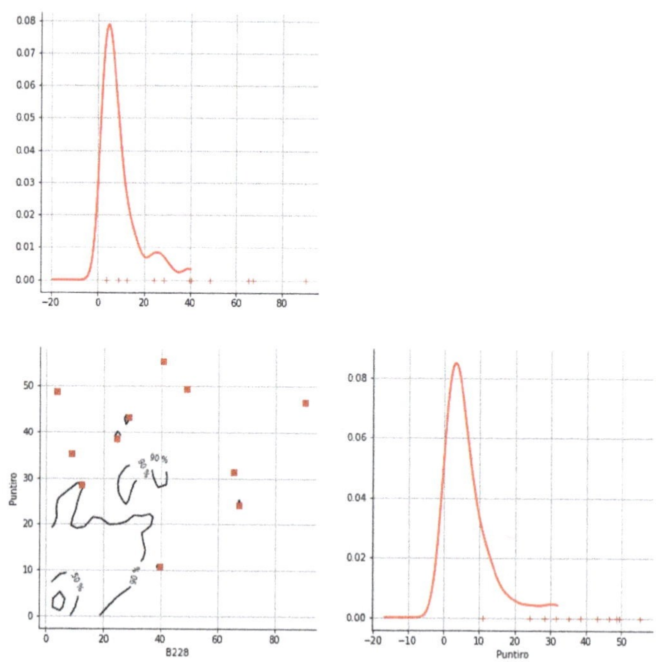

Figure A7. Bivariate HDR boxplots of B228 and Puntiró rain gauge stations.

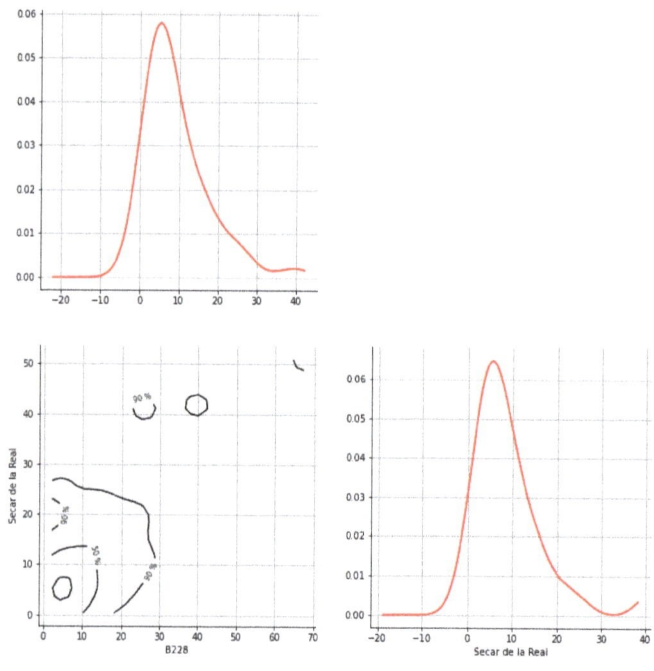

Figure A8. Bivariate HDR boxplots of B228 and Secar de la Real rain gauge stations.

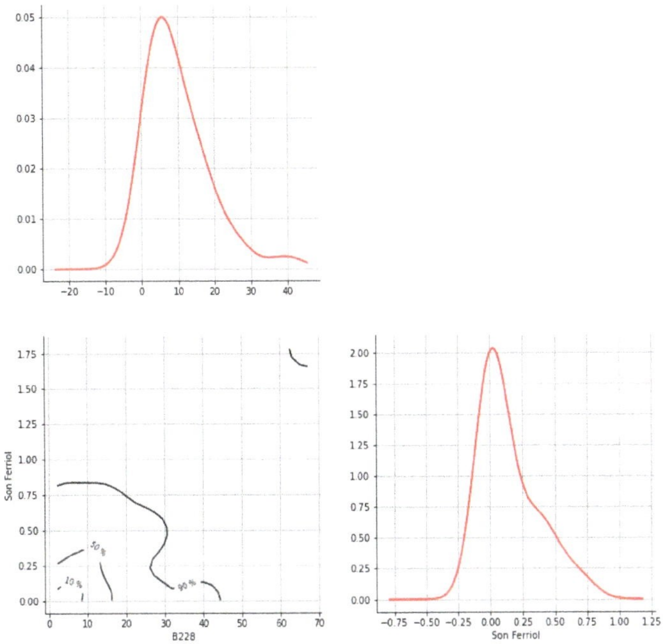

Figure A9. Bivariate HDR boxplots of B228 and Son Ferriol rain gauge stations.

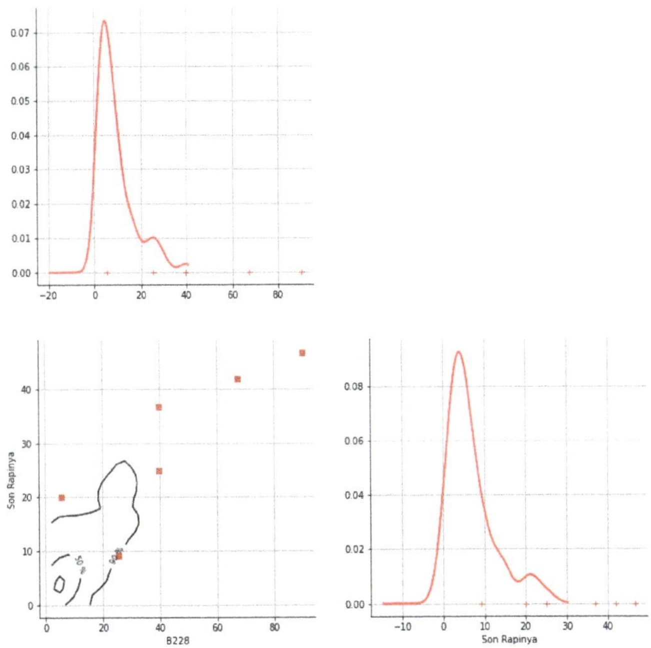

Figure A10. Bivariate HDR boxplots of B228 and Son Rapinya rain gauge stations.

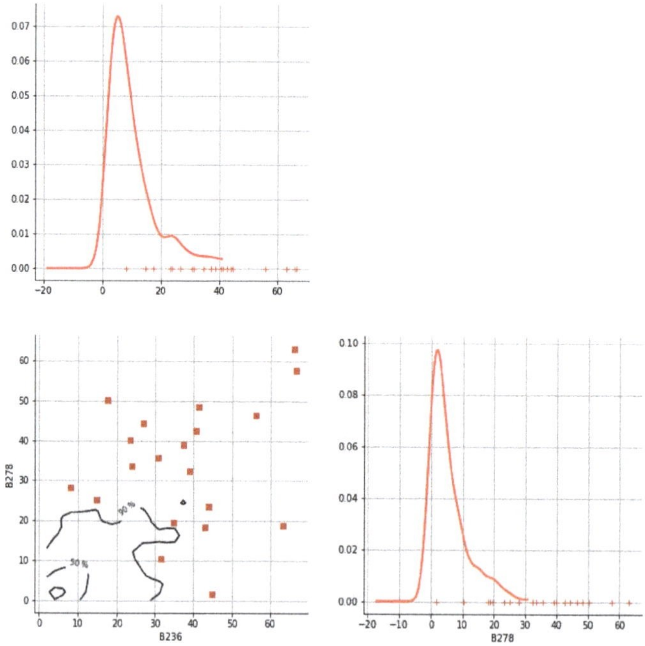

Figure A11. Bivariate HDR boxplots of B236 and B278 rain gauge stations.

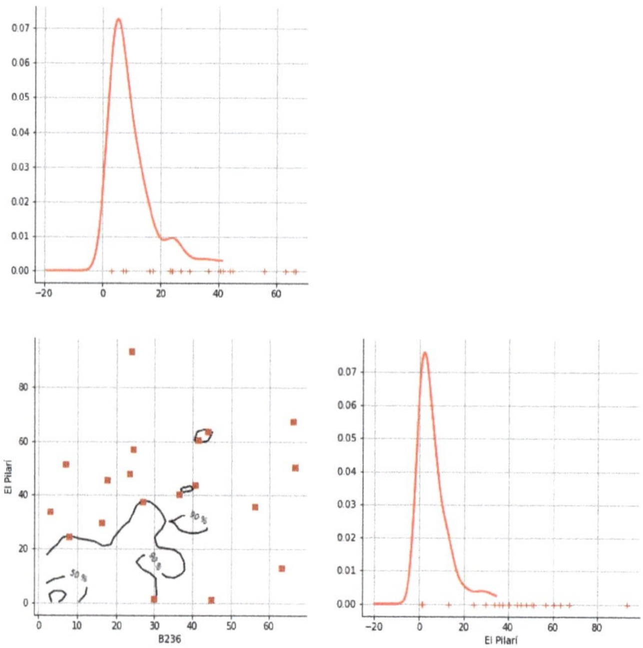

Figure A12. Bivariate HDR boxplots of B236 and El Pil·lari rain gauge stations.

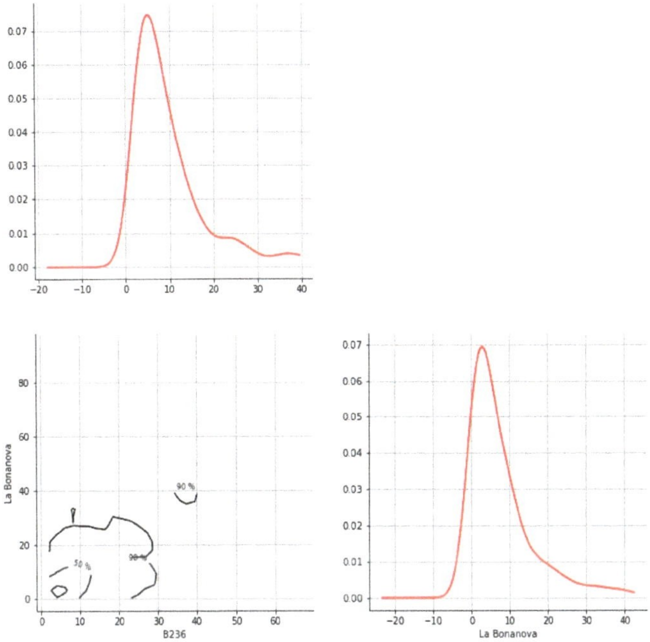

Figure A13. Bivariate HDR boxplots of B236 and La Bonanova rain gauge stations.

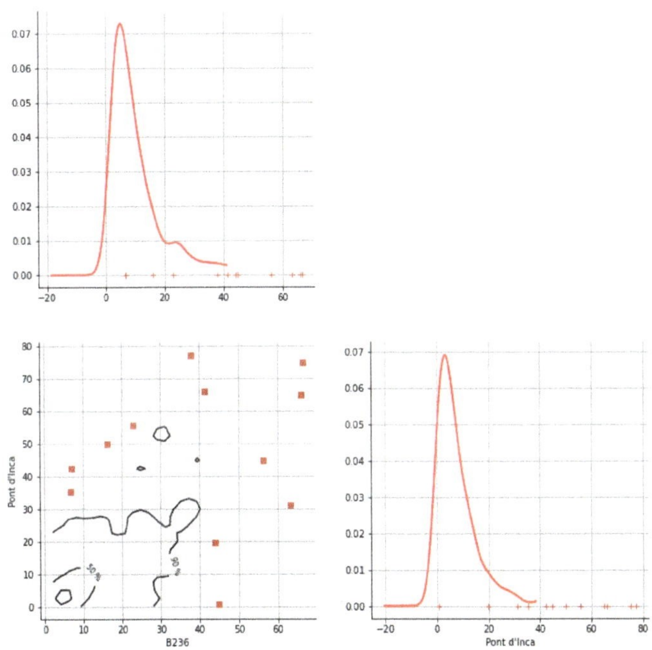

Figure A14. Bivariate HDR boxplots of B236 and Pont d'Inca rain gauge stations.

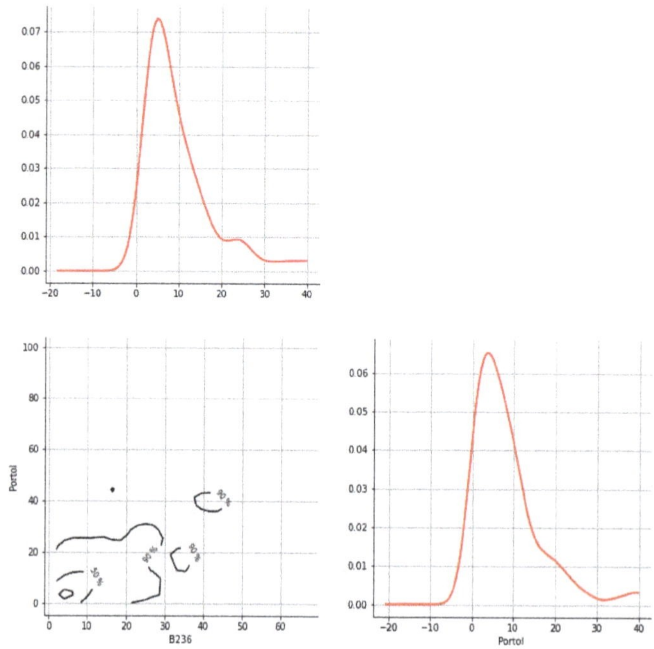

Figure A15. Bivariate HDR boxplots of B236 and Pòrtol rain gauge stations.

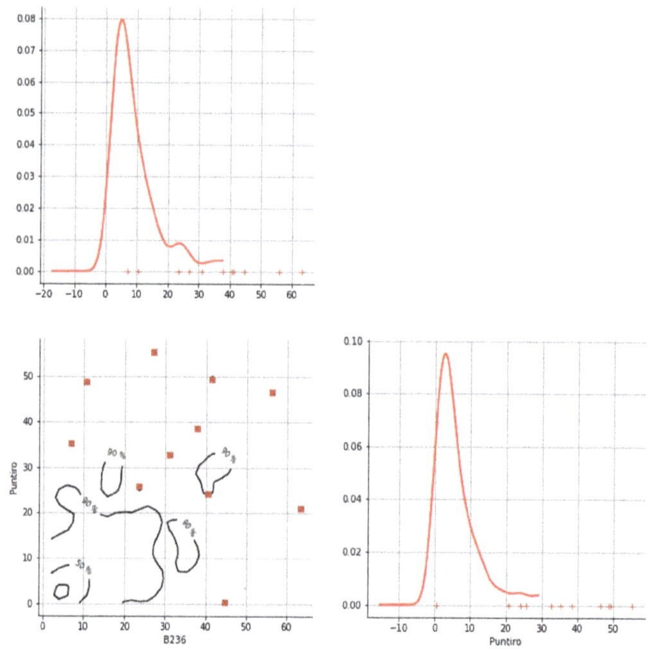

Figure A16. Bivariate HDR boxplots of B236 and Puntiró rain gauge stations.

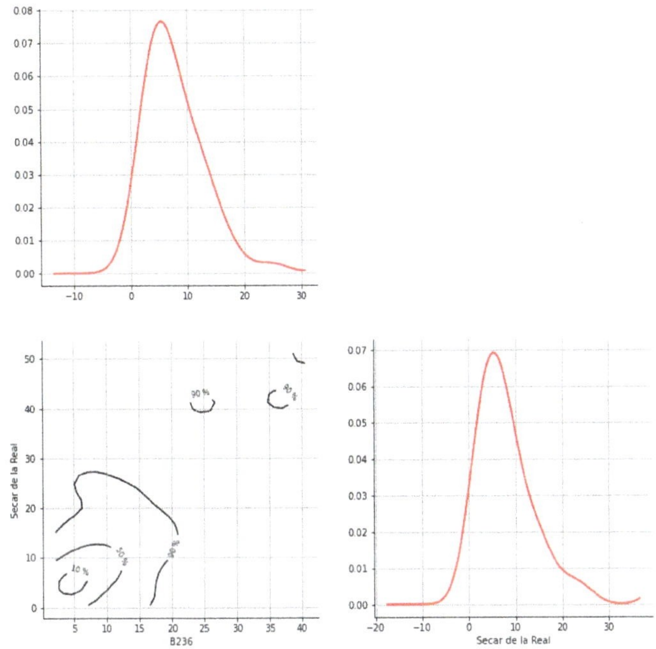

Figure A17. Bivariate HDR boxplots of B236 and Secar de la Real rain gauge stations.

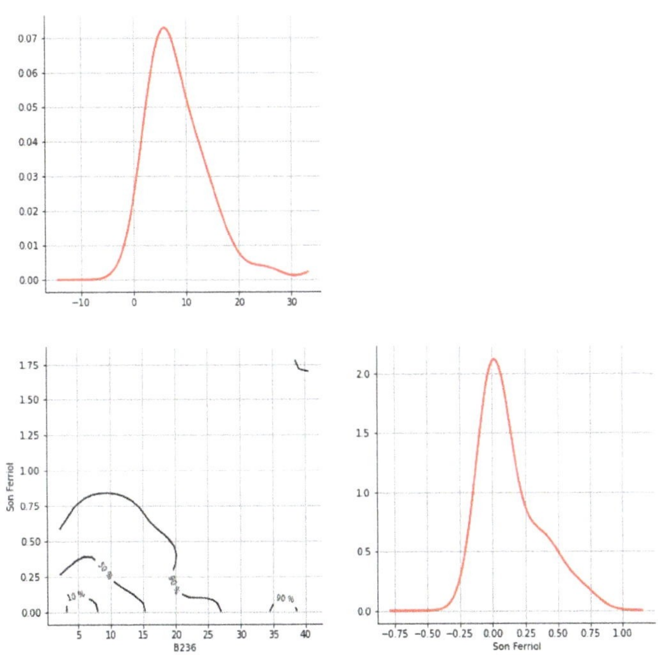

Figure A18. Bivariate HDR boxplots of B236 and Son Ferriol rain gauge stations.

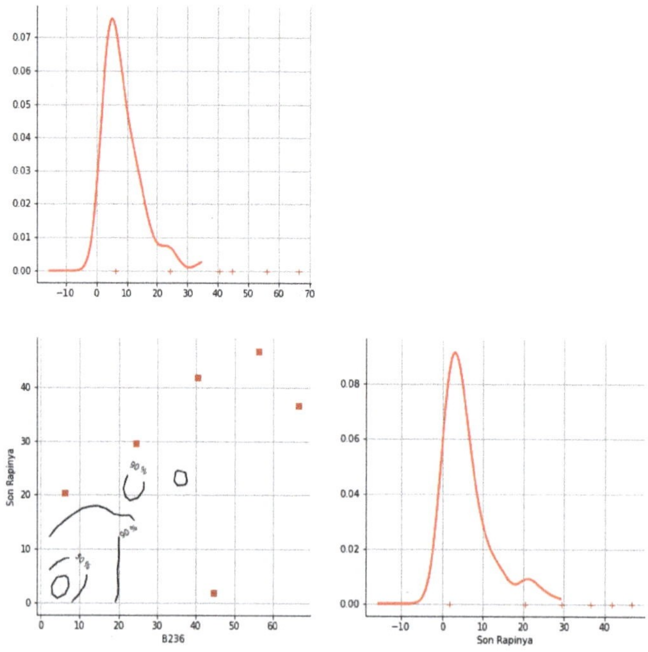

Figure A19. Bivariate HDR boxplots of B236 and Son Rapinya rain gauge stations.

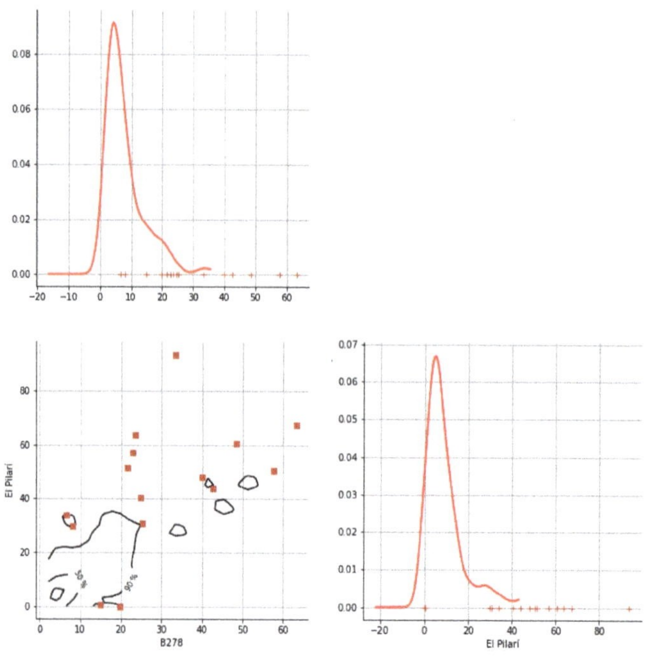

Figure A20. Bivariate HDR boxplots of B278 and El Pil·lari rain gauge stations.

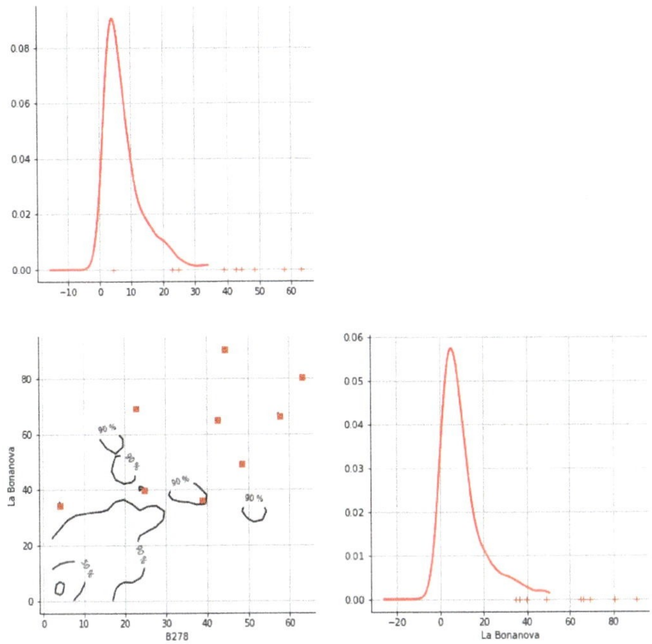

Figure A21. Bivariate HDR boxplots of B278 and La Bonanova rain gauge stations.

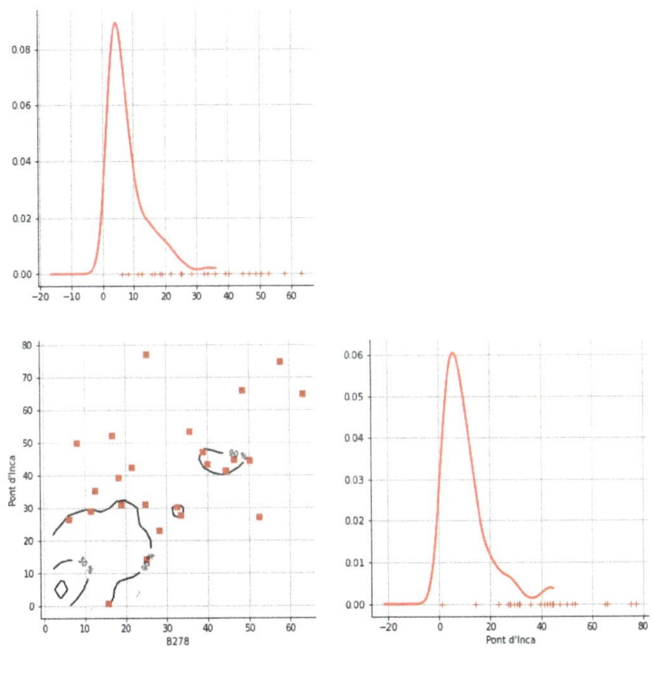

Figure A22. Bivariate HDR boxplots of B278 and Pont d'Inca rain gauge stations.

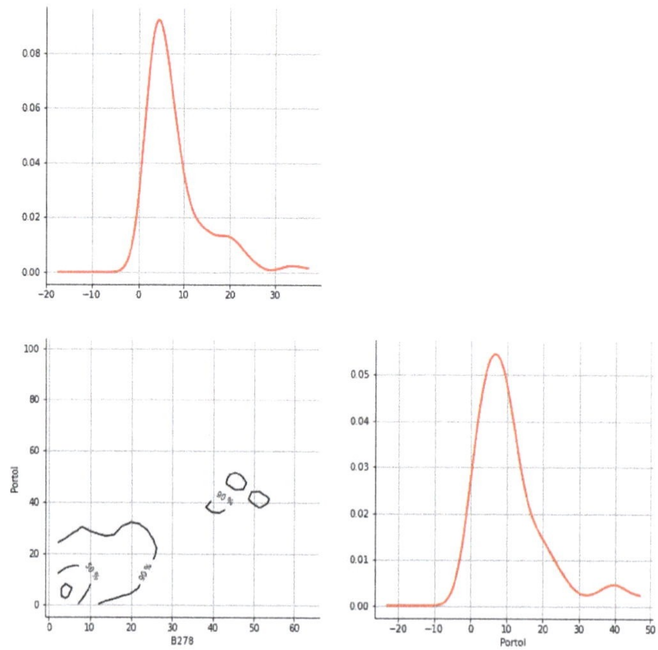

Figure A23. Bivariate HDR boxplots of B278 and Pòrtol rain gauge stations.

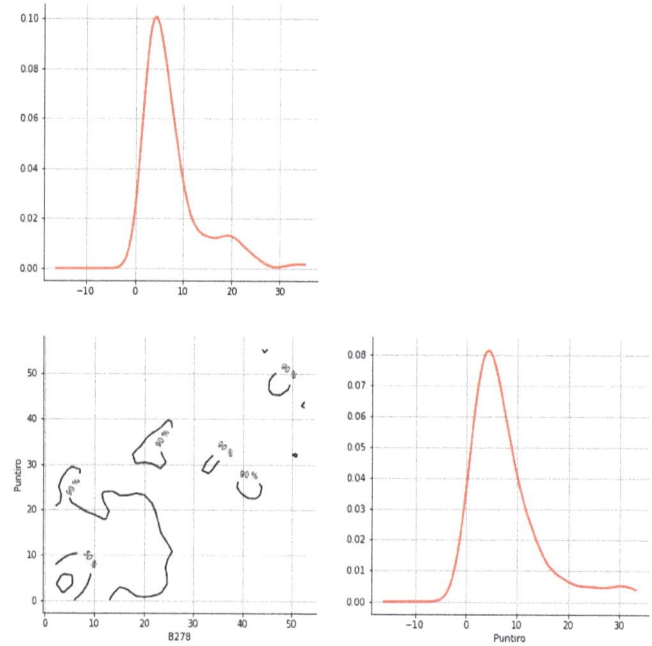

Figure A24. Bivariate HDR boxplots of B278 and Puntiró rain gauge stations.

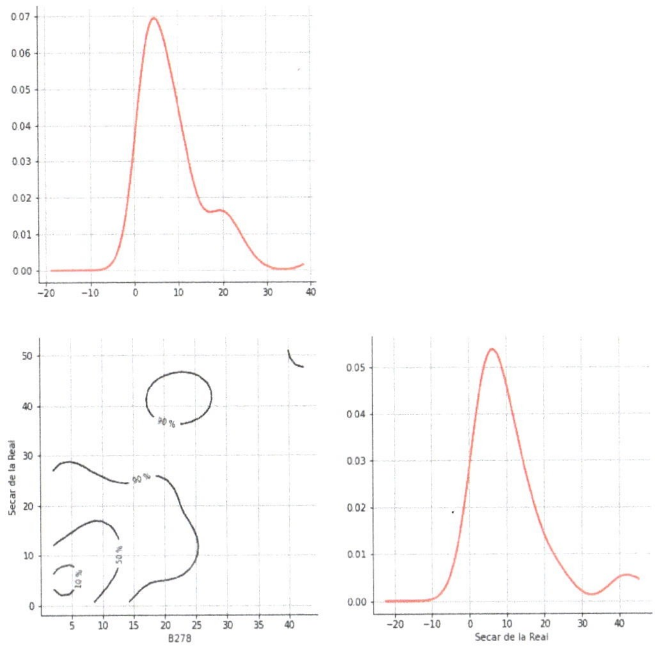

Figure A25. Bivariate HDR boxplots of B278 and Secar de la Real rain gauge stations.

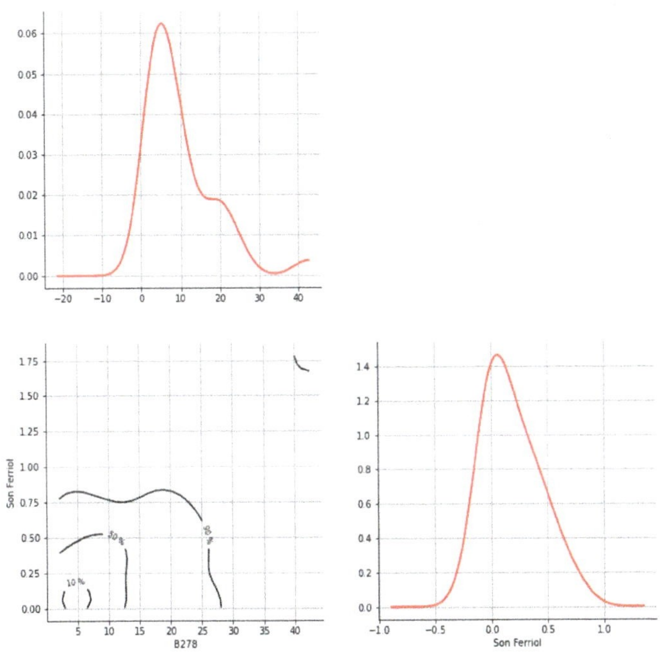

Figure A26. Bivariate HDR boxplots of B278 and Son Ferriol rain gauge stations.

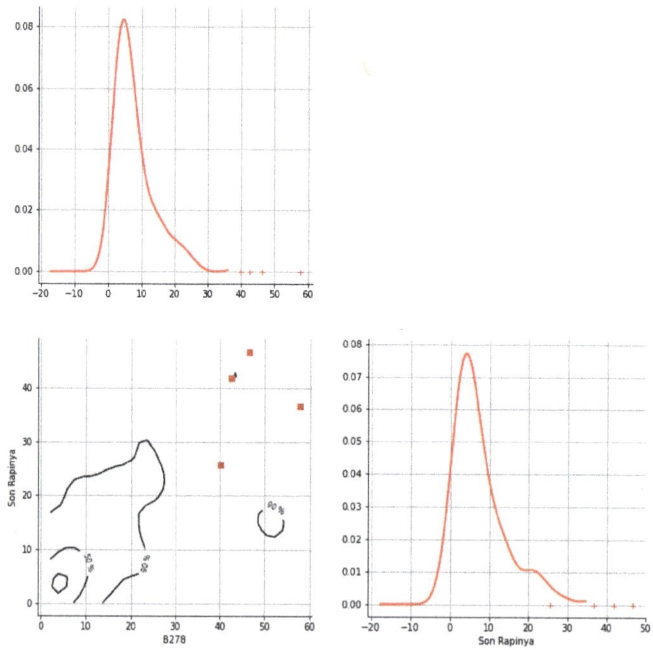

Figure A27. Bivariate HDR boxplots of B278 and Son Rapinya rain gauge stations.

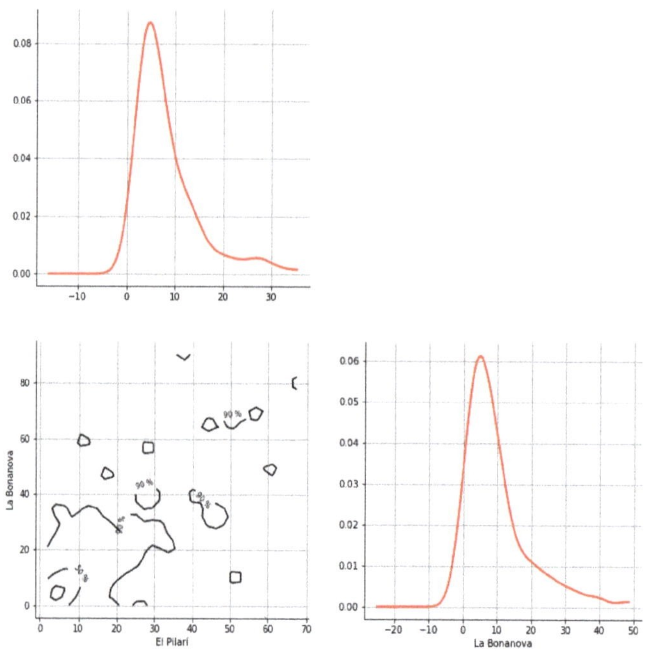

Figure A28. Bivariate HDR boxplots of El Pil·lari and La Bonanova rain gauge stations.

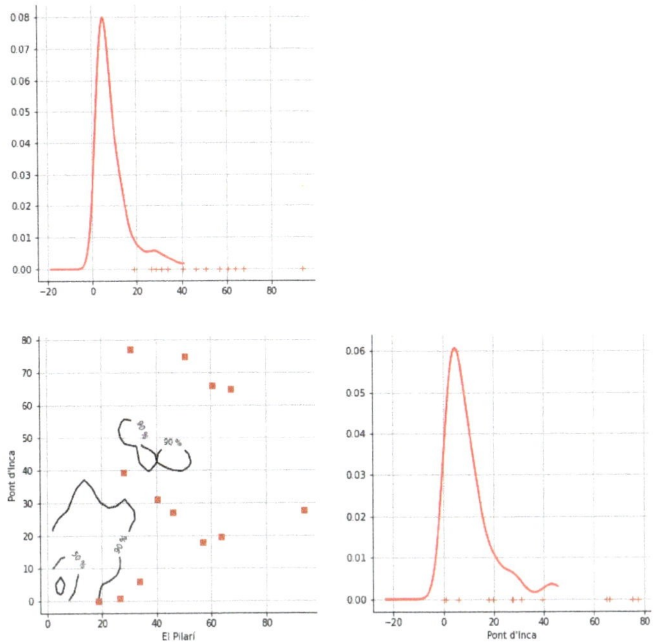

Figure A29. Bivariate HDR boxplots of El Pil·lari and Pont d'Inca rain gauge stations.

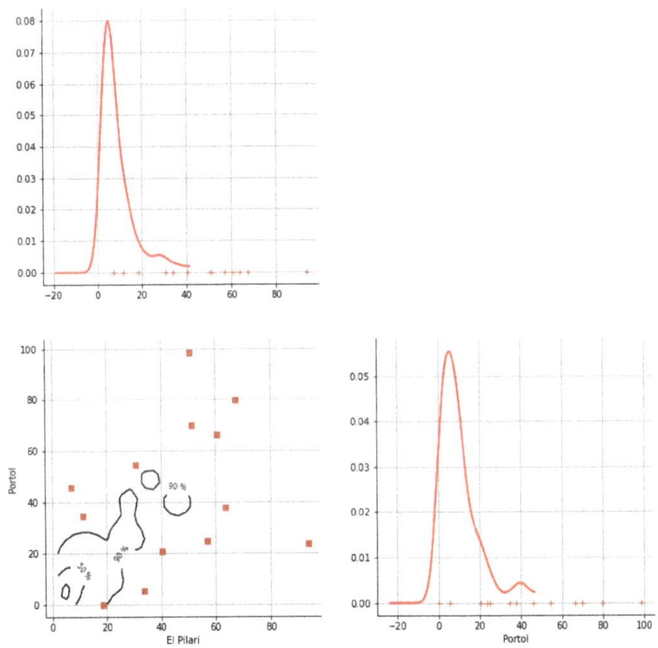

Figure A30. Bivariate HDR boxplots of El Pil·lari and Pòrtol rain gauge stations.

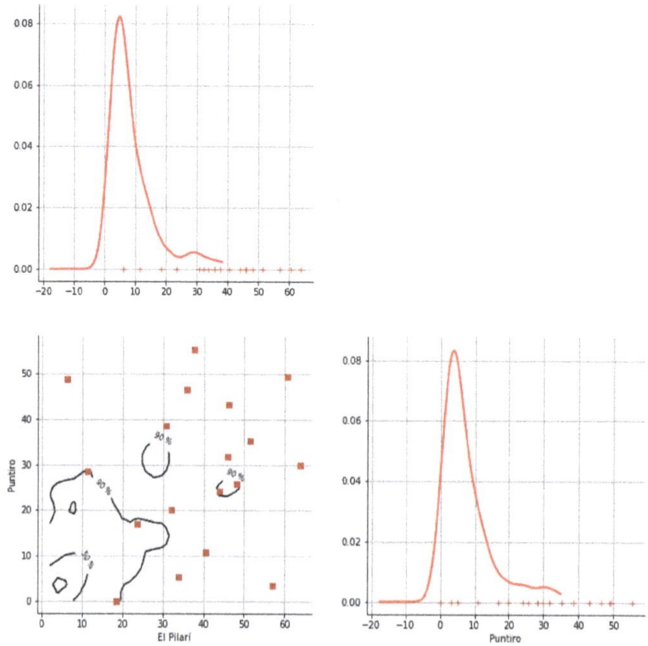

Figure A31. Bivariate HDR boxplots of El Pil·lari and Puntiró rain gauge stations.

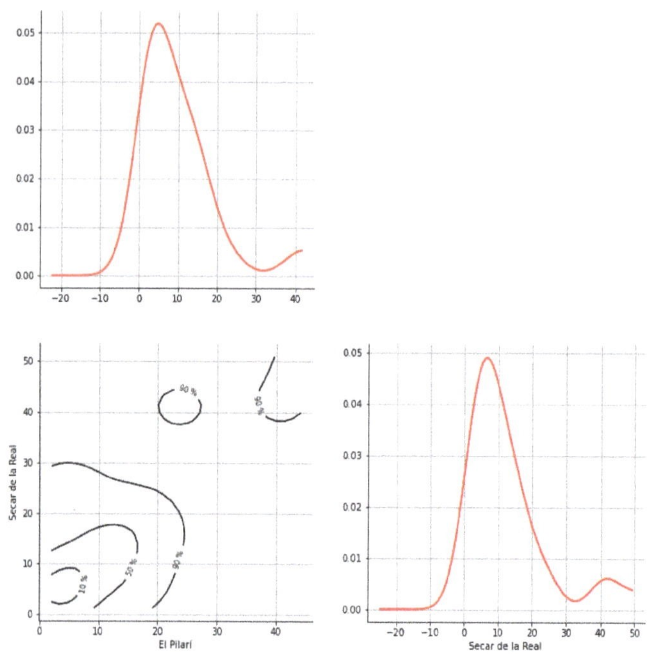

Figure A32. Bivariate HDR boxplots of El Pil·lari and Secar de la Real rain gauge stations.

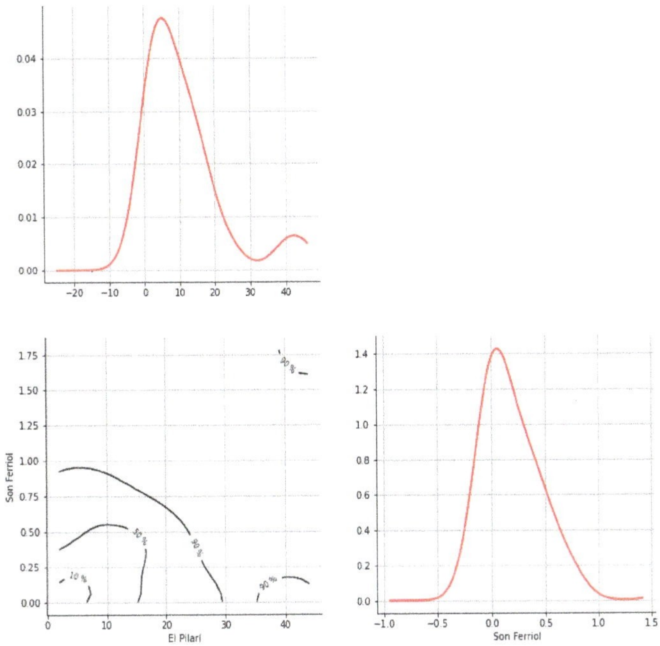

Figure A33. Bivariate HDR boxplots of El Pil·lari and Son Ferriol rain gauge stations.

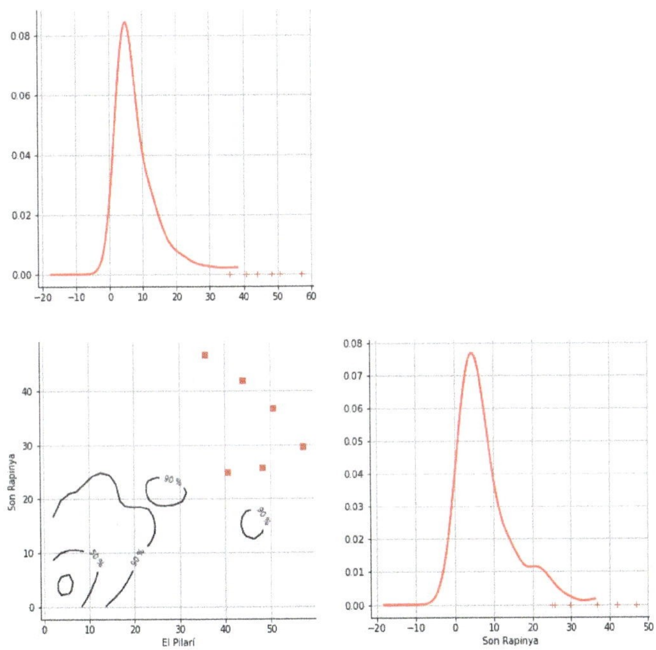

Figure A34. Bivariate HDR boxplots of El Pil·lari and Son Rapinya rain gauge stations.

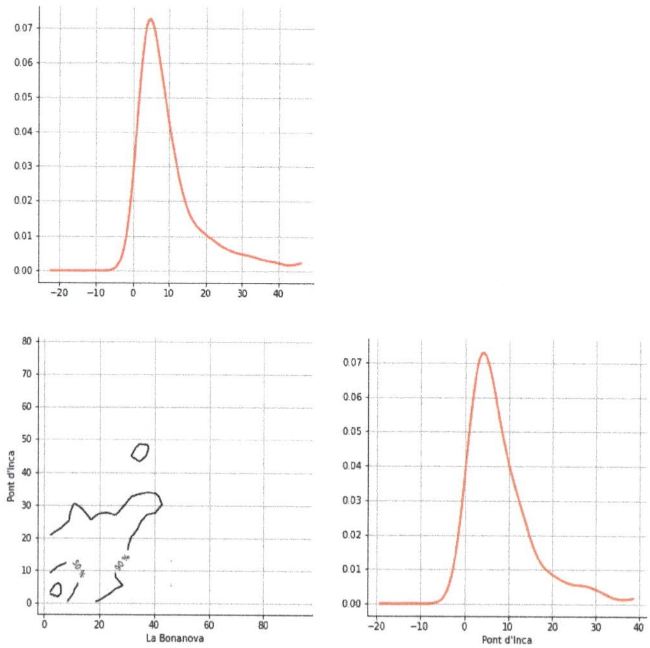

Figure A35. Bivariate HDR boxplots of La Bonanova and Pont d'Inca rain gauge stations.

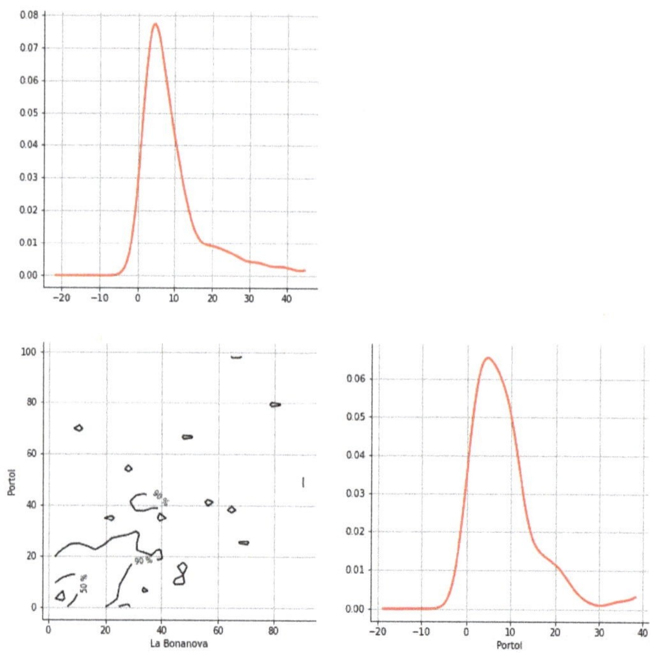

Figure A36. Bivariate HDR boxplots of La Bonanova and Pòrtol rain gauge stations.

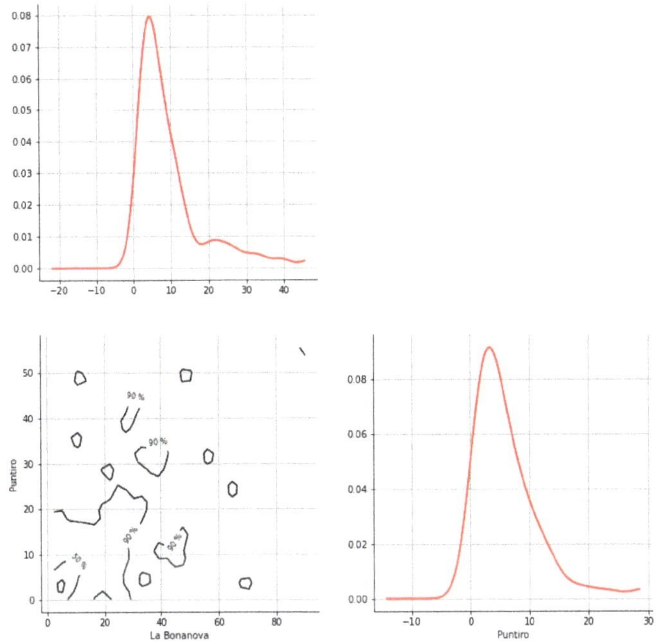

Figure A37. Bivariate HDR boxplots of La Bonanova and Puntiró rain gauge stations.

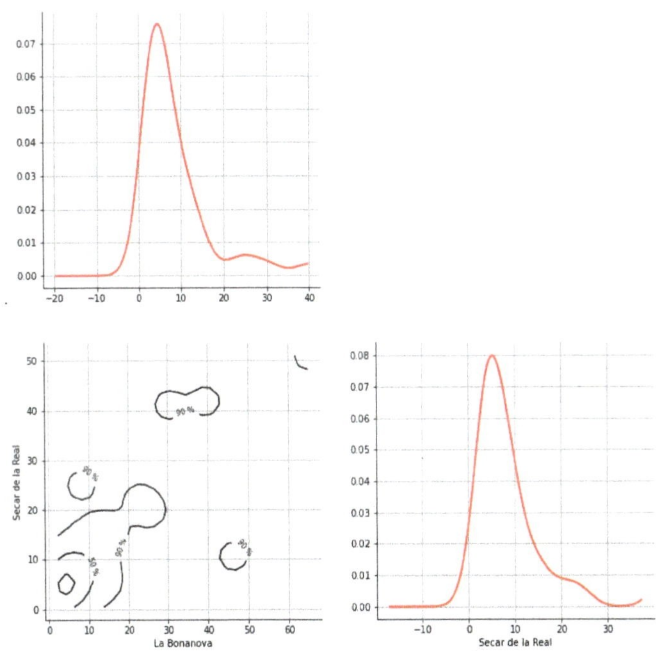

Figure A38. Bivariate HDR boxplots of La Bonanova and Secar de la Real rain gauge stations.

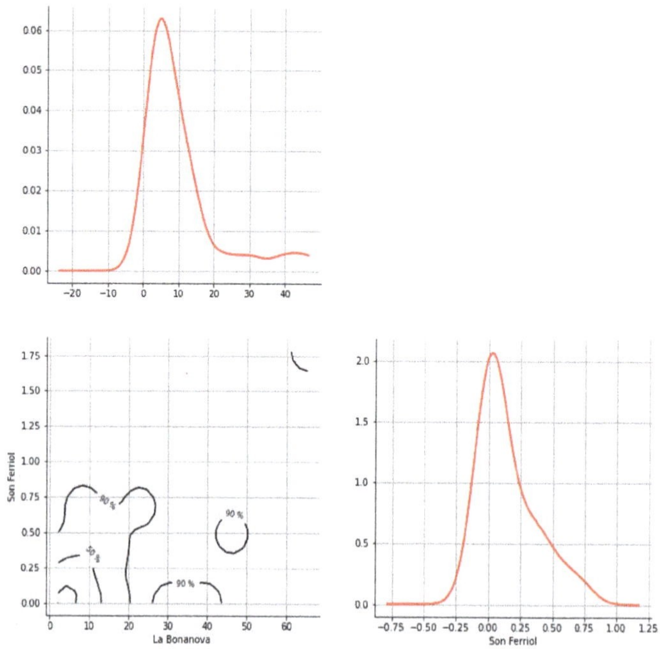

Figure A39. Bivariate HDR boxplots of La Bonanova and Son Ferriol rain gauge stations.

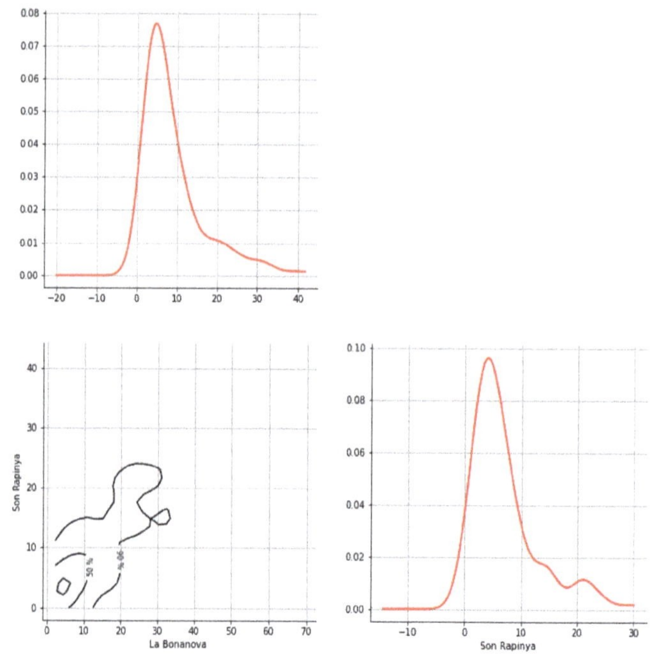

Figure A40. Bivariate HDR boxplots of La Bonanova and Son Rapinya rain gauge stations.

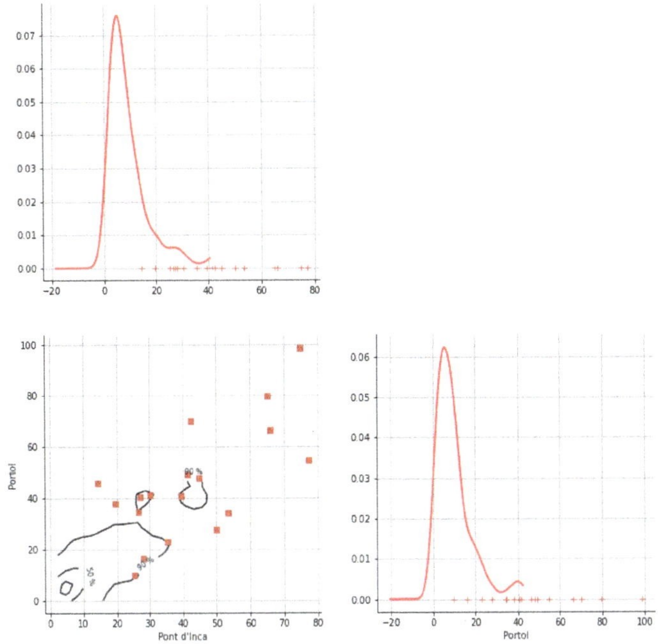

Figure A41. Bivariate HDR boxplots of Pont d'Inca and Pòrtol rain gauge stations.

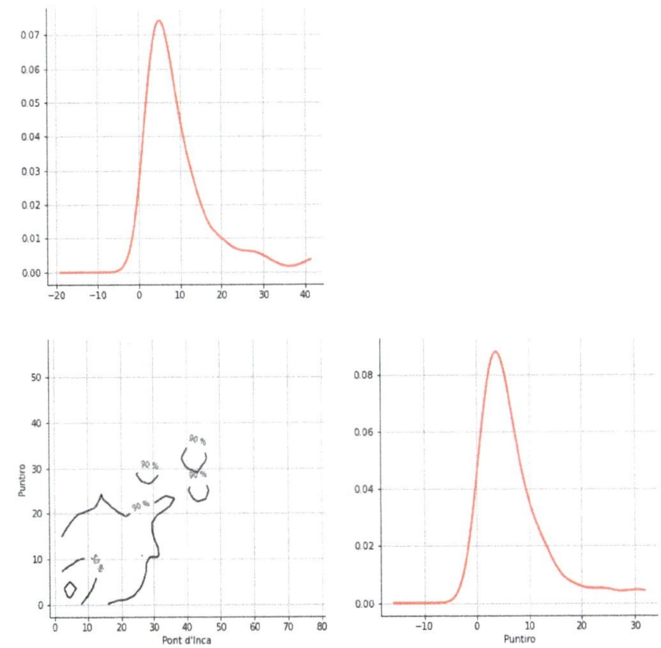

Figure A42. Bivariate HDR boxplots of Pont d'Inca and Puntiró rain gauge stations.

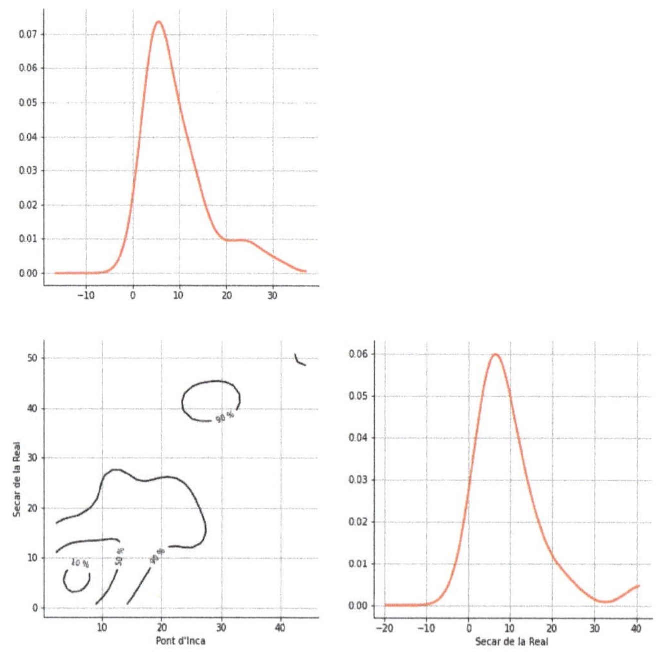

Figure A43. Bivariate HDR boxplots of Pont d'Inca and Secar de la Real rain gauge stations.

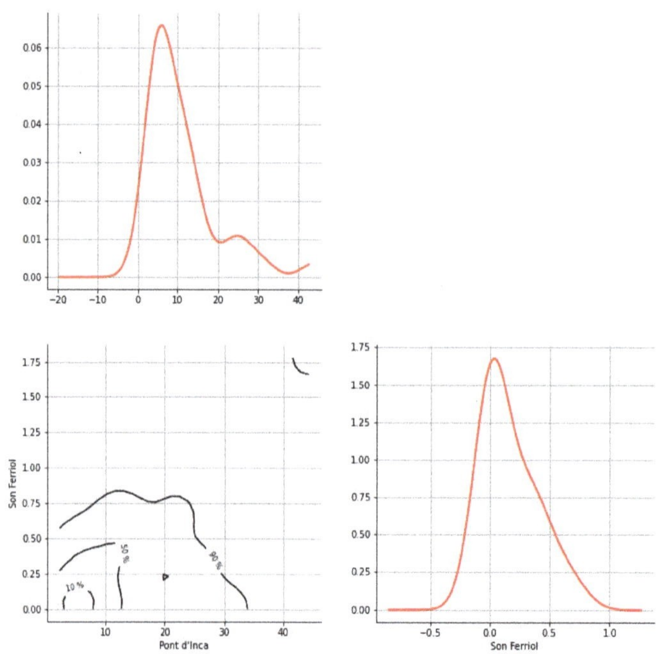

Figure A44. Bivariate HDR boxplots of Pont d'Inca and Son Ferriol rain gauge stations.

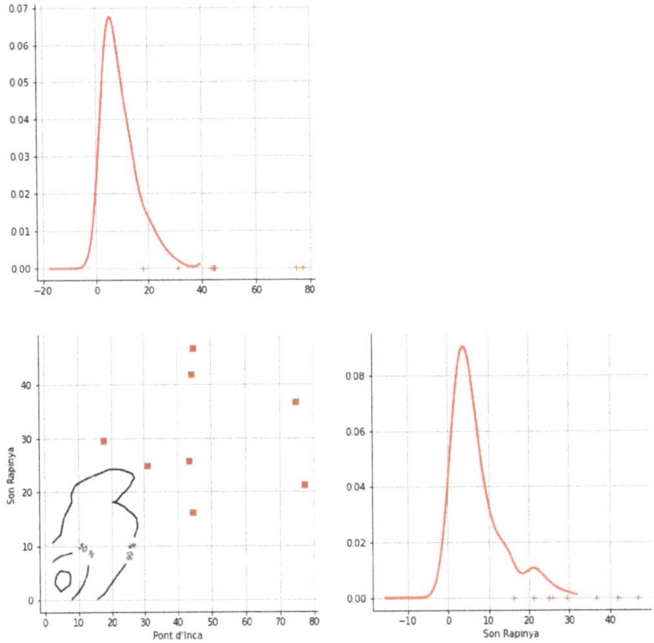

Figure A45. Bivariate HDR boxplots of Pont d'Inca and Son Rapinya rain gauge stations.

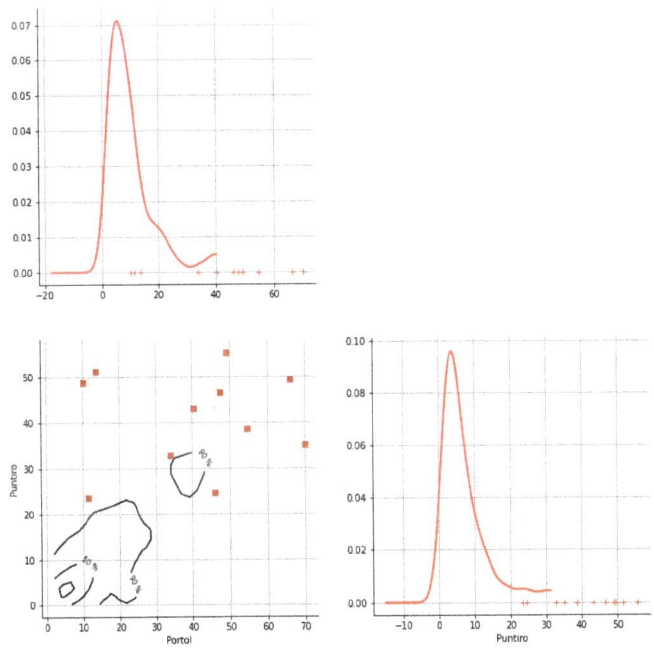

Figure A46. Bivariate HDR boxplots of Pòrtol and Puntiró rain gauge stations.

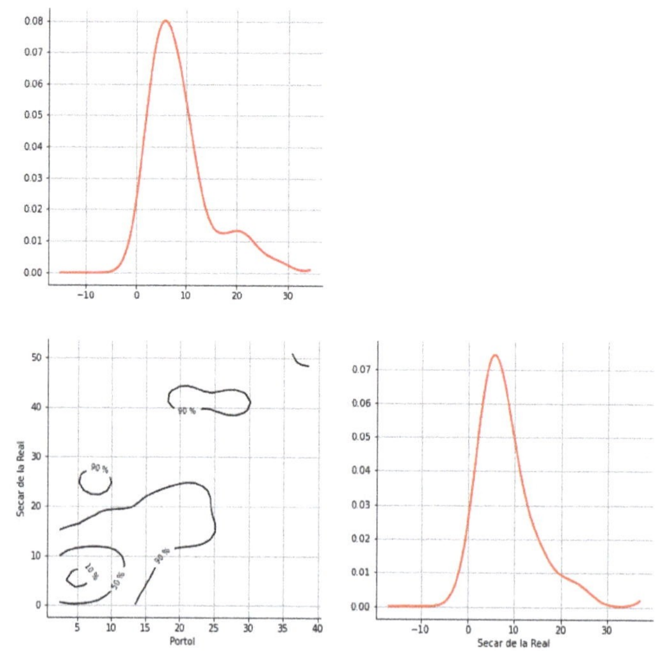

Figure A47. Bivariate HDR boxplots of Pòrtol and Secar de la Real rain gauge stations.

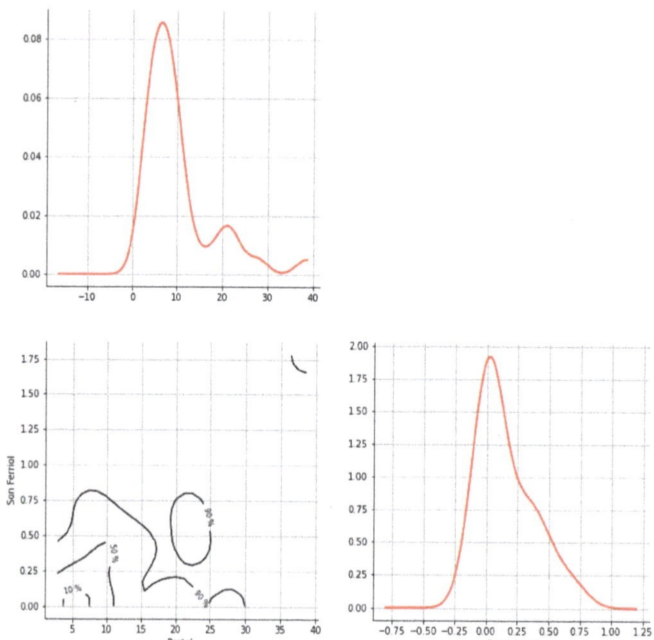

Figure A48. Bivariate HDR boxplots of Pòrtol and Son Ferriol rain gauge stations.

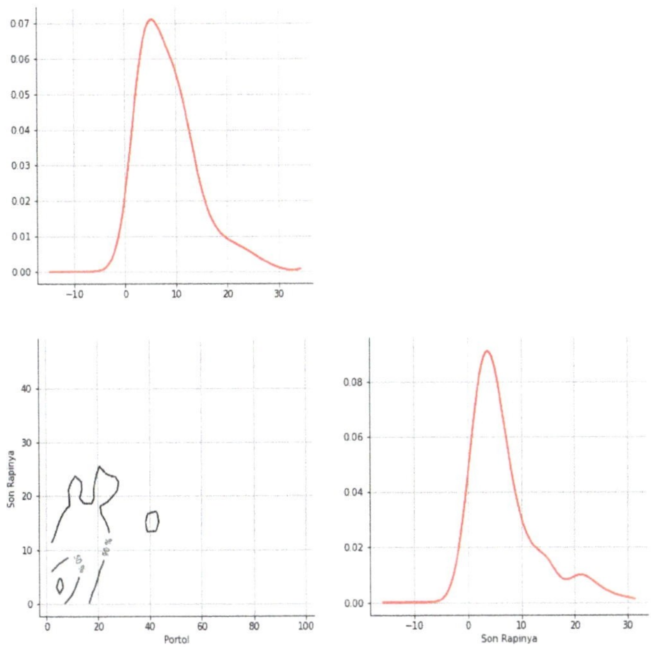

Figure A49. Bivariate HDR boxplots of Pòrtol and Son Rapinya rain gauge stations.

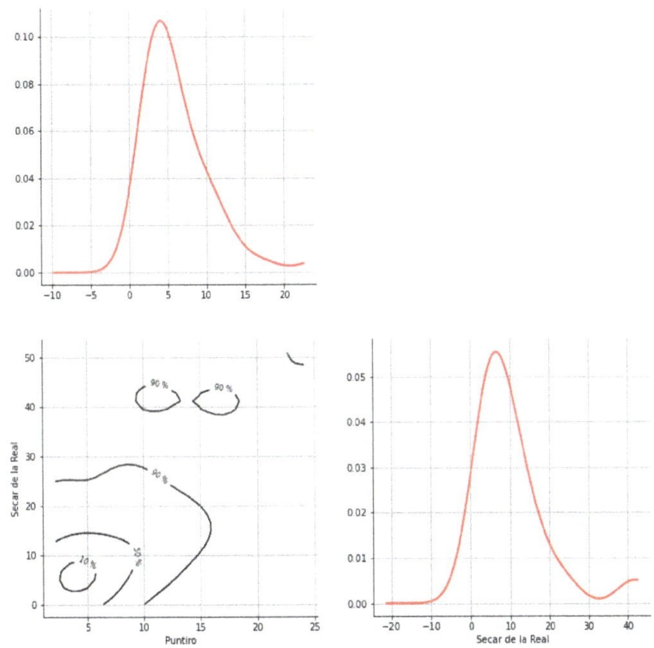

Figure A50. Bivariate HDR boxplots of Puntiró and Secar de la Real rain gauge stations.

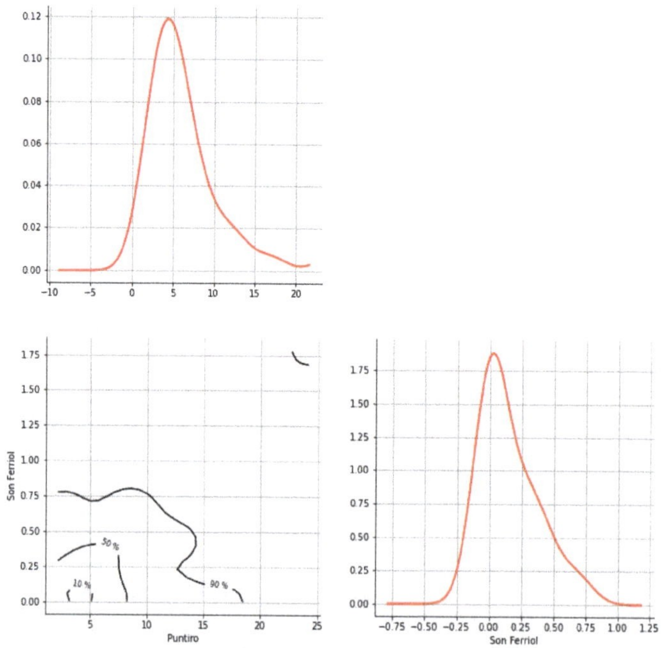

Figure A51. Bivariate HDR boxplots of Puntiró and Son Ferriol rain gauge stations.

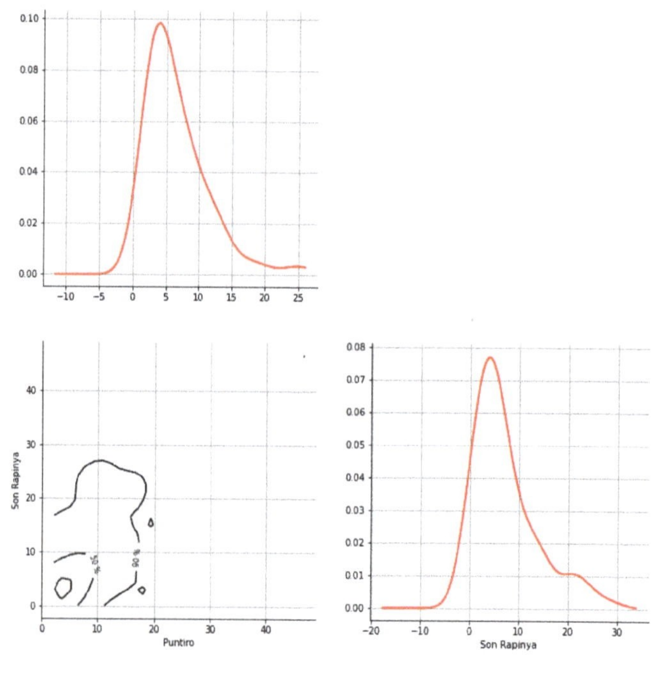

Figure A52. Bivariate HDR boxplots of Puntiró and Son Rapinya rain gauge stations.

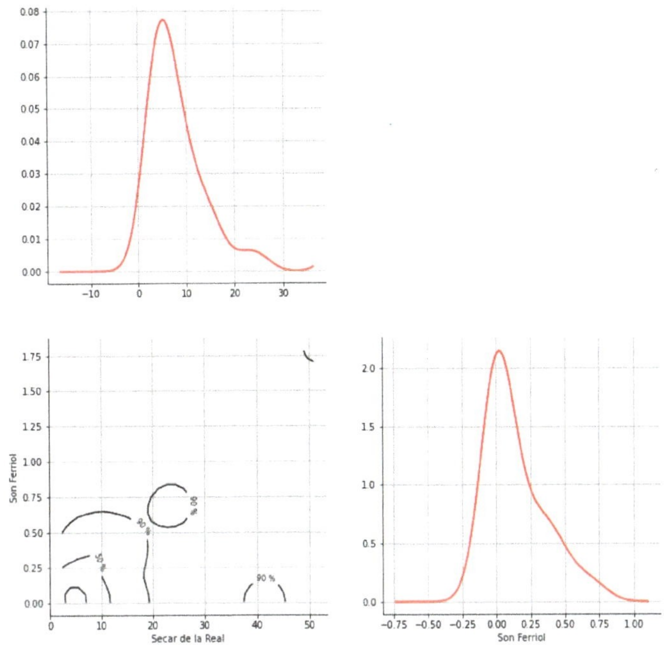

Figure A53. Bivariate HDR boxplots of Secar de la Real and Son Ferriol rain gauge stations.

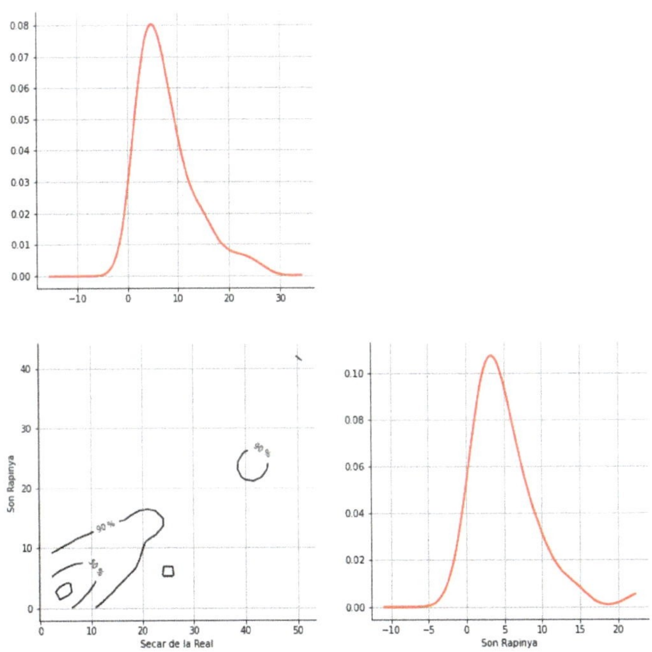

Figure A54. Bivariate HDR boxplots of Secar de la Real and Son Rapinya rain gauge stations.

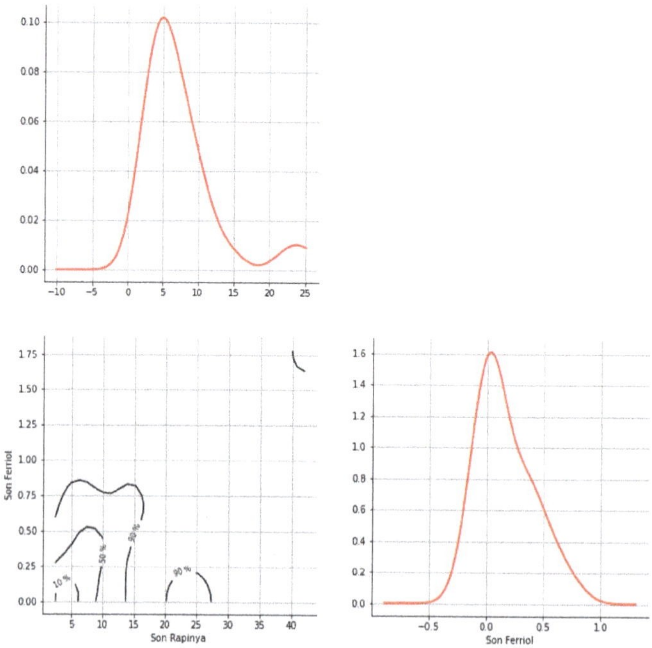

Figure A55. Bivariate HDR boxplots of Son Ferriol and Son Rapinya rain gauge stations.

Appendix B. Detection of Occurrence of Rainfall Performance

Table A1. Detection of occurrence of rainfall performance.

Combinations	POD	FAR	CSI	BS
El Pil·lari - La Bonanova - Pòrtol - Son Rapinya - Pont d'Inca - Puntiró - Secar de la Real	0.94	0.18	0.79	1.15
El Pil·lari - La Bonanova - Pòrtol - Son Rapinya - Pont d'Inca - Puntiró - B228	0.95	0.18	0.79	1.16
El Pil·lari - La Bonanova - Pòrtol - Son Rapinya - Pont d'Inca - Puntiró - B236	0.94	0.18	0.78	1.15
El Pil·lari - La Bonanova - Pòrtol - Son Rapinya - Pont d'Inca - Puntiró - B278	0.95	0.19	0.78	1.16
El Pil·lari - La Bonanova - Pòrtol - Son Rapinya - Pont d'Inca - Secar de la Real - B228	0.92	0.17	0.78	1.12
El Pil·lari - La Bonanova - Pòrtol - Son Rapinya - Pont d'Inca - Secar de la Real - B236	0.92	0.18	0.77	1.13
El Pil·lari - La Bonanova - Pòrtol - Son Rapinya - Pont d'Inca - Secar de la Real - B278	0.93	0.19	0.76	1.15
El Pil·lari - La Bonanova - Pòrtol - Son Rapinya - Pont d'Inca - B228 - B236	0.93	0.19	0.77	1.14
El Pil·lari - La Bonanova - Pòrtol - Son Rapinya - Pont d'Inca - B228 - B278	0.93	0.18	0.77	1.14
El Pil·lari - La Bonanova - Pòrtol - Son Rapinya - Pont d'Inca - B236 - B278	0.92	0.21	0.74	1.16
El Pil·lari - La Bonanova - Pòrtol - Son Rapinya - Puntiró - Secar de la Real - B228	0.94	0.15	0.81	1.10
El Pil·lari - La Bonanova - Pòrtol - Son Rapinya - Puntiró - Secar de la Real - B236	0.94	0.17	0.79	1.13
El Pil·lari - La Bonanova - Pòrtol - Son Rapinya - Puntiró - Secar de la Real - B278	0.95	0.17	0.79	1.15
El Pil·lari - La Bonanova - Pòrtol - Son Rapinya - Puntiró - B228 - B236	0.95	0.17	0.80	1.14
El Pil·lari - La Bonanova - Pòrtol - Son Rapinya - Puntiró - B228 - B278	0.95	0.17	0.80	1.14
El Pil·lari - La Bonanova - Pòrtol - Son Rapinya - Puntiró - B236 - B278	0.94	0.19	0.77	1.17
El Pil·lari - La Bonanova - Pòrtol - Son Rapinya - Secar de la Real - B228 - B236	0.93	0.15	0.80	1.10
El Pil·lari - La Bonanova - Pòrtol - Son Rapinya - Secar de la Real - B228 - B278	0.93	0.17	0.79	1.12
El Pil·lari - La Bonanova - Pòrtol - Son Rapinya - Secar de la Real - B236 - B278	0.93	0.19	0.77	1.14
El Pil·lari - La Bonanova - Pòrtol - Son Rapinya - B228 - B236 - B278	0.93	0.19	0.77	1.14
El Pil·lari - La Bonanova - Pòrtol - Pont d'Inca - Puntiró - Secar de la Real - B228	0.95	0.17	0.80	1.14
El Pil·lari - La Bonanova - Pòrtol - Pont d'Inca - Puntiró - Secar de la Real - B236	0.95	0.19	0.77	1.17
El Pil·lari - La Bonanova - Pòrtol - Pont d'Inca - Puntiró - Secar de la Real - B278	0.95	0.20	0.77	1.18
El Pil·lari - La Bonanova - Pòrtol - Pont d'Inca - Puntiró - B228 - B236	0.94	0.17	0.79	1.14
El Pil·lari - La Bonanova - Pòrtol - Pont d'Inca - Puntiró - B228 - B278	0.95	0.17	0.79	1.14
El Pil·lari - La Bonanova - Pòrtol - Pont d'Inca - Puntiró - B236 - B278	0.93	0.21	0.74	1.18
El Pil·lari - La Bonanova - Pòrtol - Pont d'Inca - Secar de la Real - B228 - B236	0.93	0.17	0.78	1.12
El Pil·lari - La Bonanova - Pòrtol - Pont d'Inca - Secar de la Real - B228 - B278	0.93	0.19	0.77	1.14
El Pil·lari - La Bonanova - Pòrtol - Pont d'Inca - Secar de la Real - B236 - B278	0.92	0.21	0.74	1.17
El Pil·lari - La Bonanova - Pòrtol - Pont d'Inca - B228 - B236 - B278	0.92	0.20	0.75	1.15
El Pil·lari - La Bonanova - Pòrtol - Puntiró - Secar de la Real - B228 - B236	0.95	0.15	0.81	1.13

Table A1. *Cont.*

Combinations	POD	FAR	CSI	BS
El Pil·lari - La Bonanova - Pòrtol - Puntiró - Secar de la Real - B228 - B278	0.95	0.16	0.80	1.13
El Pil·lari - La Bonanova - Pòrtol - Puntiró - Secar de la Real - B236 - B278	0.93	0.19	0.76	1.15
El Pil·lari - La Bonanova - Pòrtol - Puntiró - B228 - B236 - B278	0.94	0.18	0.78	1.15
El Pil·lari - La Bonanova - Pòrtol - Secar de la Real - B228 - B236 - B278	0.94	0.17	0.79	1.14
El Pil·lari - La Bonanova - Son Rapinya - Pont d'Inca - Puntiró - Secar de la Real - B228	0.93	0.16	0.79	1.11
El Pil·lari - La Bonanova - Son Rapinya - Pont d'Inca - Puntiró - Secar de la Real - B236	0.93	0.18	0.77	1.13
El Pil·lari - La Bonanova - Son Rapinya - Pont d'Inca - Puntiró - Secar de la Real - B278	0.93	0.19	0.77	1.14
El Pil·lari - La Bonanova - Son Rapinya - Pont d'Inca - Puntiró - B228 - B236	0.93	0.19	0.76	1.15
El Pil·lari - La Bonanova - Son Rapinya - Pont d'Inca - Puntiró - B228 - B278	0.93	0.18	0.77	1.13
El Pil·lari - La Bonanova - Son Rapinya - Pont d'Inca - Puntiró - B236 - B278	0.92	0.20	0.74	1.16
El Pil·lari - La Bonanova - Son Rapinya - Pont d'Inca - Secar de la Real - B228 - B236	0.94	0.16	0.80	1.11
El Pil·lari - La Bonanova - Son Rapinya - Pont d'Inca - Secar de la Real - B228 - B278	0.94	0.17	0.79	1.13
El Pil·lari - La Bonanova - Son Rapinya - Pont d'Inca - Secar de la Real - B236 - B278	0.93	0.19	0.76	1.15
El Pil·lari - La Bonanova - Son Rapinya - Pont d'Inca - B228 - B236 - B278	0.93	0.19	0.77	1.15
El Pil·lari - La Bonanova - Son Rapinya - Puntiró - Secar de la Real - B228 - B236	0.93	0.15	0.80	1.10
El Pil·lari - La Bonanova - Son Rapinya - Puntiró - Secar de la Real - B228 - B278	0.93	0.16	0.79	1.11
El Pil·lari - La Bonanova - Son Rapinya - Puntiró - Secar de la Real - B236 - B278	0.92	0.19	0.76	1.13
El Pil·lari - La Bonanova - Son Rapinya - Puntiró - B228 - B236 - B278	0.93	0.18	0.77	1.13
El Pil·lari - La Bonanova - Son Rapinya - Secar de la Real - B228 - B236 - B278	0.93	0.16	0.79	1.12
El Pil·lari - La Bonanova - Pont d'Inca - Puntiró - Secar de la Real - B228 - B236	0.93	0.17	0.78	1.13
El Pil·lari - La Bonanova - Pont d'Inca - Puntiró - Secar de la Real - B228 - B278	0.93	0.18	0.77	1.12
El Pil·lari - La Bonanova - Pont d'Inca - Puntiró - Secar de la Real - B236 - B278	0.92	0.21	0.74	1.16
El Pil·lari - La Bonanova - Pont d'Inca - Puntiró - B228 - B236 - B278	0.93	0.20	0.76	1.16
El Pil·lari - La Bonanova - Pont d'Inca - Secar de la Real - B228 - B236 - B278	0.94	0.17	0.78	1.13
El Pil·lari - La Bonanova - Puntiró - Secar de la Real - B228 - B236 - B278	0.94	0.17	0.79	1.12
El Pil·lari - Pòrtol - Son Rapinya - Pont d'Inca - Puntiró - Secar de la Real - B228	0.94	0.17	0.79	1.13
El Pil·lari - Pòrtol - Son Rapinya - Pont d'Inca - Puntiró - Secar de la Real - B236	0.95	0.18	0.78	1.16
El Pil·lari - Pòrtol - Son Rapinya - Pont d'Inca - Puntiró - Secar de la Real - B278	0.95	0.19	0.78	1.18
El Pil·lari - Pòrtol - Son Rapinya - Pont d'Inca - Puntiró - B228 - B236	0.93	0.18	0.77	1.14
El Pil·lari - Pòrtol - Son Rapinya - Pont d'Inca - Puntiró - B228 - B278	0.94	0.18	0.78	1.15
El Pil·lari - Pòrtol - Son Rapinya - Pont d'Inca - Puntiró - B236 - B278	0.94	0.20	0.76	1.17
El Pil·lari - Pòrtol - Son Rapinya - Pont d'Inca - Secar de la Real - B228 - B236	0.92	0.18	0.76	1.12
El Pil·lari - Pòrtol - Son Rapinya - Pont d'Inca - Secar de la Real - B228 - B278	0.92	0.19	0.76	1.14
El Pil·lari - Pòrtol - Son Rapinya - Pont d'Inca - Secar de la Real - B236 - B278	0.93	0.21	0.75	1.17
El Pil·lari - Pòrtol - Son Rapinya - Pont d'Inca - B228 - B236 - B278	0.91	0.20	0.74	1.14
El Pil·lari - Pòrtol - Son Rapinya - Puntiró - Secar de la Real - B228 - B236	0.93	0.16	0.79	1.11
El Pil·lari - Pòrtol - Son Rapinya - Puntiró - Secar de la Real - B228 - B278	0.95	0.17	0.79	1.13
El Pil·lari - Pòrtol - Son Rapinya - Puntiró - Secar de la Real - B236 - B278	0.94	0.19	0.77	1.15
El Pil·lari - Pòrtol - Son Rapinya - Puntiró - B228 - B236 - B278	0.93	0.19	0.77	1.15
El Pil·lari - Pòrtol - Son Rapinya - Secar de la Real - B228 - B236 - B278	0.93	0.19	0.76	1.14
El Pil·lari - Pòrtol - Pont d'Inca - Puntiró - Secar de la Real - B228 - B236	0.94	0.19	0.77	1.16
El Pil·lari - Pòrtol - Pont d'Inca - Puntiró - Secar de la Real - B228 - B278	0.95	0.19	0.78	1.16
El Pil·lari - Pòrtol - Pont d'Inca - Puntiró - Secar de la Real - B236 - B278	0.93	0.20	0.76	1.16
El Pil·lari - Pòrtol - Pont d'Inca - Puntiró - B228 - B236 - B278	0.92	0.20	0.75	1.16
El Pil·lari - Pòrtol - Pont d'Inca - Secar de la Real - B228 - B236 - B278	0.92	0.21	0.74	1.16
El Pil·lari - Pòrtol - Puntiró - Secar de la Real - B228 - B236 - B278	0.94	0.18	0.77	1.14
El Pil·lari - Son Rapinya - Pont d'Inca - Puntiró - Secar de la Real - B228 - B236	0.92	0.18	0.76	1.13
El Pil·lari - Son Rapinya - Pont d'Inca - Puntiró - Secar de la Real - B228 - B278	0.92	0.18	0.77	1.12
El Pil·lari - Son Rapinya - Pont d'Inca - Puntiró - Secar de la Real - B236 - B278	0.93	0.20	0.75	1.16
El Pil·lari - Son Rapinya - Pont d'Inca - Puntiró - B228 - B236 - B278	0.91	0.20	0.74	1.14
El Pil·lari - Son Rapinya - Pont d'Inca - Secar de la Real - B228 - B236 - B278	0.92	0.19	0.76	1.14
El Pil·lari - Son Rapinya - Puntiró - Secar de la Real - B228 - B236 - B278	0.92	0.18	0.76	1.12
El Pil·lari - Pont d'Inca - Puntiró - Secar de la Real - B228 - B236 - B278	0.92	0.20	0.75	1.15
La Bonanova - Pòrtol - Son Rapinya - Pont d'Inca - Puntiró - Secar de la Real - B228	0.95	0.14	0.83	1.11
La Bonanova - Pòrtol - Son Rapinya - Pont d'Inca - Puntiró - Secar de la Real - B236	0.95	0.15	0.81	1.12
La Bonanova - Pòrtol - Son Rapinya - Pont d'Inca - Puntiró - Secar de la Real - B278	0.95	0.17	0.80	1.14
La Bonanova - Pòrtol - Son Rapinya - Pont d'Inca - Puntiró - B228 - B236	0.96	0.15	0.81	1.13
La Bonanova - Pòrtol - Son Rapinya - Pont d'Inca - Puntiró - B228 - B278	0.95	0.15	0.81	1.13
La Bonanova - Pòrtol - Son Rapinya - Pont d'Inca - Puntiró - B236 - B278	0.94	0.17	0.79	1.14
La Bonanova - Pòrtol - Son Rapinya - Pont d'Inca - Secar de la Real - B228 - B236	0.95	0.15	0.81	1.11
La Bonanova - Pòrtol - Son Rapinya - Pont d'Inca - Secar de la Real - B228 - B278	0.93	0.16	0.79	1.10
La Bonanova - Pòrtol - Son Rapinya - Pont d'Inca - Secar de la Real - B236 - B278	0.94	0.17	0.79	1.14
La Bonanova - Pòrtol - Son Rapinya - Pont d'Inca - B228 - B236 - B278	0.95	0.17	0.79	1.14
La Bonanova - Pòrtol - Son Rapinya - Puntiró - Secar de la Real - B228 - B236	0.95	0.12	0.84	1.08
La Bonanova - Pòrtol - Son Rapinya - Puntiró - Secar de la Real - B228 - B278	0.95	0.13	0.83	1.09
La Bonanova - Pòrtol - Son Rapinya - Puntiró - Secar de la Real - B236 - B278	0.95	0.15	0.81	1.11
La Bonanova - Pòrtol - Son Rapinya - Puntiró - B228 - B236 - B278	0.95	0.15	0.82	1.11
La Bonanova - Pòrtol - Son Rapinya - Secar de la Real - B228 - B236 - B278	0.94	0.15	0.81	1.11
La Bonanova - Pòrtol - Pont d'Inca - Puntiró - Secar de la Real - B228 - B236	0.96	0.14	0.83	1.11

Table A1. *Cont.*

Combinations	POD	FAR	CSI	BS
La Bonanova - Pòrtol - Pont d'Inca - Puntiró - Secar de la Real - B228 - B278	0.95	0.15	0.81	1.12
La Bonanova - Pòrtol - Pont d'Inca - Puntiró - Secar de la Real - B236 - B278	0.95	0.18	0.79	1.15
La Bonanova - Pòrtol - Pont d'Inca - Puntiró - B228 - B236 - B278	0.95	0.16	0.80	1.14
La Bonanova - Pòrtol - Pont d'Inca - Secar de la Real - B228 - B236 - B278	0.95	0.17	0.80	1.14
La Bonanova - Pòrtol - Puntiró - Secar de la Real - B228 - B236 - B278	0.96	0.14	0.83	1.11
La Bonanova - Son Rapinya - Pont d'Inca - Puntiró - Secar de la Real - B228 - B236	0.94	0.14	0.81	1.09
La Bonanova - Son Rapinya - Pont d'Inca - Puntiró - Secar de la Real - B228 - B278	0.93	0.15	0.80	1.10
La Bonanova - Son Rapinya - Pont d'Inca - Puntiró - Secar de la Real - B236 - B278	0.94	0.17	0.79	1.13
La Bonanova - Son Rapinya - Pont d'Inca - Puntiró - B228 - B236 - B278	0.94	0.17	0.78	1.13
La Bonanova - Son Rapinya - Pont d'Inca - Secar de la Real - B228 - B236 - B278	**0.95**	**0.15**	**0.81**	**1.11**
La Bonanova - Son Rapinya - Puntiró - Secar de la Real - B228 - B236 - B278	0.93	0.15	0.81	1.09
La Bonanova - Pont d'Inca - Puntiró - Secar de la Real - B228 - B236 - B278	0.94	0.16	0.79	1.12
Pòrtol - Son Rapinya - Pont d'Inca - Puntiró - Secar de la Real - B228 - B236	0.94	0.15	0.81	1.11
Pòrtol - Son Rapinya - Pont d'Inca - Puntiró - Secar de la Real - B228 - B278	0.94	0.16	0.80	1.12
Pòrtol - Son Rapinya - Pont d'Inca - Puntiró - Secar de la Real - B236 - B278	0.95	0.18	0.79	1.15
Pòrtol - Son Rapinya - Pont d'Inca - Puntiró - B228 - B236 - B278	0.94	0.17	0.79	1.13
Pòrtol - Son Rapinya - Pont d'Inca - Secar de la Real - B228 - B236 - B278	0.93	0.18	0.78	1.13
Pòrtol - Son Rapinya - Puntiró - Secar de la Real - B228 - B236 - B278	0.94	0.15	0.81	1.11
Pòrtol - Pont d'Inca - Puntiró - Secar de la Real - B228 - B236 - B278	0.94	0.17	0.79	1.14
Son Rapinya - Pont d'Inca - Puntiró - Secar de la Real - B228 - B236 - B278	0.93	0.17	0.78	1.12

Appendix C. Accuracy of Rainfall Performance

Table A2. Accuracy of rainfall performance for various combinations.

Combinations	<1 mm	$[1,5)$ mm	$[5,20)$mm	$[20,40)$ mm	>40 mm
El Pil·lari - La Bonanova - Pòrtol - Son Rapinya - Pont d'Inca - Puntiró - Secar de la Real	0.81	0.44	0.56	0.28	0.50
El Pil·lari - La Bonanova - Pòrtol - Son Rapinya - Pont d'Inca - Puntiró - B228	0.83	0.49	0.60	0.39	0.62
El Pil·lari - La Bonanova - Pòrtol - Son Rapinya - Pont d'Inca - Puntiró - B236	0.77	0.41	0.54	0.41	0.62
El Pil·lari - La Bonanova - Pòrtol - Son Rapinya - Pont d'Inca - Puntiró - B278	0.82	0.45	0.58	0.36	0.62
El Pil·lari - La Bonanova - Pòrtol - Son Rapinya - Pont d'Inca - Secar de la Real - B228	0.82	0.47	0.59	0.38	0.50
El Pil·lari - La Bonanova - Pòrtol - Son Rapinya - Pont d'Inca - Secar de la Real - B236	0.78	0.40	0.53	0.31	0.50
El Pil·lari - La Bonanova - Pòrtol - Son Rapinya - Pont d'Inca - Secar de la Real - B278	0.79	0.42	0.55	0.35	0.55
El Pil·lari - La Bonanova - Pòrtol - Son Rapinya - Pont d'Inca - B228 - B236	0.80	0.44	0.57	0.50	0.67
El Pil·lari - La Bonanova - Pòrtol - Son Rapinya - Pont d'Inca - B228 - B278	0.83	0.50	0.62	0.53	0.67
El Pil·lari - La Bonanova - Pòrtol - Son Rapinya - Pont d'Inca - B236 - B278	0.77	0.41	0.56	0.48	0.67
El Pil·lari - La Bonanova - Pòrtol - Son Rapinya - Puntiró - Secar de la Real - B228	0.83	0.47	0.58	0.29	0.45
El Pil·lari - La Bonanova - Pòrtol - Son Rapinya - Puntiró - Secar de la Real - B236	0.77	0.41	0.53	0.30	0.45
El Pil·lari - La Bonanova - Pòrtol - Son Rapinya - Puntiró - Secar de la Real - B278	0.80	0.44	0.55	0.27	0.50
El Pil·lari - La Bonanova - Pòrtol - Son Rapinya - Puntiró - B228 - B236	0.79	0.44	0.56	0.42	0.62
El Pil·lari - La Bonanova - Pòrtol - Son Rapinya - Puntiró - B228 - B278	0.82	0.48	0.59	0.42	0.62
El Pil·lari - La Bonanova - Pòrtol - Son Rapinya - Puntiró - B236 - B278	0.75	0.41	0.55	0.40	0.62
El Pil·lari - La Bonanova - Pòrtol - Son Rapinya - Secar de la Real - B228 - B236	0.79	0.44	0.58	0.38	0.50
El Pil·lari - La Bonanova - Pòrtol - Son Rapinya - Secar de la Real - B228 - B278	0.80	0.45	0.58	0.42	0.55
El Pil·lari - La Bonanova - Pòrtol - Son Rapinya - Secar de la Real - B236 - B278	0.76	0.41	0.53	0.38	0.55
El Pil·lari - La Bonanova - Pòrtol - Son Rapinya - B228 - B236 - B278	0.78	0.43	0.57	0.52	0.67
El Pil·lari - La Bonanova - Pòrtol - Pont d'Inca - Puntiró - Secar de la Real - B228	0.82	0.46	0.59	0.31	0.45
El Pil·lari - La Bonanova - Pòrtol - Pont d'Inca - Puntiró - Secar de la Real - B236	0.76	0.37	0.55	0.33	0.50
El Pil·lari - La Bonanova - Pòrtol - Pont d'Inca - Puntiró - Secar de la Real - B278	0.79	0.38	0.53	0.28	0.50
El Pil·lari - La Bonanova - Pòrtol - Pont d'Inca - Puntiró - B228 - B236	0.78	0.43	0.57	0.47	0.56
El Pil·lari - La Bonanova - Pòrtol - Pont d'Inca - Puntiró - B228 - B278	0.83	0.46	0.60	0.38	0.56
El Pil·lari - La Bonanova - Pòrtol - Pont d'Inca - Puntiró - B236 - B278	0.75	0.37	0.53	0.48	0.62
El Pil·lari - La Bonanova - Pòrtol - Pont d'Inca - Secar de la Real - B228 - B236	0.78	0.41	0.56	0.40	0.50
El Pil·lari - La Bonanova - Pòrtol - Pont d'Inca - Secar de la Real - B228 - B278	0.79	0.41	0.56	0.36	0.50
El Pil·lari - La Bonanova - Pòrtol - Pont d'Inca - Secar de la Real - B236 - B278	0.75	0.35	0.51	0.35	0.55
El Pil·lari - La Bonanova - Pòrtol - Pont d'Inca - B228 - B236 - B278	0.78	0.41	0.56	0.48	0.60
El Pil·lari - La Bonanova - Pòrtol - Puntiró - Secar de la Real - B228 - B236	0.77	0.41	0.55	0.38	0.45
El Pil·lari - La Bonanova - Pòrtol - Puntiró - Secar de la Real - B228 - B278	0.80	0.44	0.56	0.29	0.45
El Pil·lari - La Bonanova - Pòrtol - Puntiró - Secar de la Real - B236 - B278	0.74	0.38	0.52	0.34	0.50
El Pil·lari - La Bonanova - Pòrtol - Puntiró - B228 - B236 - B278	0.77	0.41	0.55	0.41	0.56
El Pil·lari - La Bonanova - Pòrtol - Secar de la Real - B228 - B236 - B278	0.77	0.41	0.56	0.38	0.50
El Pil·lari - La Bonanova - Son Rapinya - Pont d'Inca - Puntiró - Secar de la Real - B228	0.84	0.50	0.59	0.34	0.50
El Pil·lari - La Bonanova - Son Rapinya - Pont d'Inca - Puntiró - Secar de la Real - B236	0.77	0.41	0.52	0.27	0.50
El Pil·lari - La Bonanova - Son Rapinya - Pont d'Inca - Puntiró - Secar de la Real - B278	0.80	0.46	0.56	0.33	0.55
El Pil·lari - La Bonanova - Son Rapinya - Pont d'Inca - Puntiró - B228 - B236	0.79	0.45	0.56	0.47	0.67
El Pil·lari - La Bonanova - Son Rapinya - Pont d'Inca - Puntiró - B228 - B278	0.83	0.51	0.61	0.50	0.67

Table A2. *Cont.*

Combinations	<1 mm	[1, 5) mm	[5, 20) mm	[20, 40) mm	>40 mm
El Pil·lari - La Bonanova - Son Rapinya - Pont d'Inca - Puntiró - B236 - B278	0.76	0.43	0.54	0.46	0.67
El Pil·lari - La Bonanova - Son Rapinya - Pont d'Inca - Secar de la Real - B228 - B236	0.81	0.49	0.59	0.39	0.58
El Pil·lari - La Bonanova - Son Rapinya - Pont d'Inca - Secar de la Real - B228 - B278	0.84	0.50	0.61	0.42	0.64
El Pil·lari - La Bonanova - Son Rapinya - Pont d'Inca - Secar de la Real - B236 - B278	0.78	0.44	0.54	0.39	0.64
El Pil·lari - La Bonanova - Son Rapinya - Pont d'Inca - B228 - B236 - B278	0.81	0.47	0.58	0.55	0.78
El Pil·lari - La Bonanova - Son Rapinya - Puntiró - Secar de la Real - B228 - B236	0.79	0.45	0.57	0.34	0.50
El Pil·lari - La Bonanova - Son Rapinya - Pont d'Inca - Puntiró - B228 - B278	0.83	0.51	0.61	0.50	0.67
El Pil·lari - La Bonanova - Son Rapinya - Pont d'Inca - Puntiró - B236 - B278	0.76	0.43	0.54	0.46	0.67
El Pil·lari - La Bonanova - Son Rapinya - Pont d'Inca - Secar de la Real - B228 - B236	0.81	0.49	0.59	0.39	0.58
El Pil·lari - La Bonanova - Son Rapinya - Pont d'Inca - Secar de la Real - B228 - B278	0.84	0.50	0.61	0.42	0.64
El Pil·lari - La Bonanova - Son Rapinya - Pont d'Inca - Secar de la Real - B236 - B278	0.78	0.44	0.54	0.39	0.64
El Pil·lari - La Bonanova - Son Rapinya - Pont d'Inca - B228 - B236 - B278	0.81	0.47	0.58	0.55	0.78
El Pil·lari - La Bonanova - Son Rapinya - Puntiró - Secar de la Real - B228 - B236	0.79	0.45	0.57	0.34	0.50
El Pil·lari - La Bonanova - Son Rapinya - Puntiró - Secar de la Real - B228 - B278	0.81	0.49	0.60	0.39	0.55
El Pil·lari - La Bonanova - Son Rapinya - Puntiró - Secar de la Real - B236 - B278	0.76	0.41	0.52	0.33	0.55
El Pil·lari - La Bonanova - Son Rapinya - Puntiró - B228 - B236 - B278	0.77	0.45	0.57	0.48	0.67
El Pil·lari - La Bonanova - Son Rapinya - Secar de la Real - B228 - B236 - B278	0.80	0.48	0.58	0.45	0.64
El Pil·lari - La Bonanova - Pont d'Inca - Puntiró - Secar de la Real - B228 - B236	0.78	0.43	0.56	0.36	0.50
El Pil·lari - La Bonanova - Pont d'Inca - Puntiró - Secar de la Real - B228 - B278	0.79	0.44	0.56	0.34	0.50
El Pil·lari - La Bonanova - Pont d'Inca - Puntiró - Secar de la Real - B236 - B278	0.74	0.37	0.50	0.31	0.55
El Pil·lari - La Bonanova - Pont d'Inca - Puntiró - B228 - B236 - B278	0.77	0.41	0.55	0.46	0.60
El Pil·lari - La Bonanova - Pont d'Inca - Secar de la Real - B228 - B236 - B278	0.78	0.43	0.55	0.42	0.64
El Pil·lari - La Bonanova - Puntiró - Secar de la Real - B228 - B236 - B278	0.77	0.42	0.55	0.35	0.50
El Pil·lari - Pòrtol - Son Rapinya - Pont d'Inca - Puntiró - Secar de la Real - B228	0.81	0.43	0.57	0.29	0.56
El Pil·lari - Pòrtol - Son Rapinya - Pont d'Inca - Puntiró - Secar de la Real - B236	0.76	0.38	0.54	0.36	0.44
El Pil·lari - Pòrtol - Son Rapinya - Pont d'Inca - Puntiró - Secar de la Real - B278	0.81	0.43	0.57	0.29	0.44
El Pil·lari - Pòrtol - Son Rapinya - Pont d'Inca - Puntiró - B228 - B236	0.78	0.40	0.55	0.50	0.71
El Pil·lari - Pòrtol - Son Rapinya - Pont d'Inca - Puntiró - B228 - B278	0.82	0.44	0.58	0.40	0.71
El Pil·lari - Pòrtol - Son Rapinya - Pont d'Inca - Puntiró - B236 - B278	0.77	0.40	0.56	0.52	0.33
El Pil·lari - Pòrtol - Son Rapinya - Pont d'Inca - Secar de la Real - B228 - B236	0.78	0.40	0.55	0.36	0.55
El Pil·lari - Pòrtol - Son Rapinya - Pont d'Inca - Secar de la Real - B228 - B278	0.79	0.41	0.57	0.39	0.60
El Pil·lari - Pòrtol - Son Rapinya - Pont d'Inca - Secar de la Real - B236 - B278	0.77	0.38	0.54	0.38	0.60
El Pil·lari - Pòrtol - Son Rapinya - Pont d'Inca - B228 - B236 - B278	0.77	0.40	0.56	0.57	0.75
El Pil·lari - Pòrtol - Son Rapinya - Puntiró - Secar de la Real - B228 - B236	0.77	0.40	0.54	0.33	0.50
El Pil·lari - Pòrtol - Son Rapinya - Puntiró - Secar de la Real - B228 - B278	0.81	0.43	0.55	0.28	0.56
El Pil·lari - Pòrtol - Son Rapinya - Puntiró - Secar de la Real - B236 - B278	0.75	0.38	0.53	0.34	0.38
El Pil·lari - Pòrtol - Son Rapinya - Puntiró - B228 - B236 - B278	0.75	0.39	0.54	0.45	0.71
El Pil·lari - Pòrtol - Son Rapinya - Secar de la Real - B228 - B236 - B278	0.76	0.39	0.54	0.39	0.60
El Pil·lari - Pòrtol - Pont d'Inca - Puntiró - Secar de la Real - B228 - B236	0.75	0.37	0.55	0.38	0.56
El Pil·lari - Pòrtol - Pont d'Inca - Puntiró - Secar de la Real - B228 - B278	0.79	0.39	0.55	0.31	0.56
El Pil·lari - Pòrtol - Pont d'Inca - Puntiró - Secar de la Real - B236 - B278	0.75	0.36	0.52	0.36	0.44
El Pil·lari - Pòrtol - Pont d'Inca - Puntiró - B228 - B236 - B278	0.75	0.36	0.55	0.50	0.71
El Pil·lari - Pòrtol - Pont d'Inca - Secar de la Real - B228 - B236 - B278	0.74	0.35	0.53	0.38	0.60
El Pil·lari - Pòrtol - Puntiró - Secar de la Real - B228 - B236 - B278	0.73	0.36	0.53	0.38	0.56
El Pil·lari - Son Rapinya - Pont d'Inca - Puntiró - Secar de la Real - B228 - B236	0.77	0.40	0.53	0.29	0.55
El Pil·lari - Son Rapinya - Pont d'Inca - Puntiró - Secar de la Real - B228 - B278	0.81	0.45	0.55	0.35	0.60
El Pil·lari - Son Rapinya - Pont d'Inca - Puntiró - Secar de la Real - B236 - B278	0.76	0.39	0.52	0.35	0.60
El Pil·lari - Son Rapinya - Pont d'Inca - Puntiró - B228 - B236 - B278	0.76	0.40	0.55	0.52	0.75
El Pil·lari - Son Rapinya - Pont d'Inca - Secar de la Real - B228 - B236 - B278	0.78	0.43	0.55	0.41	0.70
El Pil·lari - Son Rapinya - Puntiró - Secar de la Real - B228 - B236 - B278	0.76	0.40	0.53	0.36	0.60
El Pil·lari - Pont d'Inca - Puntiró - Secar de la Real - B228 - B236 - B278	0.74	0.36	0.53	0.36	0.60
La Bonanova - Pòrtol - Son Rapinya - Pont d'Inca - Puntiró - Secar de la Real - B228	0.85	0.49	0.59	0.35	0.56
La Bonanova - Pòrtol - Son Rapinya - Pont d'Inca - Puntiró - Secar de la Real - B236	0.81	0.43	0.57	0.33	0.56
La Bonanova - Pòrtol - Son Rapinya - Pont d'Inca - Puntiró - Secar de la Real - B278	0.82	0.44	0.56	0.31	0.56
La Bonanova - Pòrtol - Son Rapinya - Pont d'Inca - Puntiró - B228 - B236	0.82	0.46	0.59	0.48	0.71
La Bonanova - Pòrtol - Son Rapinya - Pont d'Inca - Puntiró - B228 - B278	0.84	0.49	0.61	0.42	0.71
La Bonanova - Pòrtol - Son Rapinya - Pont d'Inca - Puntiró - B236 - B278	0.80	0.45	0.58	0.41	0.71
La Bonanova - Pòrtol - Son Rapinya - Pont d'Inca - Secar de la Real - B228 - B236	0.84	0.48	0.60	0.38	0.55
La Bonanova - Pòrtol - Son Rapinya - Pont d'Inca - Secar de la Real - B228 - B278	0.83	0.48	0.59	0.43	0.60
La Bonanova - Pòrtol - Son Rapinya - Pont d'Inca - Secar de la Real - B236 - B278	0.81	0.44	0.57	0.39	0.60
La Bonanova - Pòrtol - Son Rapinya - Pont d'Inca - B228 - B236 - B278	0.83	0.47	0.60	0.52	0.75
El Pil·lari - Son Rapinya - Pont d'Inca - Puntiró - Secar de la Real - B228 - B236	0.77	0.40	0.53	0.29	0.55
El Pil·lari - Son Rapinya - Pont d'Inca - Puntiró - Secar de la Real - B228 - B278	0.81	0.45	0.55	0.35	0.60
El Pil·lari - Son Rapinya - Pont d'Inca - Puntiró - Secar de la Real - B236 - B278	0.76	0.39	0.52	0.35	0.60
El Pil·lari - Son Rapinya - Pont d'Inca - Puntiró - B228 - B236 - B278	0.76	0.40	0.55	0.52	0.75
El Pil·lari - Son Rapinya - Pont d'Inca - Secar de la Real - B228 - B236 - B278	0.78	0.43	0.55	0.41	0.70
El Pil·lari - Son Rapinya - Puntiró - Secar de la Real - B228 - B236 - B278	0.76	0.40	0.53	0.36	0.60
El Pil·lari - Pont d'Inca - Puntiró - Secar de la Real - B228 - B236 - B278	0.74	0.36	0.53	0.36	0.60

Table A2. *Cont.*

Combinations	<1 mm	[1,5) mm	[5,20) mm	[20,40) mm	>40 mm
La Bonanova - Pòrtol - Son Rapinya - Pont d'Inca - Puntiró - Secar de la Real - B228	0.85	0.49	0.59	0.35	0.56
La Bonanova - Pòrtol - Son Rapinya - Pont d'Inca - Puntiró - Secar de la Real - B236	0.81	0.43	0.57	0.33	0.56
La Bonanova - Pòrtol - Son Rapinya - Pont d'Inca - Puntiró - Secar de la Real - B278	0.82	0.44	0.56	0.31	0.56
La Bonanova - Pòrtol - Son Rapinya - Pont d'Inca - Puntiró - B228 - B236	0.82	0.46	0.59	0.48	0.71
La Bonanova - Pòrtol - Son Rapinya - Pont d'Inca - Puntiró - B228 - B278	0.84	0.49	0.61	0.42	0.71
La Bonanova - Pòrtol - Son Rapinya - Pont d'Inca - Puntiró - B236 - B278	0.80	0.45	0.58	0.41	0.71
La Bonanova - Pòrtol - Son Rapinya - Pont d'Inca - Secar de la Real - B228 - B236	0.84	0.48	0.60	0.38	0.55
La Bonanova - Pòrtol - Son Rapinya - Pont d'Inca - Secar de la Real - B228 - B278	0.83	0.48	0.59	0.43	0.60
La Bonanova - Pòrtol - Son Rapinya - Pont d'Inca - Secar de la Real - B236 - B278	0.81	0.44	0.57	0.39	0.60
La Bonanova - Pòrtol - Son Rapinya - Pont d'Inca - B228 - B236 - B278	0.83	0.47	0.60	0.52	0.75
La Bonanova - Pòrtol - Son Rapinya - Puntiró - Secar de la Real - B228 - B236	0.83	0.46	0.59	0.34	0.50
La Bonanova - Pòrtol - Son Rapinya - Puntiró - Secar de la Real - B228 - B278	0.84	0.47	0.58	0.33	0.56
La Bonanova - Pòrtol - Son Rapinya - Puntiró - Secar de la Real - B236 - B278	0.79	0.42	0.56	0.31	0.44
La Bonanova - Pòrtol - Son Rapinya - Puntiró - B228 - B236 - B278	0.81	0.47	0.60	0.44	0.71
La Bonanova - Pòrtol - Son Rapinya - Secar de la Real - B228 - B236 - B278	0.81	0.46	0.59	0.44	0.60
La Bonanova - Pòrtol - Pont d'Inca - Puntiró - Secar de la Real - B228 - B236	0.83	0.44	0.58	0.38	0.50
La Bonanova - Pòrtol - Pont d'Inca - Puntiró - Secar de la Real - B228 - B278	0.82	0.43	0.56	0.32	0.50
La Bonanova - Pòrtol - Pont d'Inca - Puntiró - Secar de la Real - B236 - B278	0.78	0.38	0.54	0.34	0.56
La Bonanova - Pòrtol - Pont d'Inca - Puntiró - B228 - B236 - B278	0.81	0.43	0.58	0.48	0.62
La Bonanova - Pòrtol - Pont d'Inca - Secar de la Real - B228 - B236 - B278	0.81	0.42	0.57	0.39	0.55
La Bonanova - Pòrtol - Puntiró - Secar de la Real - B228 - B236 - B278	0.79	0.43	0.57	0.34	0.50
La Bonanova - Son Rapinya - Pont d'Inca - Puntiró - Secar de la Real - B228 - B236	0.82	0.48	0.60	0.35	0.55
La Bonanova - Son Rapinya - Pont d'Inca - Puntiró - Secar de la Real - B228 - B278	0.84	0.49	0.60	0.39	0.60
La Bonanova - Son Rapinya - Pont d'Inca - Puntiró - Secar de la Real - B236 - B278	0.79	0.43	0.55	0.34	0.60
La Bonanova - Son Rapinya - Pont d'Inca - Puntiró - B228 - B236 - B278	0.81	0.45	0.59	0.50	0.75
La Bonanova - Son Rapinya - Pont d'Inca - Secar de la Real - B228 - B236 - B278	**0.84**	**0.51**	**0.62**	**0.44**	**0.70**
La Bonanova - Son Rapinya - Puntiró - Secar de la Real - B228 - B236 - B278	0.80	0.46	0.61	0.41	0.60
La Bonanova - Pont d'Inca - Puntiró - Secar de la Real - B228 - B236 - B278	0.80	0.43	0.57	0.35	0.55
Pòrtol - Son Rapinya - Pont d'Inca - Puntiró - Secar de la Real - B228 - B236	0.81	0.43	0.58	0.37	0.62
Pòrtol - Son Rapinya - Pont d'Inca - Puntiró - Secar de la Real - B228 - B278	0.83	0.43	0.56	0.32	0.62
Pòrtol - Son Rapinya - Pont d'Inca - Puntiró - Secar de la Real - B236 - B278	0.80	0.41	0.56	0.39	0.57
Pòrtol - Son Rapinya - Pont d'Inca - Puntiró - B228 - B236 - B278	0.80	0.43	0.58	0.50	0.83
Pòrtol - Son Rapinya - Pont d'Inca - Secar de la Real - B228 - B236 - B278	0.81	0.42	0.57	0.40	0.67
Pòrtol - Son Rapinya - Puntiró - Secar de la Real - B228 - B236 - B278	0.78	0.40	0.56	0.34	0.62
Pòrtol - Pont d'Inca - Puntiró - Secar de la Real - B228 - B236 - B278	0.77	0.37	0.55	0.38	0.62
Son Rapinya - Pont d'Inca - Puntiró - Secar de la Real - B228 - B236 - B278	0.78	0.41	0.57	0.37	0.67

Appendix D. Accuracy of Rainfall Performance II

Table A3. Accuracy of rainfall performance for various combinations.

Combinations	RMSE	NSE	PBIAS
El Pil·lari - La Bonanova - Pòrtol - Son Rapinya - Pont d'Inca - Puntiró - Secar de la Real	3.23	0.79	0.50
El Pil·lari - La Bonanova - Pòrtol - Son Rapinya - Pont d'Inca - Puntiró - B228	2.83	0.84	−5.21
El Pil·lari - La Bonanova - Pòrtol - Son Rapinya - Pont d'Inca - Puntiró - B236	3.19	0.78	−0.31
El Pil·lari - La Bonanova - Pòrtol - Son Rapinya - Pont d'Inca - Puntiró - B278	3.11	0.79	0.01
El Pil·lari - La Bonanova - Pòrtol - Son Rapinya - Pont d'Inca - Secar de la Real - B228	2.86	0.86	−2.90
El Pil·lari - La Bonanova - Pòrtol - Son Rapinya - Pont d'Inca - Secar de la Real - B236	3.17	0.81	1.01
El Pil·lari - La Bonanova - Pòrtol - Son Rapinya - Pont d'Inca - Secar de la Real - B278	3.21	0.81	−1.30
El Pil·lari - La Bonanova - Pòrtol - Son Rapinya - Pont d'Inca - B228 - B236	2.78	0.86	−4.37
El Pil·lari - La Bonanova - Pòrtol - Son Rapinya - Pont d'Inca - B228 - B278	2.69	0.87	−5.19
El Pil·lari - La Bonanova - Pòrtol - Son Rapinya - Pont d'Inca - B236 - B278	3.03	0.81	−0.02
El Pil·lari - La Bonanova - Pòrtol - Son Rapinya - Puntiró - Secar de la Real - B228	3.03	0.83	−2.31
El Pil·lari - La Bonanova - Pòrtol - Son Rapinya - Puntiró - Secar de la Real - B236	3.37	0.77	1.71
El Pil·lari - La Bonanova - Pòrtol - Son Rapinya - Puntiró - Secar de la Real - B278	3.43	0.76	−1.11
El Pil·lari - La Bonanova - Pòrtol - Son Rapinya - Puntiró - B228 - B236	2.97	0.82	−5.20
El Pil·lari - La Bonanova - Pòrtol - Son Rapinya - Puntiró - B228 - B278	2.88	0.83	−6.09
El Pil·lari - La Bonanova - Pòrtol - Son Rapinya - Puntiró - B236 - B278	3.24	0.76	−0.11
El Pil·lari - La Bonanova - Pòrtol - Son Rapinya - Secar de la Real - B228 - B236	2.98	0.85	−1.44
El Pil·lari - La Bonanova - Pòrtol - Son Rapinya - Secar de la Real - B228 - B278	3.07	0.83	−4.64
El Pil·lari - La Bonanova - Pòrtol - Son Rapinya - Secar de la Real - B236 - B278	3.37	0.78	−0.45
El Pil·lari - La Bonanova - Pòrtol - Son Rapinya - B228 - B236 - B278	2.83	0.85	−5.06
El Pil·lari - La Bonanova - Pòrtol - Pont d'Inca - Puntiró - Secar de la Real - B228	2.76	0.86	−2.55
El Pil·lari - La Bonanova - Pòrtol - Pont d'Inca - Puntiró - Secar de la Real - B236	3.06	0.82	0.56
El Pil·lari - La Bonanova - Pòrtol - Pont d'Inca - Puntiró - Secar de la Real - B278	3.23	0.79	−1.42

Table A3. *Cont.*

Combinations	RMSE	NSE	PBIAS
El Pil·lari - La Bonanova - Pòrtol - Pont d'Inca - Puntiró - B228 - B236	2.74	0.86	−2.20
El Pil·lari - La Bonanova - Pòrtol - Pont d'Inca - Puntiró - B228 - B278	2.71	0.86	−4.64
El Pil·lari - La Bonanova - Pòrtol - Pont d'Inca - Puntiró - B236 - B278	3.11	0.79	0.10
El Pil·lari - La Bonanova - Pòrtol - Pont d'Inca - Secar de la Real - B228 - B236	2.71	0.88	−1.77
El Pil·lari - La Bonanova - Pòrtol - Pont d'Inca - Secar de la Real - B228 - B278	2.91	0.86	−4.53
El Pil·lari - La Bonanova - Pòrtol - Pont d'Inca - Secar de la Real - B236 - B278	3.18	0.82	−1.02
El Pil·lari - La Bonanova - Pòrtol - Pont d'Inca - B228 - B236 - B278	2.67	0.87	−4.04
El Pil·lari - La Bonanova - Pòrtol - Puntiró - Secar de la Real - B228 - B236	2.91	0.85	−1.05
El Pil·lari - La Bonanova - Pòrtol - Puntiró - Secar de la Real - B228 - B278	3.13	0.82	−4.04
El Pil·lari - La Bonanova - Pòrtol - Puntiró - Secar de la Real - B236 - B278	3.42	0.77	−0.37
El Pil·lari - La Bonanova - Pòrtol - Puntiró - B228 - B236 - B278	2.82	0.84	−4.41
El Pil·lari - La Bonanova - Pòrtol - Secar de la Real - B228 - B236 - B278	3.07	0.84	−3.29
El Pil·lari - La Bonanova - Son Rapinya - Pont d'Inca - Puntiró - Secar de la Real - B228	3.09	0.83	−5.79
El Pil·lari - La Bonanova - Son Rapinya - Pont d'Inca - Puntiró - Secar de la Real - B236	3.38	0.77	−1.39
El Pil·lari - La Bonanova - Son Rapinya - Pont d'Inca - Puntiró - Secar de la Real - B278	3.39	0.77	−3.84
El Pil·lari - La Bonanova - Son Rapinya - Pont d'Inca - Puntiró - B228 - B236	3.00	0.83	−7.02
El Pil·lari - La Bonanova - Son Rapinya - Pont d'Inca - Puntiró - B228 - B278	2.91	0.83	−7.79
El Pil·lari - La Bonanova - Son Rapinya - Pont d'Inca - Puntiró - B236 - B278	3.23	0.77	−1.97
El Pil·lari - La Bonanova - Son Rapinya - Pont d'Inca - Secar de la Real - B228 - B236	2.86	0.86	−1.79
El Pil·lari - La Bonanova - Son Rapinya - Pont d'Inca - Secar de la Real - B228 - B278	2.92	0.85	−4.99
El Pil·lari - La Bonanova - Son Rapinya - Pont d'Inca - Secar de la Real - B236 - B278	3.25	0.80	−0.70
El Pil·lari - La Bonanova - Son Rapinya - Pont d'Inca - B228 - B236 - B278	2.73	0.86	−4.69
El Pil·lari - La Bonanova - Son Rapinya - Puntiró - Secar de la Real - B228 - B236	3.19	0.81	−4.18
El Pil·lari - La Bonanova - Son Rapinya - Puntiró - Secar de la Real - B228 - B278	3.26	0.80	−7.81
El Pil·lari - La Bonanova - Son Rapinya - Puntiró - Secar de la Real - B236 - B278	3.55	0.74	−2.96
El Pil·lari - La Bonanova - Son Rapinya - Puntiró - B228 - B236 - B278	3.02	0.82	−7.76
El Pil·lari - La Bonanova - Son Rapinya - Secar de la Real - B228 - B236 - B278	3.07	0.84	−3.72
El Pil·lari - La Bonanova - Pont d'Inca - Puntiró - Secar de la Real - B228 - B236	2.96	0.85	−4.36
El Pil·lari - La Bonanova - Pont d'Inca - Puntiró - Secar de la Real - B228 - B278	3.13	0.82	−7.34
El Pil·lari - La Bonanova - Pont d'Inca - Puntiró - Secar de la Real - B236 - B278	3.39	0.78	−3.28
El Pil·lari - La Bonanova - Pont d'Inca - Puntiró - B228 - B236 - B278	2.90	0.84	−6.46
El Pil·lari - La Bonanova - Pont d'Inca - Secar de la Real - B228 - B236 - B278	2.92	0.86	−3.46
El Pil·lari - La Bonanova - Puntiró - Secar de la Real - B228 - B236 - B278	3.29	0.80	−5.91
El Pil·lari - Pòrtol - Son Rapinya - Pont d'Inca - Puntiró - Secar de la Real - B228	3.06	0.81	−0.30
El Pil·lari - Pòrtol - Son Rapinya - Pont d'Inca - Puntiró - Secar de la Real - B236	2.96	0.80	−1.89
El Pil·lari - Pòrtol - Son Rapinya - Pont d'Inca - Puntiró - Secar de la Real - B278	2.98	0.79	−4.18
El Pil·lari - Pòrtol - Son Rapinya - Pont d'Inca - Puntiró - B228 - B236	2.94	0.80	−0.92
El Pil·lari - Pòrtol - Son Rapinya - Pont d'Inca - Puntiró - B228 - B278	2.86	0.81	−0.51
El Pil·lari - Pòrtol - Son Rapinya - Pont d'Inca - Puntiró - B236 - B278	2.66	0.82	−2.50
El Pil·lari - Pòrtol - Son Rapinya - Pont d'Inca - Secar de la Real - B228 - B236	3.01	0.83	0.25
El Pil·lari - Pòrtol - Son Rapinya - Pont d'Inca - Secar de la Real - B228 - B278	3.03	0.82	−2.01
El Pil·lari - Pòrtol - Son Rapinya - Pont d'Inca - Secar de la Real - B236 - B278	2.94	0.81	−3.36
El Pil·lari - Pòrtol - Son Rapinya - Pont d'Inca - B228 - B236 - B278	2.79	0.84	−0.55
El Pil·lari - Pòrtol - Son Rapinya - Puntiró - Secar de la Real - B228 - B236	3.21	0.78	0.90
El Pil·lari - Pòrtol - Son Rapinya - Puntiró - Secar de la Real - B228 - B278	3.26	0.77	−1.85
El Pil·lari - Pòrtol - Son Rapinya - Puntiró - Secar de la Real - B236 - B278	3.16	0.76	−3.77
El Pil·lari - Pòrtol - Son Rapinya - Puntiró - B228 - B236 - B278	2.99	0.79	−0.45
El Pil·lari - Pòrtol - Son Rapinya - Secar de la Real - B228 - B236 - B278	3.21	0.80	−1.18
El Pil·lari - Pòrtol - Pont d'Inca - Puntiró - Secar de la Real - B228 - B236	2.93	0.83	0.25
El Pil·lari - Pòrtol - Pont d'Inca - Puntiró - Secar de la Real - B228 - B278	3.09	0.80	−1.86
El Pil·lari - Pòrtol - Pont d'Inca - Puntiró - Secar de la Real - B236 - B278	3.01	0.79	−1.79
El Pil·lari - Pòrtol - Pont d'Inca - Puntiró - B228 - B236 - B278	2.92	0.81	−0.69
El Pil·lari - Pòrtol - Pont d'Inca - Secar de la Real - B228 - B236 - B278	3.05	0.82	−1.42
El Pil·lari - Pòrtol - Puntiró - Secar de la Real - B228 - B236 - B278	3.30	0.77	−0.79
El Pil·lari - Son Rapinya - Pont d'Inca - Puntiró - Secar de la Real - B228 - B236	3.22	0.79	−2.21
El Pil·lari - Son Rapinya - Pont d'Inca - Puntiró - Secar de la Real - B228 - B278	3.22	0.78	−4.63
El Pil·lari - Son Rapinya - Pont d'Inca - Puntiró - Secar de la Real - B236 - B278	3.13	0.77	−5.99
El Pil·lari - Son Rapinya - Pont d'Inca - Puntiró - B228 - B236 - B278	3.01	0.79	−2.54
El Pil·lari - Son Rapinya - Pont d'Inca - Secar de la Real - B228 - B236 - B278	3.08	0.82	−1.47
El Pil·lari - Son Rapinya - Puntiró - Secar de la Real - B228 - B236 - B278	3.39	0.76	−3.74
El Pil·lari - Pont d'Inca - Puntiró - Secar de la Real - B228 - B236 - B278	3.26	0.78	−3.74
La Bonanova - Pòrtol - Son Rapinya - Pont d'Inca - Puntiró - Secar de la Real - B228	2.84	0.85	−1.29
La Bonanova - Pòrtol - Son Rapinya - Pont d'Inca - Puntiró - Secar de la Real - B236	3.18	0.80	2.00
La Bonanova - Pòrtol - Son Rapinya - Pont d'Inca - Puntiró - Secar de la Real - B278	3.23	0.79	−0.26
La Bonanova - Pòrtol - Son Rapinya - Pont d'Inca - Puntiró - B228 - B236	2.77	0.85	−3.51
La Bonanova - Pòrtol - Son Rapinya - Pont d'Inca - Puntiró - B228 - B278	2.71	0.85	−4.24
La Bonanova - Pòrtol - Son Rapinya - Pont d'Inca - Puntiró - B236 - B278	3.06	0.79	0.89

Table A3. *Cont.*

Combinations	RMSE	NSE	PBIAS
La Bonanova - Pòrtol - Son Rapinya - Pont d'Inca - Secar de la Real - B228 - B236	2.77	0.87	−0.75
La Bonanova - Pòrtol - Son Rapinya - Pont d'Inca - Secar de la Real - B228 - B278	2.83	0.86	−2.59
La Bonanova - Pòrtol - Son Rapinya - Pont d'Inca - Secar de la Real - B236 - B278	3.14	0.81	0.39
La Bonanova - Pòrtol - Son Rapinya - Pont d'Inca - B228 - B236 - B278	2.62	0.87	−3.49
La Bonanova - Pòrtol - Son Rapinya - Puntiró - Secar de la Real - B228 - B236	2.95	0.84	−0.28
La Bonanova - Pòrtol - Son Rapinya - Puntiró - Secar de la Real - B228 - B278	3.05	0.82	−3.26
La Bonanova - Pòrtol - Son Rapinya - Puntiró - Secar de la Real - B236 - B278	3.36	0.77	0.59
La Bonanova - Pòrtol - Son Rapinya - Puntiró - B228 - B236 - B278	2.81	0.84	−4.14
La Bonanova - Pòrtol - Son Rapinya - Secar de la Real - B228 - B236 - B278	2.97	0.84	−2.61
La Bonanova - Pòrtol - Pont d'Inca - Puntiró - Secar de la Real - B228 - B236	2.68	0.87	−0.57
La Bonanova - Pòrtol - Pont d'Inca - Puntiró - Secar de la Real - B228 - B278	2.88	0.85	−2.69
La Bonanova - Pòrtol - Pont d'Inca - Puntiró - Secar de la Real - B236 - B278	3.17	0.80	0.06
La Bonanova - Pòrtol - Pont d'Inca - Puntiró - B228 - B236 - B278	2.66	0.86	−2.96
La Bonanova - Pòrtol - Pont d'Inca - Secar de la Real - B228 - B236 - B278	2.81	0.87	−2.31
La Bonanova - Pòrtol - Puntiró - Secar de la Real - B228 - B236 - B278	3.04	0.83	−1.99
La Bonanova - Son Rapinya - Pont d'Inca - Puntiró - Secar de la Real - B228 - B236	3.03	0.83	−3.71
La Bonanova - Son Rapinya - Pont d'Inca - Puntiró - Secar de la Real - B228 - B278	3.11	0.82	−6.48
La Bonanova - Son Rapinya - Pont d'Inca - Puntiró - Secar de la Real - B236 - B278	3.36	0.77	−2.27
La Bonanova - Son Rapinya - Pont d'Inca - Puntiró - B228 - B236 - B278	2.87	0.84	−6.20
La Bonanova - Son Rapinya - Pont d'Inca - Secar de la Real - B228 - B236 - B278	**2.86**	**0.86**	**-2.49**
La Bonanova - Son Rapinya - Puntiró - Secar de la Real - B228 - B236 - B278	3.19	0.81	−5.51
La Bonanova - Pont d'Inca - Puntiró - Secar de la Real - B228 - B236 - B278	3.07	0.83	−5.20
Pòrtol - Son Rapinya - Pont d'Inca - Puntiró - Secar de la Real - B228 - B236	3.02	0.81	1.19
Pòrtol - Son Rapinya - Pont d'Inca - Puntiró - Secar de la Real - B228 - B278	3.08	0.80	−1.02
Pòrtol - Son Rapinya - Pont d'Inca - Puntiró - Secar de la Real - B236 - B278	2.95	0.79	−2.72
Pòrtol - Son Rapinya - Pont d'Inca - Puntiró - B228 - B236 - B278	2.83	0.81	0.33
Pòrtol - Son Rapinya - Pont d'Inca - Secar de la Real - B228 - B236 - B278	2.99	0.82	−0.35
Pòrtol - Son Rapinya - Puntiró - Secar de la Real - B228 - B236 - B278	3.21	0.78	−0.20
Pòrtol - Pont d'Inca - Puntiró - Secar de la Real - B228 - B236 - B278	3.06	0.81	−0.41
Son Rapinya - Pont d'Inca - Puntiró - Secar de la Real - B228 - B236 - B278	3.21	0.78	−3.08

Appendix E. GPA Inverse CDF on Site

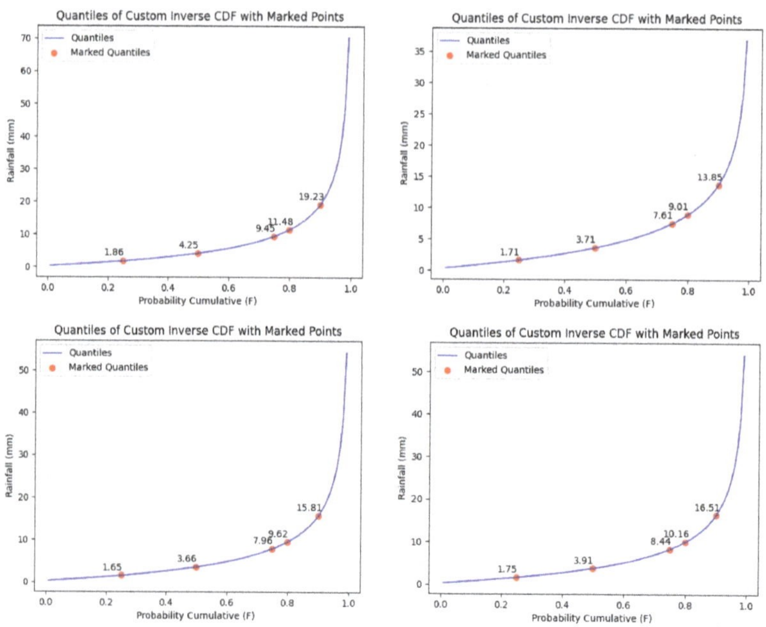

Figure A56. *Cont.*

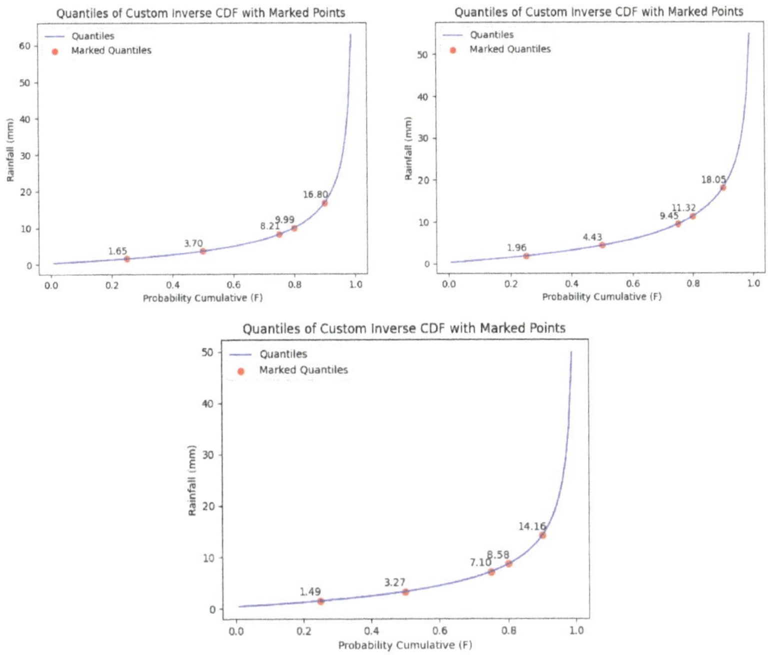

Figure A56. GPA inverse CDF on the different rainfall station sites.

References

1. Ramos, M. Divisive and hierarchical clustering techniques to analyse variability of rainfall distribution patterns in a Mediterranean region. *Atmos. Res.* **2001**, *57*, 123–138. [CrossRef]
2. IPCC. *Climate Change 2023: Synthesis Report (Full Volume) Contribution of Working Groups I, II and III to the Sixth Assessment Report of the Intergovernmental Panel on Climate Change*; Core Writing Team, Lee, H., Romero, J., Eds.; IPCC: Geneva, Switzerland, 2023; p. 184. [CrossRef]
3. Diez-Sierra, J.; del Jesus, M. A rainfall analysis and forecasting tool. *Environ. Model. Softw.* **2017**, *97*, 243–258. [CrossRef]
4. Cassalho, F.; Beskov, S.; de Mello, C.R.; de Moura, M.M. Regional flood frequency analysis using L-moments for geographically defined regions: An assessment in Brazil. *J. Flood Risk Manag.* **2019**, *12*, e12453. [CrossRef]
5. Senent-Aparicio, J.; Soto, J.; Pérez-Sánchez, J.; Garrido, J. A novel fuzzy clustering approach to regionalise watersheds with an automatic determination of optimal number of clusters. *J. Hydrol. Hydromech.* **2017**, *65*, 359–365. [CrossRef]
6. Aragón, L.; Lavado-Casimiro, W.; Montesinos, C.; Zubieta, R.; Laqui, W. Estimación de lluvias extremas mediante un enfoque de análisis regional y datos satelitales en Cusco, Perú. *Tecnol. Cienc. Agua* **2024**, *15*, 1–64. [CrossRef]
7. Smithers, J.; Schulze, R. A methodology for the estimation of short duration design storms in South Africa using a regional approach based on L-moments. *J. Hydrol.* **2001**, *241*, 42–52. [CrossRef]
8. Topçu, E.; Seçkin, N. Drought Analysis of the Seyhan Basin by Using Standardized Precipitation Index SPI and L-moments. *J. Agric. Sci.* **2016**, *22*, 196–215. [CrossRef]
9. Rahman, A.S.; Rahman, A.; Zaman, M.A.; Haddad, K.; Ahsan, A.; Imteaz, M. A study on selection of probability distributions for at-site flood frequency analysis in Australia. *Nat. Hazards* **2013**, *69*, 1803–1813. [CrossRef]
10. Serrano-Notivoli, R.; de Luis, M.; Beguería, S. An R package for daily precipitation climate series reconstruction. *Environ. Model. Softw.* **2017**, *89*, 190–195. [CrossRef]
11. Hosking, J.R.M.; Wallis, J.R. *Regional Frequency Analysis: An Approach Based on L-Moments*; Cambridge University Press: Cambridge, UK, 1997.
12. Malekinezhad, H.; Zare-Garizi, A. Regional frequency analysis of daily rainfall extremes using L-moments approach. *Atmósfera* **2014**, *27*, 411–427. [CrossRef]
13. Kjeldsen, T.R.; Smithers, J.C.; Schulze, R.E. Regional flood frequency analysis in the KwaZulu-Natal province, South Africa using the index-flood method. *J. Hydrol.* **2002**, *255*, 194–211. [CrossRef]

14. Vivekanandan, N. Flood frequency analysis using method of moments and L-moments of probability distributions. *Cogent Eng.* **2015**, *2*, 1018704. [CrossRef]
15. Blanco-Gómez, P.; Ortiz-Vallespi, J.; Estrany-Planas, P.; Orihuela-Martínez, J. Rainfall Frequency Analysis (RainFA) Package [Software]. 2023. Available online: https://github.com/vielca/RainFA (accessed on 7 May 2025).
16. Blanco-Gómez, P.; Estrany-Planas, P.; Jiménez-García, J.L. Cumulative Rainfall Radar Recalibration with Rain Gauge Data Using the Colour Pattern Regression Algorithm QGIS Plugin. *Remote Sens.* **2024**, *16*, 3496. [CrossRef]
17. Estrany-Planas, P.; Blanco-Gómez, P.; Vilarrasa, V. Aplicación del análisis regional de frecuencias a las series diarias de precipitación usando la metodología de los L-moments en la ciudad de Palma para la caracterización de la lluvia de diseño. In Proceedings of the VII Jornadas de Ingeniería del Agua. La resiliencia de las Infraestructuras Hidráulicas ante el Cambio Climático, 18–19 October, Cartagena, Spain; Universidad Politécnica de Cartagena: Cartagena, Spain, 2023; pp. 359–372. [CrossRef]
18. Council of European Communities. Council Directive 91/271/EEC of 21 May 1991 Concerning Urban Waste-Water Treatment. 1991. Available online: http://data.europa.eu/eli/dir/1991/271/oj (accessed on 7 May 2025).
19. European Parliament and Council of the European Union. Directive (EU) 2024/3019 of the European Parliament and of the Council Concerning Urban Wastewater Treatment (Recast). 2024. Available online: https://eur-lex.europa.eu/legal-content/EN/TXT/?uri=celex:52022PC0541 (accessed on 7 May 2025).
20. Ministerio de la Presidencia, Relaciones con las Cortes y Memoria Democrática. Real Decreto 665/2023, de 18 de julio, por el que se Modifica el Reglamento del Dominio Público Hidráulico, Aprobado por Real Decreto 849/1986, de 11 de Abril; el Reglamento de la Administración Pública del Agua, Aprobado por Real Decreto 927/1988, de 29 de julio; y el Real Decreto 9/2005, de 14 de enero, por el que se Establece la Relación de Actividades Potencialmente Contaminantes del suelo y los Criterios y estáNdares para la Declaración de suelos Contaminados. 2023. Available online: https://www.boe.es/diario_boe/txt.php?id=BOE-A-2023-18806 (accessed on 7 May 2025).
21. Mann, H.B. Nonparametric Tests Against Trend. *Econometrica* **1945**, *13*, 245–259. [CrossRef]
22. Kendall, M.G. *Rank Correlation Methods*, 4th ed.; Charles Griffin: London, UK, 1975; pp. 245–259.
23. The Pandas Development Team. Pandas-dev/Pandas: Pandas 2023. Available online: https://zenodo.org/records/7979740 (accessed on 7 May 2025).
24. Wright, D.B.; Mantilla, R.; Peters-Lidard, C.D. A remote sensing-based tool for assessing rainfall-driven hazards. *Environ. Model. Softw.* **2017**, *90*, 34–54. [CrossRef]
25. Hael, M.A.; Yongsheng, Y.; Saleh, B.I. Visualization of rainfall data using functional data analysis. *SN Appl. Sci.* **2020**, *2*, 461. [CrossRef]
26. Hyndman, R.T.; Shang, H.L. Rainbow plots, bagplots, and boxplots for functional data. *J. Comput. Graph. Stat.* **2010**, *19*, 29–45. [CrossRef]
27. Tukey, J.W. *Exploratory Data Analysis*; Addison Wesley: Reading, PA, USA, 1977.
28. Wickham, H.; Stryjewski, L. 40 Years of Boxplots. 2011. Available online: http://vita.had.co.nz/papers/boxplots.pdf (accessed on 7 May 2025).
29. Hyndman, R.T. Computing and Graphing Highest Density Regions. *Am. Stat.* **1996**, *52*, 120–126. [CrossRef]
30. Searcy, J.K.; Hardison, C.H. Double-Mass Curves. *Geol. Surv. Water Supply Pap.* **1960**, *1541-B*, 31–65.
31. Partal, T.; Kahya, E. Trend analysis in Turkish precipitation data. *Hydrol. Process.* **2006**, *20*, 2011–2026. [CrossRef]
32. Salmi, T.; Määttä, A.; Anttila, P.; Ruoho-Airola, T.; Amnell, T. *Detecting Trends of Annual Values of Atmospheric Pollutants by the Mann-Kendall Test and Sen's Slope Estimates—The Excel Template Application MAKESENS*; User Manual; Air Quality, Finish Meteorological Institute: Helsinki, Finland, 2002.
33. Citakoglu, H.; Minarecioglu, N. Trend analysis and change point determination for hydro-meteorological and groundwater data of Kizilirmak basin. *ATheoretical Appl. Climatol.* **2021**, *145*, 1275–1292. [CrossRef]
34. Blanco-Gómez, P.; Jimeno-Sáez, P.; Senent-Aparicio, J.; Pérez-Sánchez, J. Impact of Climate Change on Water Balance Components and Droughts in the Guajoyo River Basin (El Salvador). *Water* **2019**, *11*, 2360. [CrossRef]
35. Hussain, M.; Mahmud, I. pyMannKendall: A python package for non parametric Mann Kendall family of trend tests. *J. Open Source Softw.* **2019**, *4*, 1556. [CrossRef]
36. Das, S. Extreme rainfall estimation at ungauged sites: Comparison between region-of-influence approach of regional analysis and spatial interpolation technique. *Int. J. Climatol.* **2019**, *39*, 407–423. [CrossRef]
37. Agencia Estatal de Meteorología (AEMET). Valores Climatológicos. 2024. Available online: https://www.aemet.es/es/serviciosclimaticos/datosclimatologicos/valoresclimatologicos (accessed on 7 May 2025).
38. Borcard, D.; Gillet, F.; Legendre, P. *Numerical Ecology with R*; Springer International Publishing AG: Berlin/Heidelberg, Germany, 2011. [CrossRef]
39. Ward, J.H. Hierarchical grouping to optimize an objective function. *J. Am. Statist. Assoc.* **1963**, *58*, 236–244. [CrossRef]
40. Rousseeuw, P.J. Silhouettes: A Graphical Aid to the Interpretation and Validation of Cluster Analysis. *Comput. Appl. Math.* **1987**, *20*, 53–65. [CrossRef]

41. Arbelaitz, O.; Gurrutxaga, I.; Muguerza, J.; Pérez, J.M.; Perona, I. An extensive comparative study of cluster validity indices. *Pattern Recognit.* **2013**, *46*, 243–256. [CrossRef]
42. Ramachandra Rao, A.; Srinivas, V. Regionalization of watersheds by hybrid-cluster analysis. *J. Hydrol.* **2006**, *318*, 37–56. [CrossRef]
43. Aktürk, G.; Çıtakoğlu, H.; Demir, V.; Beden, N. Meteorological Drought Analysis and Regional Frequency Analysis in the Kızılırmak Basin: Creating a Framework for Sustainable Water Resources Management. *Water* **2024**, *16*, 2124. [CrossRef]
44. Millington, N.; Das, S.; Simonovic, S. The Comparison of GEV, Log-Pearson Type 3 and Gumbel Distributions in the Upper Thames River Watershed Under Global Climate Models. 2011. Available online: https://ir.lib.uwo.ca/wrrr/40 (accessed on 2 June 2025).
45. Liu, Z.N.; Yu, X.Y.; Jia, L.F.; Wang, Y.S.; Song, Y.C.; Meng, H.D. The influence of distance weight on the inverse distance weighted method for ore-grade estimation. *Sci. Rep.* **2021**, *11*, 2689. [CrossRef]
46. Achilleos, G. The Inverse Distance Weighted interpolation method and error propagation mechanism—creating a DEM from an analogue topographical map. *J. Spat. Sci.* **2011**, *56*, 283–304. [CrossRef]
47. Sahu, R.; Verma, M.; Ahmad, I. *Regional Frequency Analysis Using L-Moment Methodology—A Review*; Recent Trends in Civil Engineering; Springer: Singapore, 2021; Volume 77, pp. 811–832. [CrossRef]
48. Badyalina, B.; Mokhtar, N.A.; Mat Jan, N.A.; Hassim, N.H.; Yusop, H. Flood Frequency Analysis using L-Moment For Segamat River. *Matematika* **2021**, *37*, 47–62. [CrossRef]
49. Zambrano-Bigiarini, M.; Nauditt, A.; Birkel, C.; Verbist, K.; Ribbe, L. Temporal and spatial evaluation of satellite-based rainfall estimates across the complex topographical and climatic gradients of Chile. *Hydrol. Earth Syst. Sci.* **2017**, *21*, 1295–1320. [CrossRef]
50. Jimeno-Sáez, P.; Blanco-Gómez, P.; Pérez-Sánchez, J.; Cecilia, J.M.; Senent-Aparicio, J. Impact Assessment of Gridded Precipitation Products on Streamflow Simulations over a Poorly Gauged Basin in El Salvador. *Water* **2021**, *13*, 2497. [CrossRef]
51. Shapiro, S.S.; Wilk, M.B. An Analysis of Variance Test for Normality (Complete Samples). *Biometrika* **1965**, *52*, 591–611. [CrossRef]
52. Gelabert, M.; Rodríguez-Perea, A. Unidades morfológicas del llano de Palma (Mallorca). In Geomorfología en España; Sociedad Espanola de Geomorfologia: Logrono, Spain, 1994; pp. 403–411.
53. Torrens Calleja, J.M.; Rosselló Geli, J.; Grimalt Gelabert, M. Recopilación de Información Vinculada a Temporales de Viento, Precipitaciones Torrenciales e Inundaciones en la Ciudad de Palma de Mallorca Entre los Años 2000 y 2015. In Proceedings of the X Congreso Internacional AEC: Clima, Sociedad, Riesgos y Ordenación del Territorio, Alicante, Spain, 5–8 October 2016; pp. 417–425.
54. Torrens Calleja, J.M. Percepción de las precipitaciones en el Municipio de Palma de Mallorca (Illes Balears) entre 1980–2010; In *Análisis Espacial y Representación Geográfica: Innovación y Aplicación*; Realidades del Medio fíSico; Universidad de Zaragoza, Departamento de Geografía y Ordenación del Territorio: Zaragoza, Spain, 2015; pp. 1891–1900.
55. Govern de les Illes Balears. Servei de Mapa de la Pluviometria de Règim Extrem de les Illes Balears. Pluviometria de Règim Extrem (Període de Retorn: 10 Anys). 2024. Available online: https://ideib.caib.es/geoserveis/rest/services/public/GOIB_Meteo_IB/MapServer/5 (accessed on 7 May 2025).
56. Gelabert, M. *Repartiment de les Precipitacions Màximes a Mallorca*; Treballs de Geografía; Universitat de les Illes Balears: Palma, Spain, 1989; Volume 41, pp. 7–18.
57. Balearsmeteo. Balears Meteo, Network of Amateur Weather Stations in the Balearic Islands. 2024. Available online: http://balearsmeteo.com/ (accessed on 7 May 2025).
58. Sevruk, B. Adjustment of tipping-bucket precipitation gauge measurements. *Atmos. Res.* **1996**, *42*, 237–246. [CrossRef]
59. Wagner, P.D.; Fiener, P.; Wilken, F.; Kumar, S.; Schneider, K. Comparison and evaluation of spatial interpolation schemes for daily rainfall in data scarce regions. *J. Hydrol.* **2012**, *464–465*, 388–400. [CrossRef]
60. Das, S.; Millington, N.; Simonovic, S. Distribution choice for the assessment of design rainfall for the city of London (Ontario, Canada) under climate change. *Can. J. Civ. Eng.* **2013**, *40*, 121–129. [CrossRef]
61. Beersma, J.J.; Buishand, T.A. Drought in the Netherlands – Regional frequency analysis versus time series simulation. *J. Hydrol.* **2007**, *347*, 332–346. [CrossRef]
62. Lopes Martins, L.; Souza, J.; Sobierajski, G.; Blain, G. Is it possible to apply the regional frequency analysis to daily extreme air temperature data? *Bragantia* **2022**, *81*, e4322. [CrossRef]

hydrology

MDPI

Article

Refining Rainfall Derived from Satellite Radar for Estimating Inflows at Lam Pao Dam, Thailand

Nathaporn Areerachakul [1,*], Jaya Kandasamy [2,*], Saravanamuthu Vigneswaran [2] and Kittitanapat Bandhonopparat [3]

1 Faculty of Engineering and Industrial Technology, Suansunandha Rajhabhat University, Dusit, Bangkok 10300, Thailand
2 Faculty of Engineering and Information Technology, University of Technology Sydney, Brosdway, Ultimo, NSW 2007, Australia; saravanamuth.vigneswaran@uts.edu.au
3 Faculty of Science and Technology, Rajamangala University Thanyaburi, Amphoe Khlong Luang, Pathum Thani 12110, Thailand; kittanapat_ba@rmutt.ac.th
* Correspondence: nathaporn.ar@ssru.ac.th (N.A.); jaya.kandasamy@uts.edu.au (J.K.)

Abstract: This project aimed to evaluate the use of meteorological satellite-derived rainfall data to estimate water inflows to dams. In this study, the Lam Pao Dam in the Chi Basin, Thailand, was used as a case study. Rainfall data were obtained using the PERSIANN technique. To improve accuracy, satellite-derived rainfall estimates were adjusted using ground-based rainfall measurements from stations located near and within the catchment area, applying the 1-DVAR method. The Kriging method was employed to estimate the spatial distribution of rainfall over the catchment area. This approach resulted in a Probability of Detection (POD) of 0.92 and a Threat Score (TS) of 0.72 for rainfall estimates in the Chi Basin. Rainfall data from the Weather Research and Forecasting (WRF) numerical models were used as inputs for the HEC-HMS model to simulate water inflows into the dam. To refine rainfall estimates, various microphysics schemes were tested, including WSM3, WSM5, WSM6, Thompson, and Thompson Aerosol-Aware. Among these, the Thomson Aerosol-Aware scheme demonstrated the highest accuracy, achieving an average POD of 0.96, indicating highly reliable rainfall predictions for the Lam Pao Dam catchment. The findings underscore the potential benefits of using satellite-derived meteorological data for rainfall estimation, particularly where installing and maintaining ground-based measurement stations is difficult, e.g., forests/mountainous areas. This research contributes to a better understanding of satellite-derived rainfall patterns and their influence on catchment hydrology for enhanced water resource analysis.

Keywords: rainfall estimation; satellite; WRF; HEC-HMS; Lam Pao Dam

1. Introduction

Hydrological modeling has been widely applied in Thailand using tools such as SWAT and HEC-HMS. These models rely heavily on accurate data inputs to understand catchment hydrodynamics and to develop effective strategies for water resource management, flood mitigation, drainage planning, water supply, and irrigation systems. This research focuses on simulating and predicting inflows to dams for the purposes of improved water resource investigations. The Lam Pao Dam, a key agricultural reservoir in northeastern Thailand, plays a critical role in supporting irrigation, particularly during dry seasons. Constructed in 1968, the dam has an estimated storage capacity of 1980 million cubic meters. It supplies water to approximately 54,080 hectares during the rainy season and 4240 hectares during the dry season [1].

The study incorporates radar-derived rainfall data to a HEC-HMS hydrological model of the Lam Pao Basin to evaluate its effectiveness for water resource purposes. Variability in rainfall significantly influences river levels and the overall water balance. Additionally, changes in runoff behavior have been exacerbated by deforestation within the catchment area. Research conducted in the Chi River Basin from 1951 to 2003 reported a 20% reduction in forest cover over a span of 52 years [2]. To assess water availability across different sub-regions of the Chi Basin—such as the Lumsapung Basin and Mae Sruai Dam—hydrological models employing net water balance methods have been utilized. These models typically account for inflows (Qin), precipitation (Pt), evaporation (Et), and outflows (Qout) to estimate variations in reservoir storage (ΔS). Additional studies [3–5] have sought to improve hydrological modeling techniques to enhance water resource management practices. The integration of hydrological and meteorological data has proven effective in improving streamflow estimation across numerous river basins and catchments in Thailand [6–8].

Radar systems operate by emitting electromagnetic waves—similar to those used in mobile phones and wireless networks—which are transmitted in brief pulses. These pulses reflect off objects in their path, with a portion of the signal returning to the radar. When the pulses encounter precipitation, part of the energy is scattered back, allowing the radar to detect and estimate rainfall intensity [9]. While rain gauges offer precise measurements of rainfall at specific locations, they are limited in their ability to capture spatial variability. In contrast, radar provides broad spatial coverage but measures precipitation within the atmosphere rather than the exact amount that reaches the ground [9].

To overcome the limitations of both methods, rainfall accumulation data from radar are typically presented as images that represent estimated precipitation totals over defined time intervals—such as every five minutes or hourly. These images are created by combining the high spatial resolution of radar with the point-specific accuracy of rain gauge data. Rain gauges are used to calibrate radar outputs, improving rainfall estimation in areas between gauge locations [10]. This combined approach yields more accurate precipitation estimates than using either radar or gauges independently. However, the quality of these accumulation images depends on accurate radar reflectivity, reliable rain gauge measurements, and effective correction of errors and biases [9,10].

This study incorporates satellite-based rainfall estimates using the PERSIANN (Precipitation Estimation from Remotely Sensed Information using Artificial Neural Networks) approach. Specifically, infrared-based PERSIANN rainfall data were applied to the Chi River Basin and integrated with the HEC-HMS model to evaluate hydrological conditions at the Lam Pao Dam. While ground-based rain gauge data were utilized, their spatial coverage is limited due to the Chi Basin's diverse and complex terrain. Comparing gridded satellite precipitation data with point-based gauge observations within the HEC-HMS modeling framework provides valuable insights into the influence of data source variability on hydrological model outputs.

In addition to radar and rain gauge data, rainfall information can also be derived from satellite observations and outputs from numerical weather prediction (NWP) models. One such model is the Weather Research and Forecasting (WRF) model—a state-of-the-art mesoscale NWP system designed for both research and operational flow estimation. WRF supports meteorological simulations across a wide range of spatial scales, from tens of meters to several thousand kilometers. It generates precipitation accumulation outputs based on real-time radar observations, which are calibrated with rain gauge data to reduce uncertainties. The WRF modeling framework has been applied in various regions worldwide [11–14]. In this study, WRF-derived data are integrated with the HEC-HMS hydrological model to simulate streamflow in the Lam Pao Basin.

This study aims to evaluate the effectiveness of integrating rainfall data from the WRF model and the PERSIANN algorithm with the HEC-HMS model for water resource and irrigation purposes. The primary objective is to simulate streamflow within the Lam Pao Basin using this combined modeling approach. The HEC-HMS model was employed to estimate the Lam Pao Dam's storage levels over annual periods, facilitating an analysis of long-term hydrological trends. The simulation period of up to one year was selected to assess the model's performance and consistency, with the potential for extension to longer-term applications in future research. Overall, this approach supports the evaluation of the model's reliability and its potential use as a decision-support tool for managing water resources at Lam Pao Dam.

2. Data and Methods

This study is organized into three core components. First, rainfall estimates across the catchment were obtained using the PERSIANN satellite-based technique. To improve the precision of these estimates, the 1-DVAR method was applied, incorporating observational data from local rain gauge stations. Spatial analysis of rainfall over the reservoir area was further conducted using the kriging interpolation method.

In the second phase, rainfall estimates for the catchment were produced using the WRF atmospheric model, employing five different microphysics parameterization schemes. The performance of each scheme was assessed through statistical evaluation, using observed rainfall data from ground stations for validation.

Lastly, the satellite-derived PERSIANN rainfall data were fed into the HEC-HMS hydrological model to simulate inflow into Lam Pao Dam. These simulated values were compared with observed discharge data to assess accuracy. In parallel, rainfall estimates from the WRF model were also used as input for hydrological simulations, and the outputs were evaluated using statistical validation techniques.

2.1. Study Area

The Chi Basin, where PERSIANN satellite rainfall data were collected and WRF estimated rainfall was estimated, is depicted in Figure 1. Located in northeastern Thailand, the basin spans an area of 49,131.92 km^2 and covers 14 provinces, including Chaiyaphum, Khon Kaen, Nong Bua Lamphu, Udon Thani, Maha Sarakham, Nakhon Ratchasima, Loei, Phetchabun, Kalasin, Roi Et, Yasothon, Ubon Ratchathani, Si Sa Ket, and Mukdahan.

The Lam Pao Dam, located in northern Kalasin Province, was constructed between 1963 and 1968. It has a storage capacity of 1430 million cubic meters and plays a vital role in flood prevention and agricultural water supply. The dam's catchment area extends into a large part of Udon Thani Province. Downstream, the Lam Pao River flows through Kalasin Province before joining the Chi River at the boundary between Kalasin and Roi Et Provinces.

Water usage from the dam is distributed across various sectors: irrigation (87.1%), aquaculture (9.5%), domestic consumption (1.8%), industry (1.6%), and livestock (0.1%). Water shortages typically occur during the dry season, and demand during drought conditions—particularly for aquaculture and irrigation—places significant pressure on water allocation.

The irrigated area includes approximately 50,000 hectares during the wet season and around half that in the dry season. Rice is the predominant crop, occupying about 50% of the farmland, followed by cassava, groundnuts, sugarcane, and kenaf.

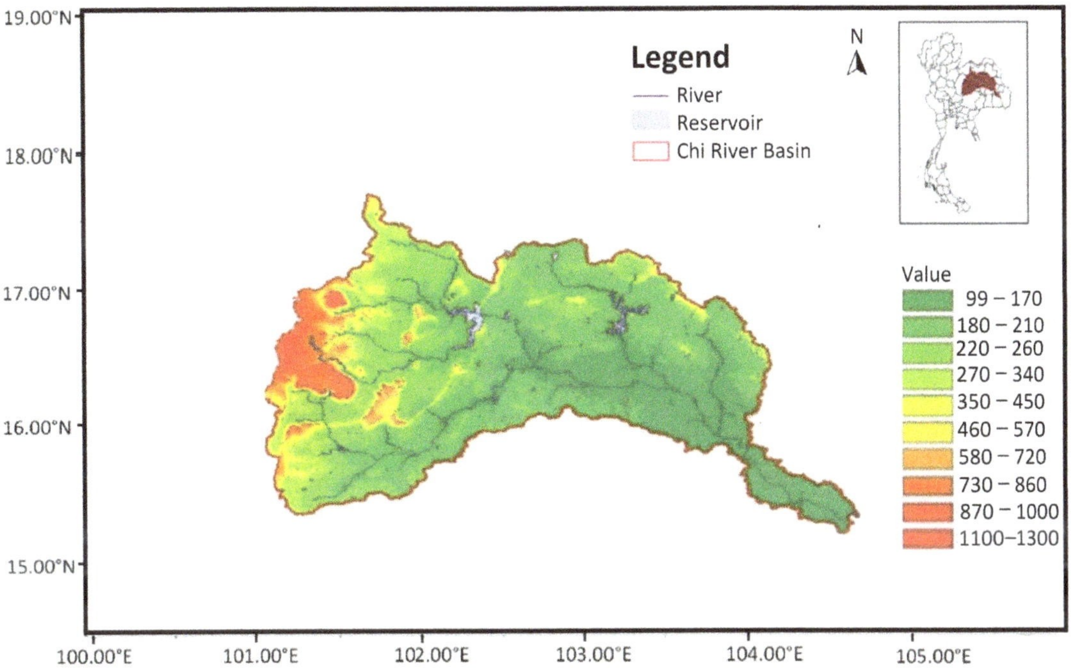

Figure 1. Map of northeastern region of Thailand showing important rivers including the Chi River.

Aquaculture is mainly practiced in major water bodies, including the Lam Pao River and Lam Pao Reservoir. In 1998, total aquaculture production reached nearly 251 tons. Additionally, commercial-scale shrimp farming within the Lam Pao Irrigation Project in Kalasin Province produced about 700 tons in the same year.

2.2. *Rainfall Data*

The rainfall data used in this study were supplied by the Thailand Deptartment of Meteorology. Rainfall estimation in this study made use of the PERSIANN dataset, which is based on satellite remote sensing data. This method relies on infrared (IR) brightness temperature observations from NCEP, collected at 15 min intervals between 2009 and 2019 [15]. In addition, Tropical Rainfall Measuring Mission (TRMM) Multi-Satellite Precipitation Analysis (TMPA) was employed, which combines data from several satellites along with ground-based rain gauge observations [16,17]. The TRMM dataset covers the latitudinal range of 50° S to 50° N and spans the years 2009 to 2018. For this study, the TRMM 3B42 Version 7 (TRMM-3B42V7) algorithm was used, offering high-resolution data at $0.25^{\circ} \times 0.25^{\circ}$ spatial and 6 h temporal scales for detailed rainfall analysis.

Statistical Verification and Evaluation Methods to Compare Satellite with Rain Gauge Stations

The 1DVAR technique was applied to generate an optimal estimate of the vertical atmospheric state profile, represented as x, by integrating observational data (y_o) with a background state estimate (x_b). This method takes into account the associated error covariance matrices for both the observational inputs and the background information.

The 1DVAR approach is mathematically framed as the minimization of the following function [1]:

$$J(x) = \frac{1}{2}(x - x_b)^T B^{-1}(x - x_b) + \frac{1}{2}(y_0 - H\lfloor x \rfloor)^T O^{-1}(y_0 - H\lfloor x \rfloor) \quad (1)$$

where y_o = elevation of observation (m); T = temperature in K; O = observation error covariance matrix (m, m); B = background error covariance matrix (n, n); and $H(x)$ = converts the n dimensional state vector defined on the 1D-Var levels and surface into a corresponding profile of refractivity given on the m refractivity levels.

To evaluate the model's performance, the coefficient of determination (R^2) was utilized. The correlation coefficient (CC) was also used to quantify the strength and direction of the relationship between two variables—typically either two observed data series or components of a multivariate dataset. The CC ranges from −1.0 to 1.0, with values approaching 1 indicating a strong positive correlation, meaning the estimated values closely align with the observed trends [18].

$$CC = \frac{\sum_{i=1}^{N}(Y_i - \overline{Y})(O_i - \overline{O})}{\sqrt{\sum_{i=1}^{N}\left(Y_i - \overline{Y}\right)^2 \sum_{i=1}^{N}\left(O_i - \overline{O}\right)^2}} \quad (2)$$

where Y_i = gauge rainfall measurement, $\overline{Y}$ = average gauge rainfall measurement, O_i = estimated rainfall, and $\overline{O}$ = estimated rainfall average.

Table 1 illustrates how the relationship between satellite data (estimate) and rain gauge station data (observed) is classified. In this study, "Yes" refers to rainfall amounts greater than or equal to 0.1 mm, while "No" refers to rainfall amounts less than 0.1 mm. When comparing rainfall quantities from the rain gauge station (observed) and those estimated from satellite data (estimate), the variables in Table 1 are classified based on the following criteria:

- Hit = the rainfall quantity of the satellite (estimate) and rain gauge station (observed) ≥ 0.1;
- False alarm = the rainfall quantity of the satellite (estimate) ≥ 0.1 and rain gauge station (observed) < 0.1;
- Miss = the rainfall quantity of the satellite (estimate) < 0.1 and the rate of the rain gauge station (observed) > 0.1;
- Correct rejection = the satellite (estimate) and rain gauge station (observed) < 0.1.

Table 1. Binary contingency table for whether there is rain.

Estimate	Observed		Total
	Yes	No	
Yes	a (hit)	b (false alarm)	a + b
No	c (miss)	d (correct rejection)	c + d
Total	a + c	b + d	a + b + c + d = n

The scores in Table 1 are divided into four indices, as outlined in Equations (3)–(6). The Probability of Detection (POD) is the ratio of hits (a) to the total number of observed events (a + c). The False Alarm Ratio (FAR) represents the number of false alarms (b) compared to the total number of "yes" estimates (a + b). The Probability of False Detection (POFD) is the ratio of false alarms (b) to the total number of "no" estimates (b + d). The Threat Score (TS) is the ratio of hits (a) to the total number of observed "yes" (rain) events plus false alarms (a + b + c).

1. Probability of Detection (POD) [19]

$$POD = \frac{a}{a+c} \quad (3)$$

2. False Alarm Ratio (FAR) [20]

$$FAR = \frac{b}{a+b} \quad (4)$$

3. Probability of False Detection (POFD) [19]

$$POFD = \frac{b}{b+d} \quad (5)$$

4. Threat Score (TS) [21]

$$TS = \frac{b}{a+b+c} \quad (6)$$

The perfect estimate has a POD equal to 1. Similarly, a perfect estimate occurs where the FAR and POFD equal to 0. The TS gives the rate of all well-predicted locations with respect to the reference, where the reference or the simulation are above the threshold. The POD, FAR, and TS generally need to be examined together with MAE and RMSE goodness of fit criteria.

2.3. WRF Model Rainfall Estimates

Rainfall across the Chi Basin in northeastern Thailand was monitored using meteorological satellite data. The Thai Meteorological Department (TMD) utilized the WRF model, driven by the National Center for Environmental Prediction (NCEP) Global model, to generate rainfall estimates. For calibration purposes, rainfall data from 10 nearby stations (listed in Table 2) were collected over a 10-year span from 2008 to 2017. A statistical summary of the data is given in Table S3. The WRF-simulated rainfall was then compared to observed values from these rain gauge stations and adjusted accordingly to improve model accuracy.

Table 2. Rainfall station used in this study.

Code	Station	Latitude	Longitude
48354	Udon Thani	17.383	102.800
48356	Sakon Nakhon	17.150	104.133
48355	Sakon Nakhon (1)	17.125	104.061
48360	Nong Bua Lumphu	17.233	102.429
48381	Khon Kaen	16.461	102.790
48384	Tha Pra (1)	16.333	102.817
48382	Kosum Phisai (2)	16.247	103.068
48390	Kamalasai (2)	16.333	103.588
48405	Roi Et	16.050	103.683
48404	Roi Et (1)	16.073	103.608

Among the various factors influencing numerical weather prediction models, cloud microphysics plays a critical role as a major source of uncertainty. Five different WRF microphysics schemes used in this study are listed in Table 3.

Table 3. Microphysics (MP) schemes used in WRF model.

MP Scheme	Scheme	Reference
3	WSM3	[22,23]
4	WSM5	[23]
6	WSM6	[24]
8	Thompson	[25]
28	Thompson Aerosol-Aware	[25–27]

2.4. HEC-HMS Model of Lam Pao Dam

The Lam Pao Dam, located in the Chi Basin in northern Kalasin Province, was constructed between 1963 and 1968. Its primary function is agricultural irrigation. The Lam Pao Reservoir effectively divides the catchment into two sections (see Figure S1). Most of the catchment area upstream of the dam is mountainous. To the north-northeast, the Phu Phan Mountain Range extends, with part of it designated as a national park, and beyond that lies Sakhon Nakhon Province. To the northwest of the catchment is Udon Thani. The river upstream of the dam originates from Nong Han Kumphawapi Lake in Udon Thani Province, flowing through Kalasin Province before converging with the Chi River at the boundary between Kalasin and Roi Et Provinces.

For this study, the HEC-HMS model was employed to simulate specific rainfall events and assess inflows into the Lam Pao Dam. Flow data measured at the dam were used to validate the model's output. These flow data were supplied by the Thailand Department of Irrigation. Information regarding the model setup, description, and calibration (see Figure S2) can be found in the Supplementary Information (SI).

3. Results and Discussion

3.1. Comparison and Validation of PERSIANNs Satellite Data

This section presents a comparison between rainfall data from ground-based rain gauge stations and satellite-derived meteorological data from PERSIANN, along with the application of bias correction using the 1-DVAR method. The PERSIANN data from 2009 to 2018, covering the rainy season months of June, July, August, and September, were selected for analysis. The correlation coefficient (R^2) values, both pre- and post-bias correction, for the ten-year period are shown in Table 4.

Table 4. Correlation coefficient during June–September (R^2) of rainfall gauge data compared with PERSIANN satellite.

Years	June		July		August		September	
	R^{2+}	R^{2*}	R^{2+}	R^{2*}	R^{2+}	R^{2*}	R^{2+}	R^{2*}
2009	0.04	0.79	0.23	0.72	0.12	0.69	0.10	0.60
2010	0.35	0.19	0.19	0.49	0.01	0.27	0.24	0.05
2011	0.73	0.94	0.40	0.73	0.33	0.83	0.01	0.49
2012	0.34	0.87	0.56	0.78	0.44	0.76	0.01	0.25
2013	0.21	0.52	0.41	0.78	0.27	0.70	0.21	0.58
2014	0.48	0.69	0.46	0.70	0.00	0.82	0.65	0.90
2015	0.02	0.78	0.54	0.83	0.18	0.58	0.05	0.48
2016	0.01	0.37	0.30	0.66	0.44	0.81	0.63	0.83
2017	0.41	0.82	0.58	0.80	0.38	0.73	0.18	0.63
2018	0.28	0.86	0.52	0.81	0.63	0.88	0.04	0.32
Avg	0.28	0.68	0.42	0.73	0.28	0.71	0.21	0.51

+ before bias correction; * after bias correction.

Satellite imagery, including data from rain gauge stations, PERSIANN satellite observations, and the images after data assimilation with the 1-DVAR technique, are presented in Figure S3. In July 2018, rainfall recorded by the rain gauge stations ranged from 200 to 340 mm, with the heaviest precipitation occurring in the central, eastern, and northeastern parts of the Chi Basin. The PERSIANN data estimated rainfall between 240 and 400 mm, covering areas from the north to the east and northeast of the basin. The correlation coefficient with the rain gauge data was $R^2 = 0.54$. Following data assimilation with 1-DVAR, the satellite data showed improved agreement with the rain gauge data, resulting in a higher R^2 of 0.83. In August 2018, rain gauge data varied from 200 to 300 mm, while the satellite data ranged from 220 to 340 mm, yielding an initial R^2 of 0.33. After applying 1-DVAR assimilation, the satellite rainfall data aligned much more closely with the ground measurements, improving the R^2 to 0.83. For September 2018 (see Figure S3), the rain gauge data ranged from 200 to 420 mm, whereas the satellite estimates ranged from 200 to 320 mm, with an R^2 of 0.66. After 1-DVAR assimilation, the satellite data adjusted to a range of 220 to 340 mm, resulting in an R^2 of 0.90. September showed the most significant improvement, with the best correlation between the rain gauge data and the post-assimilation satellite estimates.

In July 2018, the Probability of Detection (POD) reached its highest value of 0.924, while the False Alarm Ratio (FAR) was 0.15, as shown in Figure S3. In August 2018, the POD remained high at 0.892, which also corresponds to the data presented in Figure S3. For September 2018, the POD score was 0.75, and the FAR was 0.37, demonstrating the best match between the rain gauge data and the post-assimilation satellite data using the 1-DVAR technique, as illustrated in Figure S3.

3.2. WRF-Derived Rainfall

The results from the Numerical Weather Prediction (NWP) approach using the WRF model are presented. The WRF rainfall estimates for the period from 2013 to 2018 were utilized to predict rainfall in the Lam Pao Dam catchment.

An analysis of the data in Table 5 for June 2013 indicates that Scheme 8 outperformed the other four schemes, showing the best fit overall. Scheme 8 achieved the highest values for Probability of Detection (POD) and Threat Score (TS), and the lowest values for False Alarm Ratio (FAR) and Probability of False Detection (POFD), as shown in Table 5, although Scheme 28 performed similarly, though slightly less effectively.

Table 5. The event of hits, false alarms, missing events, correct negatives, PODs, FARs, POFDs, and TSs for June 2013, June 2014, and June 2018. Values with the highest POD and TS and lowest FAR and POFD are highlighted.

		June 2013				
Scheme	3	4	6	8	28	Avg
Hits	275	283	300	304	309	
False Alarm	75	73	60	48	50	
Missing Event	26	14	21	13	22	
Correct negative	44	50	39	55	39	
POD	0.914	0.953	0.935	0.959	0.934	0.939
FAR	0.214	0.205	0.167	0.136	0.139	0.172
POFD	0.630	0.593	0.606	0.466	0.562	0.572
TS	0.731	0.765	0.787	0.833	0.811	0.786

Table 5. *Cont.*

	June 2014					
Scheme	**3**	**4**	**6**	**8**	**28**	**Avg**
Hits	199	249	253	256	258	
False Alarm	63	104	104	106	103	
Missing Event	56	34	30	27	25	
Correct negative	32	19	19	17	20	
POD	0.780	0.880	0.894	0.905	0.912	0.874
FAR	0.240	0.295	0.291	0.293	0.285	0.281
POFD	0.663	0.846	0.846	0.862	0.837	0.811
TS	0.626	0.643	0.654	0.658	0.668	0.650
	June 2018					
Scheme	**3**	**4**	**6**	**8**	**28**	**Avg**
Hits	194	245	241	238	243	
False Alarm	79	110	112	104	105	
Missing Event	36	21	22	24	24	
Correct negative	41	44	45	54	48	
POD	0.843	0.921	0.916	0.908	0.910	0.900
FAR	0.298	0.310	0.317	0.304	0.304	0.304
POFD	0.658	0.714	0.713	0.658	0.686	0.686
TS	0.628	0.652	0.643	0.653	0.653	0.645

In June 2014, Scheme 28 achieved the highest POD value of 0.912 and the highest TS value of 0.668. Meanwhile, Scheme 3 had the lowest FAR value of 0.285 and the lowest POFD value of 0.837.

For June 2018, results varied across the different schemes. Scheme 4 had the highest POD value, while Scheme 3 exhibited the lowest FAR. Schemes 8 and 28 showed slightly higher FAR values. Both Scheme 3 and Scheme 20 had the lowest POFD values, while Scheme 28 attained the highest TS value.

Overall, Scheme 28 demonstrated the most robust performance, as indicated by its high POD, FAR, POFD, and TS values, highlighted in Table 5. Schemes 3 and 8 also performed well but slightly behind Scheme 28 in terms of effectiveness. Figure 2 illustrates this by showing the spatial distribution of rainfall derived from rain gauge data (Figure 2a) and the mapping results from kriging spatial methods for Schemes 3, 8, and 28 (Figure 2b–d). Scheme 28 (Figure 2d) most closely mirrored the spatial map from the rain gauge stations, especially in terms of rainfall intensity variation from west to east, ranging from 200 to 320 mm.

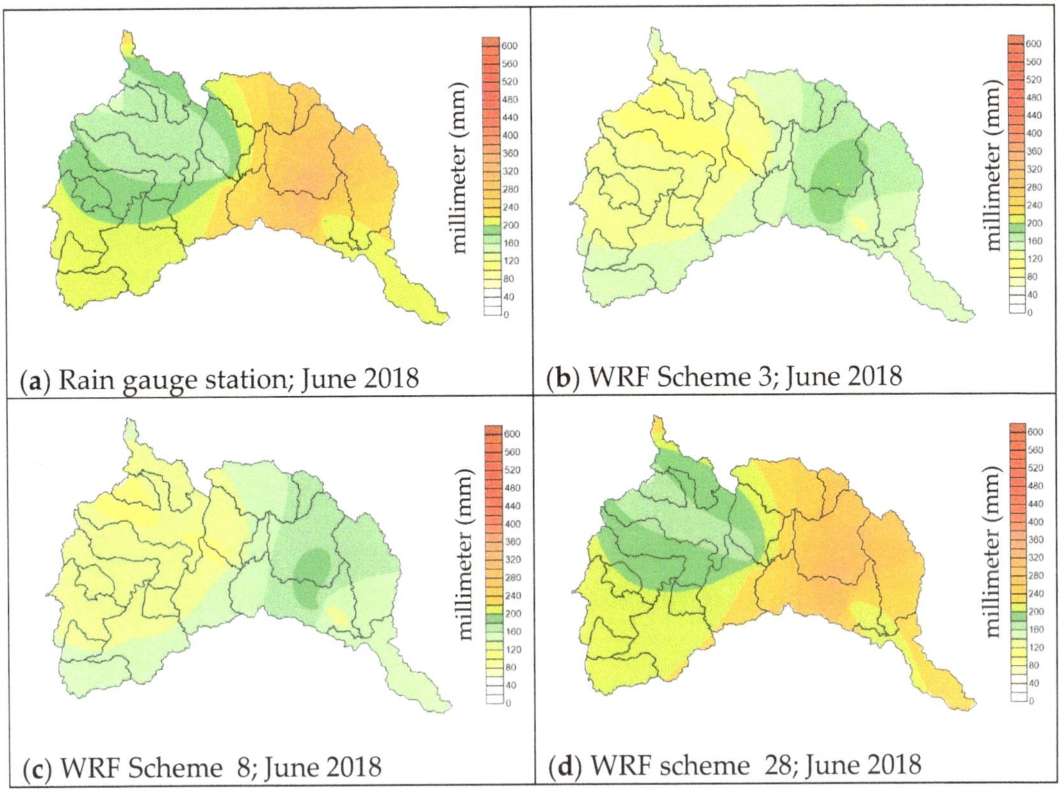

Figure 2. Rainfall distribution after kriging methods for June 2018.

3.3. Hydrology Modeling Coupling with PERSIANN Satellite and WRF Data

The PERSIANN satellite-derived rainfall data were integrated into the HEC-HMS model, and the resulting simulations were compared with observed inflow data from the Lam Pao Dam (Figure 3). For comparison, a separate model using rain gauge data was also developed and evaluated.

In 2010 (Figure 3a,b), the model utilizing satellite-derived rainfall closely aligned with the results from the rain gauge-based model, indicating good agreement between the two data sources. Table S4 shows that the inflow to Lam Pao Dam was within 10%. However, in 2011 (Figure 3c,d), a noticeable discrepancy emerged between the results obtained from satellite data and those based on rain gauge observations. Both models, however, failed to fully replicate the observed inflows, particularly during September, suggesting limitations in capturing peak flow events. The estimated rainfall over the Lam Pao Dam catchment area in 2010 was slightly underestimated compared to the measured rainfall, highlighting the need for further refinement (Table S4).

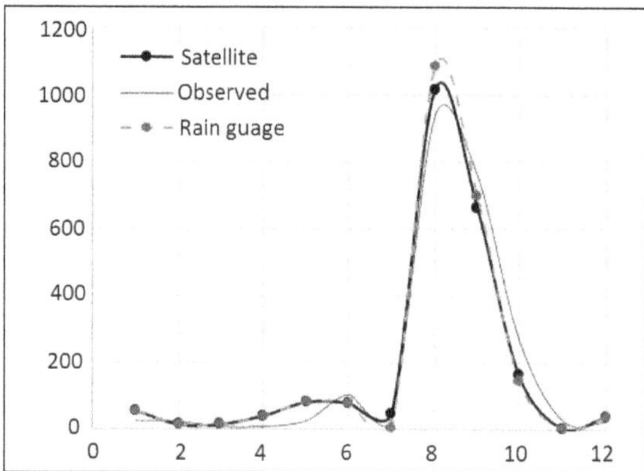

(**a**) 2010: Inflow to Lam Pao Dam (1 × 10^6 m^3) vs. month (in number); 1 × 10^6 m^3/m = 100,000 cubic meters per month.

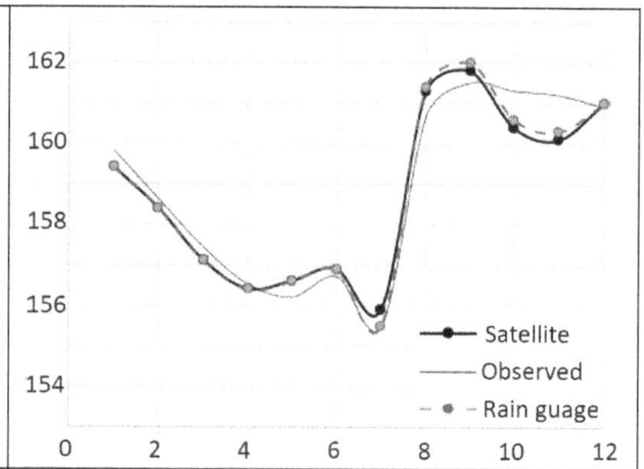

(**b**) 2010: Water level of Lam Pao Dam (m above datum) vs. month (in number).

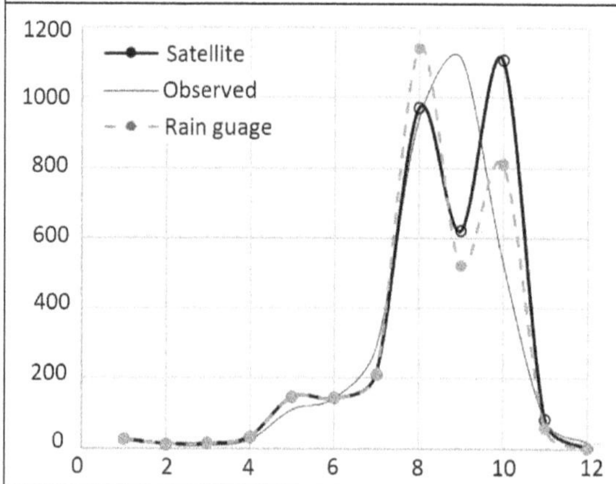

(**c**) 2011: Inflow to Lam Pao Dam (1 × 10^6 m^3) vs. month (in number); 1 × 10^6 m^3/m = 100,000 cubic meters per month.

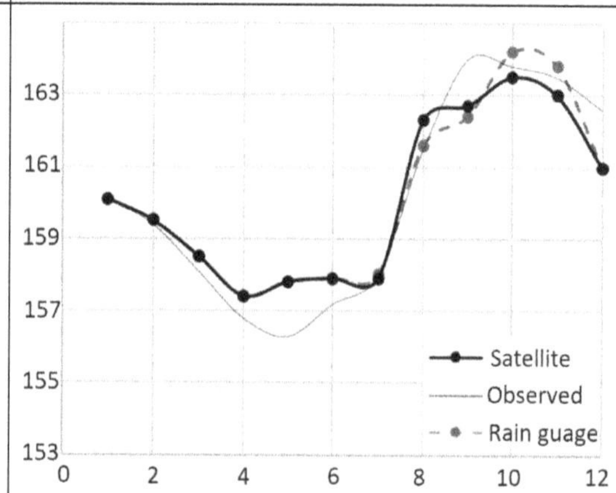

(**d**) 2011: Water level of Lam Pao Dam (m above datum) vs. month (in number).

Figure 3. Comparison between observed data with HEC-HMS model results using PERSIANN satellite station data. Inflow and water levels estimated by HEC-HMS from rain gauge data given for comparison.

These findings underscore the potential for improving the simulation of catchment flows into the Lam Pao Dam. A more detailed assessment of satellite rainfall data accuracy is required, along with appropriate corrections to better align with ground-based measurements. Moreover, the acquisition of higher-resolution Digital Terrain Model (DTM) data is critical for enhancing the spatial accuracy of hydrological modeling. By employing advanced HEC-HMS capabilities and integrating both radar and satellite-derived rainfall data, the overall precision and reliability of the hydrological simulations can be significantly improved. A sensitivity analysis on input parameters (satellite and WRF) could be carried out to estimate the uncertainty in rainfall inputs and assess how this affects HEC-HMS outputs and, by extension, their application in water management planning.

WRF data for the years 2013 and 2014 were available, and the comparison, shown in Figure S4, focuses specifically on the months of June to September, where WRF data were available. No comparison with the HEC-HMS based on rain gauge data was possible due to lack of data in those years. The limited comparison shows that the rain estimates in the water catchment area of the Lam Pao Dam provided higher estimates than observed data.

This comparison provides insight into the potential of satellite data to serve as a viable alternative to ground-based rain gauge measurements. A robust conclusion on the possibility of replacing ground-based gauges with satellite data would require long-term data across multiple locations. This is necessary to capture the diversity of rainfall mechanisms that occur over time and across different geographical settings. Reaching such a conclusion would require the accumulation of evidence from many studies conducted over time and in varied regions [28,29]. Accordingly, our study is a contribution toward that broader objective.

4. Conclusions

In the initial phase of the study, the focus was on validating PERSIANN rainfall estimates within the Chi Basin. Validation was carried out using bias correction threshold scores and the 1D-VAR data assimilation technique. During June 2011, PERSIANN satellite-derived estimates indicated higher intensities than rain gauge data and a correlation coefficient (R^2) of 0.73. The 1D-VAR assimilation method resulted in a significantly improved fit with an R^2 of 0.94, closely aligning with the ground-based measurements. Rainfall validation POD, TS, etc., demonstrated a close agreement between the corrected PERSIANN estimates and observed rainfall data. The PERSIANN-derived rainfall data were subsequently integrated into the HEC-HMS model and gave adequate simulated estimates of water inflows into the Lam Pao Dam.

Rainfall estimation was carried out using the WRF atmospheric model, which employed five different microphysics schemes. The estimated rainfall outputs were statistically validated by comparing them with observed rainfall data from measurement stations across the Chi Basin. The performance of the microphysics schemes, reflected by an average Probability of Detection (POD) value of 0.90, indicated that the WRF model provided consistent and reasonably accurate rainfall predictions for the Lam Pao Dam catchment area when benchmarked against ground-based observations. These validated estimates, used with the HEC-HMS hydrological model, were able to simulate water inflow into the Lam Pao Dam, albeit in a limited comparison.

The results of this study provide insight into the potential of PERSIANN satellite-based rainfall estimates, particularly in regions where the installation and maintenance of ground-based measurement stations are impractical—such as forested areas or rugged terrain. Furthermore, the integration of satellite-derived rainfall data with other observational and modeling resources offers significant potential to enhance the ability to provide flow estimates for water resource studies. This approach can help mitigate economic losses, improve operational planning, and ultimately strengthen community resilience.

Supplementary Materials: The following supporting information can be downloaded at: https://www.mdpi.com/article/10.3390/hydrology12070163/s1, Figure S1: Location map of Lam Pao Dam catchment. Inset is location of Lam Pao Dam catchment relative to Thailand; Figure S2: Comparison between observed data with HEC-HMS model results using rainfall station data; Figure S3: PERSIANN forecast in Chi basin; Figure S4: Comparison between observed data with HEC-HMS model results using WRF Scheme 28; Table S1: Principal catchment characteristics of sub-basins 1 to 5; Table S2: Data sources used to develop HEC-HMS model; Table S3: Data summary of rainfall station used in this study; Table S4: Predicted inflow to Lam Pao Dam using rainfall derived from satellite and from rainfall.

Author Contributions: Conceptualization, N.A., K.B. and J.K.; methodology, K.B. and J.K.; software, K.B.; validation, K.B. and J.K.; formal analysis, K.B. and J.K.; investigation, K.B.; resources, N.A.; data curation, K.B.; writing—original draft preparation, K.B. and NA.; writing—review and editing, J.K. and S.V.; supervision, N.A. and J.K.; project administration, N.A. All authors have read and agreed to the published version of the manuscript.

Funding: This research was in part funded by the National Research Council of Thailand, grant number 106/2563, and the Agricultural Development Agency of Thailand (Public Organization).

Data Availability Statement: The data supporting the findings of this study are available upon reasonable request to the authors.

Acknowledgments: The authors would like to thank the Thai Meteorology Department for providing the data and Suansunandha Rajhabhat University for supporting this research.

Conflicts of Interest: The authors declare no conflicts of interest.

References

1. MAC. *Water Management Plan and Dry Season Crop Cultivation in Irrigation Areas 2018/19*; Water Allocation Division 1, Water Management Division, Office of Water Management and Hydrology, Ministry of Agriculture and Cooperatives (MAC): Bangkok, Thailand, 2018.
2. Boochabun, K.; Vongtanaboon, S.; Sukrarasmi, A. *Impact of Land-Use Development on the Water Balance and Flow Regime of the Chi River Basin, Thailand*; Environments in the Mekong River Basin; Springer: Tokyo, Japan, 2007. [CrossRef]
3. Legesse, D.; Vallet-Coulomb, C.; Gasse, F. Analysis of the hydrological response of a tropical terminal lake, lakeAbiyata (Main Ethiopian Rift Valley) to changes in climate and human activities. *Hydrol. Process.* **2004**, *18*, 487–504. [CrossRef]
4. Kunstmann, H.; Heckl, A.; Rimmer, A. Physically based distributed hydrological modelling of the Upper Jordan Catchment and investigation of effective model equations. *Adv. Geosci.* **2006**, *9*, 123–130. [CrossRef]
5. Tingsanchali, T.; Tanmanee, S. Assessment of hydrological safety of Mae Sruai Dam, Thailand. *Procedia Eng.* **2012**, *32*, 1198–1204. [CrossRef]
6. Maxwell, R.M.; Lundquist, J.K.; Mirocha, J.D.; Smith, S.G.; Woodward, C.S.; Tompson, A.F. Development of a coupled groundwater-atmosphere model. *Mon. Weather Rev.* **2011**, *139*, 96–116. [CrossRef]
7. Larsen, M.A.; Refsgaard, J.C.; Drews, M.; Butts, M.B.; Jensen, K.H.; Christensen, J.H.; Christensen, O.B. Results from a full coupling of the HIRHAM regional climate model and MIKE SHE hydrological model for a Danish catchment. *Hydrol. Earth Syst. Sci.* **2014**, *18*, 4733–44749. [CrossRef]
8. Larsen, M.A.; Refsgaard, J.C.; Jensen, K.H.; Butts, M.B.; Stisen, S.; Mollerup, M. Calibration of a distributed hydrology and land surface model using energy flux measurements. *Agric. For. Meteorol.* **2016**, *217*, 74–88. [CrossRef]
9. Sunmin, K. Real-time flood forecasting with weather radar and distributed hydrological model, Chapter 18. In *Water Engineering Modeling and Mathematic Tools*; Samui, P., Bonakdari, H., Deo, R., Eds.; Elsevier: Amsterdam, The Netherlands, 2021; Volume 2021, pp. 369–380. ISBN 9780128206447.
10. BOM. How Radar Works, Bureau of Meterology Austraila. 2025. Available online: http://www.bom.gov.au/australia/radar/about/what_is_radar.shtml#:~:text=Radars%20send%20out%20electromagnetic%20waves,scattered%20back%20to%20the%20radar (accessed on 18 June 2025).
11. Yucel, I.; Onen, A.; Yilmaz, K.K.; Gochis, D.J. Calibration & evaluation of a flood forecasting system; utility of numerical weather prediction model, data assimilation and satellite-based rainfall. *J. Hydrol.* **2015**, *523*, 49–66.
12. Gochis, D.; Mc Creight, J.; Yu, W.; Dugger, A.; Sampson, K.; Yates, D.; Wood, A.; Clark, M.; Rasmussen, R. Multi-Scale Water Cycle Predictions Using the Community WRF-Hydro Modeling System. NCAR, Boulder, Colorado. 2015. Available online: http://www.ral.ucar.edu/projects/wrf_hydro/ (accessed on 18 June 2025).
13. Arnault, J.; Wagner, S.; Rummler, T.; Fersch, B.; Bliefferricht, J.; Andresen, S.; Kunstmann, H. Role of runoff-infiltration partitioning and additionally resolved overland flow on land-atmosphere feedbacks: A case study with the WRF-Hydro coupled modeling system for West Africa. *J. Hydrometeorol.* **2015**, *17*, 1489–1516. [CrossRef]
14. Kerandi, N.; Arnault, J.; Laux, P.; Wagner, S.; Kitheka, J.; Kunstmann, H. Joint atmospheric-terrestrial water balances for East Africa: A WRF-Hydro case study for the upper Tana River basin. *Theor. Appl. Climatol.* **2017**, *131*, 1337–1355. [CrossRef]
15. Ashouri, H.; Kuo-Lin, H.; Soroosh, S.; Dan, K.B.; Kenneth, R.K.; Dewayne, C.L. PERSIANN-CDR: Daily Precipitation Climate Data Record from Multisatellite Observations for Hydrological and Climate Studies. *Bull. Am. Meteorol. Soc.* **2015**, *96*, 69–83. [CrossRef]

16. Huffman, G.J.; David, T.B.; Eric, J.N.; David, B.W.; Robert, F.A.; Guojun, G. The TRMM Multisatellite Precipitation Analysis (TMPA): Quasi-Global, Multiyear, Combined-Sensor Precipitation Estimates at Fine Scales. *J. Hydrometeorol.* **2007**, *8*, 38–55. [CrossRef]
17. ROM-SAF, Algorithm Theoretical Baseline Document: Level 2B and 2C 1DVar products Version 3.1. 2018. Available online: http://www.romsaf.org/product_documents.php (accessed on 15 February 2020).
18. Boddy, S.; Smith, G. *Statistical Methods in Practice: For Scientists and Technologists for Sciences and Technologies*; Wiley: Hoboken, NJ, USA, 2009; p. 248.
19. Yates, E.; Anquetin, S.; Ducrocq, V.; Creutin, J.D.; Ricard, D.; Chancibault, K. Point and areal validation of forecast precipitation field. *Meteorol. Appl.* **2006**, *13*, 1–20. [CrossRef]
20. Mason, I. A model for assessment of weather forecasts. *Aust. Meteor. Mag.* **1982**, *30*, 291–303.
21. Wilks, D.S. *Statistical Methods in the Atmospheric Sciences*, 3rd ed.; Academic Press: New York, NY, USA, 2011; p. 676.
22. Hong, S.Y.; Dudhia, J.; Chen, S.H. Chen A revised approach to ice microphysical processes for the bulk parameterization of clouds and precipitation. *Mon. Weather Rev.* **2004**, *132*, 103–120. [CrossRef]
23. Hong, S.Y.; Lim, J.O. The WRF Single-Moment 6-Class Microphysics Scheme (WSM6). *Asia-Pac. J. Atmos. Sci.* **2006**, *42*, 129–151.
24. Thompson, G.; Field, P.R.; Rasmussen, R.M.; Hall, W.D. Explicit forecasts of winter precipitation using an improved bulk microphysics scheme. Part I: Description and sensitivity analysis. *Mon. Weather Rev.* **2008**, *132*, 519–542.
25. Thompson, G.; Field, P.R.; Rasmussen, R.M.; Hall, W.D. Explicit forecasts of winter precipitation using an improved bulk microphysics scheme. Part II: Implementation of a new snow parameterization. *Mon. Weather Rev.* **2008**, *136*, 5095–5115. [CrossRef]
26. Thompson, G.; Eidhammer, T. A study of aerosol impacts on clouds and precipitation development in a large winter cyclone. *J. Atmos. Sci.* **2014**, *71*, 3636–3658. [CrossRef]
27. Kain, J.S. The Kain–Fritsch Convective Parameterization: An Update. *J. Appl. Meteorol.* **2004**, *43*, 170–181. [CrossRef]
28. Yu, C.; Hu, D.; Di, Y.; Wang, Y. Performance evaluation of IMERG precipitation products during typhoon Lekima. *J. Hydrol.* **2019**, *597*, 126307. [CrossRef]
29. Koutsoyiannis, D.; Iliopoulou, T.; Koukouvinos, A.; Malamos, N.; Mamassis, N.; Dimitriadis, P.; Tepetidis, N.; Markantonis, D. In Search of Climate Crisis in Greece Using Hydrological Data. *Water* **2023**, *15*, 1711. [CrossRef]

hydrology

MDPI

Article

Rainfall Organization and Storm Tracking in Urban Barcelona, NE Spain, Using a High-Resolution Rain Gauge Network

María del Carmen Casas-Castillo [1,*], Xavier Navarro [2] and Raül Rodríguez-Solà [3]

[1] Department of Physics, School of Industrial, Aerospace and Audiovisual Engineering of Terrassa, Universitat Politècnica de Catalunya · BarcelonaTech, Colom 1, 08222 Terrassa, Spain
[2] Department of Physics, Polytechnic School of Engineering of Vilanova i la Geltrú, Universitat Politècnica de Catalunya · BarcelonaTech, Víctor Balaguer 1, 08800 Vilanova i la Geltrú, Spain; javier.navarro@upc.edu
[3] Department of Physics, Barcelona School of Industrial Engineering, Universitat Politècnica de Catalunya · BarcelonaTech, Diagonal 647, 08028 Barcelona, Spain; raul.rodriguez@upc.edu
* Correspondence: m.carmen.casas@upc.edu

Abstract

Extreme rainfall in urban areas can cause major economic damage, a problem expected to intensify with climate change. Despite this, high-resolution studies at the city scale remain limited. This study analyzes rainfall organization and storm dynamics over Barcelona using data from a dense rain gauge network (1994–2019). The aim is to identify dominant spatial patterns and understand how storms evolve in relation to local urban and topographic features. Principal component analysis and simple scaling analysis revealed signs of a rainfall island effect, possibly linked to the urban heat island and modulated by orographic and coastal influences. Tailored rainfall indices highlighted a division between inland areas shaped by orography and coastal zones influenced by the sea. These spatial structures evolved with rainfall duration, shifting from localized contrasts at a 10 min resolution to more homogeneous distributions at daily scales. Storm tracking showed that 90% of speeds ranged from 5 to 60 km/h and intense rainfall events typically moved east–southeast toward the sea and north–northeast. Faster storms tended to follow preferred directions reflecting mesoscale circulations and possible modulations by local terrain. These findings underscore how urban morphology, local relief, and a coastal setting may shape rainfall at the city scale, in interaction with broader Mediterranean synoptic dynamics.

Keywords: storm tracking; rain gauge network; rainfall intensity; urban rain island; Barcelona

1. Introduction

Understanding the temporal and spatial distribution of intense local and mesoscale rainfall events over urban basins, along with their kinematics and evolution, is crucial for effective urban drainage modeling and flood management. In highly populated cities with complex sewerage systems, such as Barcelona (NE of the Iberian Peninsula), accurately tracking storm movement is essential for minimizing flood risks. The high variability of precipitation characteristic of the Western Mediterranean region [1–8] makes storm monitoring even more critical.

Urbanization can influence precipitation patterns by modifying land surface characteristics and atmospheric conditions, leading to changes in the behavior of convective storm cells and rainfall distribution [9,10]. Factors such as the urban heat island (UHI) effect [11], increased air pollution, and complex city skylines contribute to these changes by enhancing convective processes and cloud formation [12]. Additionally, urban-induced modifications can affect cloud dynamics and lightning activity, further impacting precipitation patterns [13].

In recent decades, the influence of climate change on precipitation extremes has attracted increasing scientific attention. The IPCC Sixth Assessment Report (AR6) affirms with high confidence that extreme precipitation events have become more frequent and intense in many regions worldwide, including much of Europe [14]. This intensification is primarily attributed to the atmosphere's enhanced capacity to hold moisture under warmer conditions, which increases the potential for convective precipitation [15]. The Mediterranean Basin, in particular, has been identified as a climate change hotspot, where both thermodynamic and dynamic factors (including changes in large-scale moisture transport) may contribute to a rise in the frequency of short-duration, high-intensity rainfall events [14,16,17]. However, observed trends in the region have not been spatially uniform. While some studies report a decrease in extreme precipitation in the western Mediterranean, others identify increases in the eastern part of the basin [18–22]. A recent comprehensive European assessment also found no clear signal of intensification of sub-daily extremes in Spain over the past six decades, underscoring the complexity of detecting consistent regional trends [23]. Looking ahead, projections indicate that extreme precipitation is likely to increase across most of Europe under global warming scenarios of up to 2 °C. Nevertheless, regional disparities persist, with the southern regions (particularly the southern Mediterranean) expected to experience little change or even decreases in extreme precipitation [24,25]. Beyond these large-scale patterns, climate change may also reshape rainfall distribution at the urban scale [26–28]. Recent studies show that urbanization can amplify local effects of climate change, intensifying atmospheric drying [29,30] and enhancing the risk of compound extreme events, including simultaneous heat and heavy rainfall [31]. These shifting patterns highlight the importance of understanding how urban morphology, land cover, and topography interact with atmospheric processes (particularly in Mediterranean regions) to shape rainfall variability at the city scale.

In the Mediterranean region, such interactions are particularly relevant, as urbanized areas frequently experience intense, localized precipitation events, which are a primary cause of floods [32]. Additionally, the region's topography, geographical location, and the presence of a warm sea create favorable conditions for the development of hazardous convective systems or the intensification of unsettled Atlantic perturbations [33]. In this context, a related phenomenon known as the urban rain island (URI) effect emerges, wherein urban areas exhibit altered rainfall accumulation [34]. The URI effect can sometimes lead to increased precipitation over urban centers due to UHI-induced convection, enhanced surface roughness, and higher aerosol concentrations acting as cloud condensation nuclei. The URI effect can thus have significant implications for urban hydrology, particularly in regions with high precipitation variability, such as the Western Mediterranean, where small-scale convective storms can lead to intense, localized flooding.

Urban precipitation is shaped by a wide range of interacting factors, including land use, topography, coastal influences, and localized atmospheric conditions. These factors create what we refer to here as complex urban environments, especially in medium to large cities. The Barcelona metropolitan area exemplifies such complexity: the interaction between urban morphology, river corridors, coastal influences, and the surrounding Collserola mountain range plays a central role in shaping storm behavior and rainfall distribution. This study aims to explore how these elements affect local convective organization and precipitation patterns in the city.

Individual convective storms, typically composed of simple convective cells with a short lifespan of around 20–30 min, have been found to generally follow the direction of the prevailing mid-level winds (700 hPa) [35]. However, these cells frequently merge, forming multicellular systems whose movement is no longer dictated solely by the predominant wind. Instead, their propagation results from a combination of individual cell motions and the continuous development of new cells within the system [36]. In regions with complex orography and coastal influence, such as Catalunya (the Spanish region where the city of Barcelona is located), wind shear is commonly observed, further modifying storm propagation by altering cell regeneration patterns and redirecting storm movement relative to the predominant mid-level flow [37]. This effect highlights the crucial role of topography in shaping local convective dynamics.

Beyond these geographic complexities, urban areas add another layer of complexity to convective dynamics. The urban heat island effect enhances atmospheric instability, favoring the initiation and intensification of convective cells [38], while urban breeze circulations and cold pool interactions can further support cell regeneration and influence storm trajectories [39]. These urban-induced processes may increase the frequency, intensity, and spatial extent of convective rainfall, particularly in downwind areas [34,40,41]. While cold pool dynamics associated with mesoscale convective systems have been studied in Catalonia [42,43], their interaction with urban-induced processes remains largely unexplored. In the case of Barcelona, the consolidated urban core presents a notably uniform morphology, which may limit the relative influence of intra-urban variability compared to mesoscale drivers such as orography and coastal proximity. In fact, studies explicitly linking urban-induced convection and rainfall in Barcelona are still limited. Nevertheless, existing research has documented a well-defined UHI across the metropolitan area, shaped by urban morphology and land use [44], and recent work by [12] has begun to explore how urbanization influences mesoscale precipitation organization in various regions, offering valuable insight into urban–convective interactions relevant to complex coastal and urban environments such as Barcelona.

As convective cells continue to organize, they may evolve into larger, long-lived structures, such as mesoscale convective systems (MCSs) and other multicellular storm complexes, which are frequently responsible for significant rainfall episodes in the study area [45]. These systems, sustained by the ongoing interaction and regeneration of convective cells, can produce widespread and intense precipitation, increasing the risk of severe flooding.

To monitor and track storms, different approaches can be used. High-resolution radar images [46,47] provide valuable information about precipitation patterns, but they have known limitations, including beam blockage, attenuation, and difficulties in detecting near-ground precipitation. Therefore, they often need to be complemented with surface rainfall data [45,48,49] to enhance accuracy and improve hydrological models. Among these ground-based methods, a widely used option is a sufficiently dense rain gauge network within the urban area of interest [50,51]. While sparsely distributed rain gauge networks struggle to capture the high spatial and temporal variability of precipitation fields, especially highly convective ones, compared to weather radars [52], this limitation can be overcome with a sufficiently dense network. This ground-based observation method is particularly valuable for large, densely populated cities, where local rainfall intensities can vary significantly [7]. Compared to other ground-based instruments such as standard rain barrels (which lack temporal resolution) or microwave disdrometers (costlier and more complex), dense tipping-bucket networks offer a practical and accurate solution for urban storm monitoring. Even when radar data is available, a sufficiently dense rain gauge network provides direct ground-based rainfall measurements, avoiding uncertainties associated with interpreting radar images at a comparable resolution.

The goal of the present study is to improve the understanding of rainfall variability over the study area, particularly for intense rainfall events, to enhance drainage system management and mitigate urban flooding. Also, to better understand this variability, we analyze the catalogue of storm movements recorded by Barcelona's rainfall network between 1994 and 2019. Additionally, the role of the URI effect is explored as a potential factor influencing precipitation distribution within the city, providing new insights into the complex interactions between urbanization and rainfall patterns.

2. Materials and Methods

2.1. The Pluviometric Network of Barcelona

The city of Barcelona, with an approximate extension of 101 km^2, is located on a small coastal plain bounded by the Mediterranean Sea to the east, the Collserola mountain range to the west (512 m in altitude at its highest point), the Llobregat River to the south, and the Besòs River to the north, with various hills scattered throughout the city. Barcelona features a dense rainfall network (Figure 1, Table 1), currently managed by the public company Barcelona Cicle de l'Aigua S.A. (BCASA), Barcelona, Spain, with the participation of the Barcelona City Council through its governing bodies. The pluviometric network, originally deployed and maintained by Clavegueram de Barcelona, S.A. (CLABSA, located in Barcelona, Spain), and now by BCASA, consists of high-resolution tipping bucket rain gauges, most of them operating continuously since 1996 [2,3,49]; for this study, data from twenty-four of them have been used.

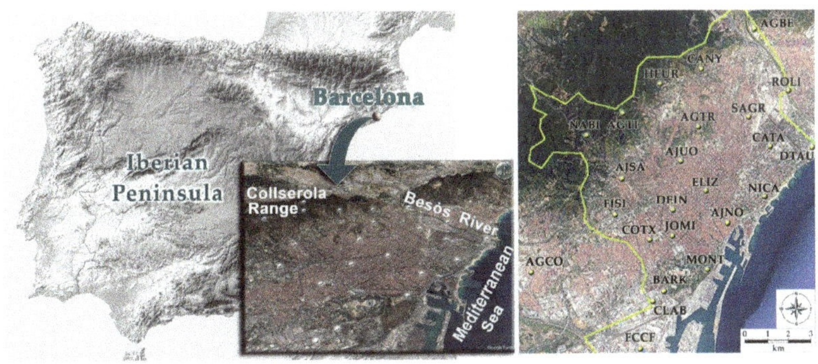

Figure 1. Pluviometric network of Barcelona. In the right panel, the yellow line indicates the municipal boundaries of the city. The figures showing satellite views of Barcelona have been adapted by the authors from Google Earth imagery © Google.

Careful attention has been given to site selection and maintenance to ensure accurate rainfall measurements, minimizing potential sources of error such as nearby obstacles, turbulence, or channeling effects. The volume resolution of the pluviometers is 0.1 mm, their integration time is 5 min, and they feature a digital recording unit equipped with an accurate calendar clock [37]. The network is fairly evenly distributed, with an average density of 0.2 gauges/km^2 and an approximate spacing of 1.8 km between rain gauges [53]. Given the relatively homogeneous morphology of the consolidated urban area, dominated by dense, mid-rise buildings and limited green space, no fine-scale analysis of intra-urban morphological variability was included, as its explanatory value was expected to be limited compared to larger-scale geographical features such as coastline and topography. The urbanized area studied lies predominantly on a flat coastal plain known as the Pla de Barcelona, with gentle slopes across most of the rain gauge network, although a few stations are located on more elevated terrain along the Collserola range (Table 1).

Table 1. Location and elevation of the pluviometers in the urban network of Barcelona.

Code	UTMx (m)	UTMy (m)	Elevation (m a.s.l.)
AGBE	432,572	4,590,189	18
AGCO	422,530	4,579,549	15
AGTI	426,690	4,586,494	450
AGTR	430,034	4,585,878	162
AJNO	431,351	4,581,678	8
AJSA	426,648	4,583,627	120
AJUO	429,270	4,584,410	86
BARK	428,485	4,578,698	9
CANY	430,181	4,588,414	148
CATA	433,279	4,585,023	7
CLAB	427,987	4,578,247	4
COTX	427,879	4,580,946	27
DEIN	428,911	4,582,279	37
DEPU	435,184	4,584,975	5
ELIZ	430,373	4,583,077	32
FCCF	427,443	4,576,149	3
FISI	426,341	4,582,030	67
HEUR	428,357	4,587,775	206
JOMI	428,880	4,581,117	30
MONT	430,431	4,579,649	146
NABI	425,042	4,585,529	293
NICA	433,035	4,582,822	3
ROLI	434,117	4,587,439	14
SAGR	432,334	4,586,332	26

Fractal analysis provides a useful framework for evaluating the spatial density and arrangement of rain gauge networks. In an ideal case, such as a fully space-filling system, the fractal dimension would approach 2.0. However, in real-world networks with non-uniform distribution, this value is typically lower. The correlation dimension [54], a specific measure used to characterize the scaling properties of discontinuous networks, has been applied to various cases. For instance, a value of 1.72 was reported for the Barcelona network in [55], in a study where the authors found that the spatial structure exhibits scale-invariant properties over spatial extents ranging from 5 to 100 km^2.

From the network registers, a total of 2678 rainfall events recorded between 1994 and 2019 were detected in at least 10 of the pluviometers, with durations longer than 25 min. From this initial set of storms, further filtering steps were introduced to refine the analysis, focusing on subsets of events of interest, for example, those reaching intensities defined by the Meteorological Service of Catalunya (SMC) [56] as heavy (20 mm in 30 min) and torrential (40 mm in 30 min) at least at one of the rain gauges. This filtering resulted in 133 intense rain episodes.

2.2. Spatial Organization of Rainfall in Barcelona

For detailed storm tracking studies over a city, having more rain gauges recording an event improves the ability to investigate storm movement and evolution. However, for other analyses, it is crucial to determine whether all gauges provide independent and relevant information. Closely positioned gauges tend to have highly correlated measurements, meaning that only one effectively contributes new information. In contrast, gauges farther apart are less correlated and more independent, provided they capture the same rainfall event. This dependence varies with the temporal scale of analysis.

To better understand the spatial distribution of independent rainfall information across the city and how it varies with temporal scale, we applied the principal component analysis (PCA) statistical method, widely used in atmospheric studies [3,57]. PCA reduces dataset dimensionality while preserving as much information as possible by transforming original variables into uncorrelated principal components, which are linear combinations of the original ones [58]. Given the long-term observational record available, PCA was chosen for its ability to identify dominant spatial patterns in rainfall variability and its suitability for analyzing datasets at multiple temporal scales. Using rainfall data recorded by the urban network over 26 years (1994–2019) for different durations, this initial analysis helps identify areas where rain gauge data are more independent at each timescale, offering insights into the spatial organization of rainfall patterns in Barcelona. In a subsequent step, we derive a weighted combination of the principal components to produce a single summary field per duration, capturing the dominant spatial structures across the network.

Beyond assessing the independence of rain gauge data, further spatial analyses were conducted to explore how rainfall patterns relate to specific scaling properties and climatic indices. In particular, we examined the spatial distribution of the simple scaling parameter [59–61], which describes the fractal scale-invariance properties of rainfall across different durations. This parameter has been shown to correlate with the local rainfall regime, generally aligning with total precipitation while also capturing the irregularity of rainfall patterns. In this study, it may offer valuable insight into rainfall variability within the urban area.

The scaling invariance of rainfall can be described [61] through a relationship that links the probability distribution of rainfall intensity for a given duration, I_t, to that of another duration, $I_{\lambda t}$, via a scaling factor expressed as a power law of the scale parameter λ. This relationship is mathematically defined in Equation (1) for the monofractal (or simple scaling) case, where β is the scaling parameter mentioned in the previous paragraph, and q represents the order of the statistical moments $\left\langle I_t^q \right\rangle$ of the probability distributions (Equation (2)).

$$\left\langle I_{\lambda t}^q \right\rangle = \lambda^{\beta q} \left\langle I_t^q \right\rangle \quad (1)$$

$$\left\langle I_t^q \right\rangle = \frac{\sum_{i=1}^{n} I_{t_i}^q}{n} \quad (2)$$

The parameter β can be determined by performing a linear regression of the logarithmic values of the statistical moments against the logarithm of t. In the case of scale invariance, the resulting straight lines have a slope of βq. To ensure the validity of the monofractal hypothesis, the statistical moments were computed using the annual maximum series at each rain gauge.

Additionally, we analyzed the spatial distribution of selected climatic indices, including the mean values over the study period (1994–2019) of PRCPTOT, defined as the total annual precipitation, and Rx1day, the maximum daily precipitation [62]. To capture sub-daily extremes relevant in urban hydrology, we also defined tailored indices: Rx4h, Rx1h, and Rx10min, representing the mean of the annual maximum precipitation values for 4 h, 1 h, and 10 min durations, respectively, over the study period (1994–2019). These indices offer a more detailed understanding of short-duration rainfall events, which are especially critical in assessing flood risks in urban areas. This analysis provides insights into how different timescales of precipitation extremes vary spatially across Barcelona. By integrating these approaches, we aim to identify spatial patterns influenced by orographic and geographic features, including the influence of the coastline and the sea, as well as other urban-related factors that may shape rainfall dynamics.

2.3. Storm-Tracking Technique

The basis of storm-tracking techniques that utilize rainfall network data to estimate storm movement across an area relies on monitoring specific characteristics of the rainfall pattern, often referred to as a reference point [63]. Common reference points include the onset of rainfall at each gauge [64], the peak intensity time, and the centroid of the storm hyetogram. Other approaches involve computing maximum spatio-temporal cross-correlation functions between rain gauges [65] or applying temporal shifts to maximize correlations between pairs of gauges [66]. However, these methods do not perform satisfactorily in cases where multiple rainfall cells are present, as they struggle to account for the complex, evolving structure of convective storms [63,67]. While storms are often composed of multiple evolving and decaying cells, many analyses based on rainfall network data rely on the simplifying assumption that storm movement can be represented by a uniform speed. More recent studies [50,51] have further developed rain gauge-based tracking methods by applying these principles to urban areas, demonstrating their effectiveness in monitoring storm motion. These approaches, based on methods proposed in [53], utilize dense rain gauge networks to capture rainfall variability and track storm propagation, offering a practical alternative for analyzing storm dynamics in regions lacking radar coverage or where simplicity and accessibility are prioritized over more complex radar-based techniques.

The storm-tracking technique chosen to monitor the rainfall events recorded by the urban network is the one proposed in [50,51]. To track these events across the network, the time centroids of the hyetograms and the highest-intensity peaks have been used as reference points (Figure 2). Other techniques, such as the lag-correlation method [65–67], were initially tested but later discarded due to unsatisfactory results, possibly caused by this method's sensitivity to the presence of multiple rain cells within the storm [63].

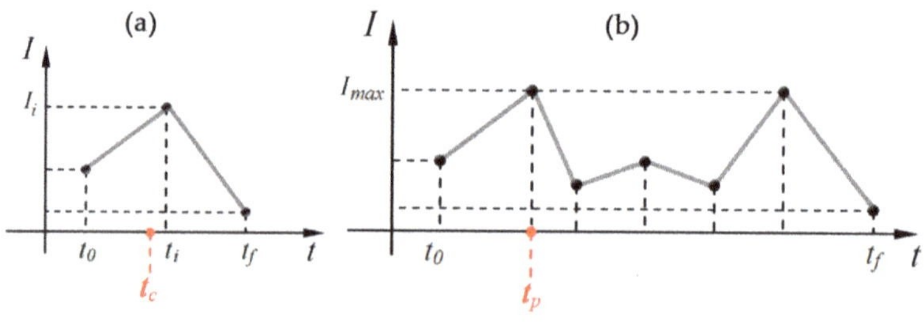

Figure 2. Examples of a hyetogram (**a**) time centroid and (**b**) first maximum intensity peak.

For each rain gauge k, the centroid of its hyetogram is a time $t_{c,k}$ calculated as Equation (3), where t_0 and t_f represent the start and end times of the storm hyetogram, and I_i are its intensity values.

$$t_{c,k} = \frac{\sum_{i=0}^{f} I_i t_i}{\sum_{i=0}^{f} I_i} \tag{3}$$

If the reference point for monitoring the storms is the maximum-intensity peak instead of the centroid, the alternative time $t_{p,k}$ is selected as the time for which the highest intensity in the hyetogram of rain gauge k occurs. If multiple peaks of equal intensity are present, the first recorded one is chosen. Throughout the rest of this paper, t_k denotes the time corresponding to the chosen reference point for tracking the storm, either the centroid $t_{c,k}$ or the first maximum peak $t_{p,k}$, depending on the selected method.

For a given storm, once the reference points for each rain gauge in the network have been determined, a set of experimental points $(x_k\ ,\ y_k\ ,\ t_k)$ is considered, with $(x_k\ ,\ y_k\ )$ the coordinates of the gauge k, and t_k the time of the storm feature to be tracked. This set of experimental points is then fitted to a plane, defined by Equation (4) by the least squares method (planar regression) [51].

$$ax + by + c = t \tag{4}$$

In particular, the speed of the movement v can be estimated by the inverse of the maximum plane slope according to Equation (5), while the direction of movement, which is assumed to be parallel to the direction of the maximum slope, is expressed as an azimuth angle (θ), measured clockwise from north (0°), and calculated using Equation (6).

$$v = \frac{1}{\sqrt{a^2 + b^2}} \tag{5}$$

$$\theta = \begin{cases} \arctan\left(\frac{a}{b}\right) \text{ if } b > 0 \text{ and } a \geq 0 \\ \arctan\left(\frac{a}{b}\right) + 180^\circ \text{ if } b < 0 \\ \arctan\left(\frac{a}{b}\right) + 360^\circ \text{ if } b > 0 \text{ and } a < 0 \end{cases} \tag{6}$$

In the special case where $b = 0$, then $\theta = 90^\circ$ (East) if $a > 0$, and $\theta = 270^\circ$ (West) if $a < 0$. If $a = 0$ and $b = 0$ the direction is undefined, corresponding to a stationary storm [51].

As an example, Figure 3 presents the planar regression for a rainfall event recorded by the urban network of Barcelona on 14 September 2009, where the hyetogram centroids for each of the 15 pluviometers that captured the storm are represented on the t-axis. The parameters of the fitted plane equation in this space (x, y, t), where longitude is represented by x and latitude by y, were determined as $a = -0.0822$ s/m, $b = -0.3712\frac{\text{s}}{\text{m}}$, and $c = 1.741 \times 10^6$ s. This corresponds to a movement speed of 2.63 m/s, that is, 9.5 km/h towards the south–southwest ($\theta = 192.5^\circ$). See also Figure 4, which presents a sequence of images from the radar operated by the SMC in Vallirana, a nearby town west of Barcelona. The images, captured at 12 min intervals, clearly show the storm moving approximately in a S–SW direction.

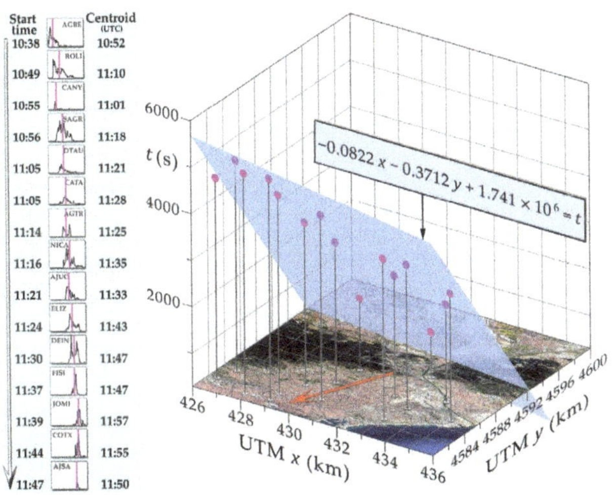

Figure 3. Example of a planar regression using the hyetogram centroids from the 15 rain gauges of the urban network of Barcelona capturing the storm occurred on 14 September 2009. The fitting gives a movement speed of 9.5 km/h towards the S–SW. On the left, the start times and centroids of the 15 hyetograms are shown. Each hyetogram includes a vertical purple line indicating the time centroid at that station, providing insight into the storm progression. On the right, the vertical axis expresses time in seconds, with t = 0 assigned to 10:30 UTC. Other reference times: t = 2000 s ≈ 11:03 UTC; t = 4000 s ≈ 11:37 UTC; t = 6000 s ≈ 12:10 UTC.

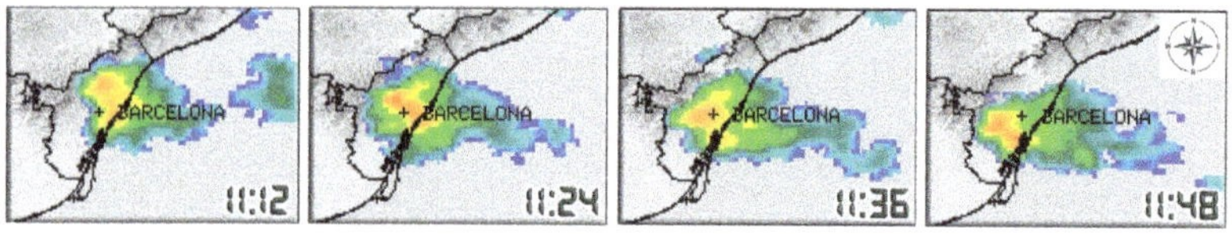

Figure 4. Sequence of radar reflectivity images from the Vallirana C-band Doppler radar (SMC), capturing the storm that occurred on 14 September 2009 in the Barcelona metropolitan area, shown at 12 min intervals.

This technique has been applied to the 2678 rainfall events recorded between 1994 and 2019 by the urban network, with the aim of compiling a catalog of the main movement directions of the storms moving over the city, as well as their range of speeds.

Equation (7) calculates the value of the significance ratio (SR) [51], a dimensionless metric ranging from 0 (perfect fit) to 1 (no fit), which quantifies the goodness of fit of the planar regression model, where n is the number of rain gauges involved in the regression. Lower SR values indicate better fits, with values under 0.85 typically considered satisfactory. In the example shown in Figure 3, the SR value is 0.31, indicating a very good fit.

$$SR = \sqrt{\frac{\sum_{k=1}^{n}(t_k - a - bx_k - cy_k)^2}{\sum_{k=1}^{n}(t_k - \bar{t}_k)^2}} \quad (7)$$

3. Results

3.1. Spatial Organization of Rainfall: The Urban Rainfall Island in Barcelona

The PCA technique was applied to the urban network records from 1994 to 2019, using as variables the rainfall quantities recorded by each pluviometer for durations ranging from 5 min to 24 h. To further explore the spatial distribution of the principal components, we interpolated their scores obtained at the rain gauges across a grid covering the study area. This allowed us to generate contour lines (isohyet-like representations), which indicate the spatial variation in the dominant modes of rainfall variability. Figure 5 shows the spatial distribution of the first three principal components of rainfall for a duration of 15 min, explaining 30%, 11%, and 9% of the total variance, respectively.

In the first principal component (PCA_1), which explains 30% of the variance, the rain gauges are grouped into concentric bands that extend across the entire map (Figure 5a). According to this pattern, there is a central area with the highest scores (above 0.25) located in the center of Barcelona, indicating that this region strongly follows the dominant spatial pattern captured by this component. Surrounding this, a second zone forms a ring around the central core, with moderate scores between 0.2 and 0.25. The third ring (with values between 0.15 and 0.20) appears further out, covering areas such as the Collserola range, the Besòs River area, and the central part of the port. Finally, the outermost zone (with scores below 0.15) corresponds to the peripheral rain gauges of the network, where the influence of this component is weakest. Since all scores in this component are positive, it suggests a common spatial mode of rainfall distribution across the city, with its intensity decreasing toward the periphery. The concentric structure of the first principal component values suggests a spatial pattern similar to the urban heat island (UHI) effect [11,68]. Just as the UHI effect leads to higher temperatures in the city center, this pattern may indicate a stronger expression of the dominant rainfall mode in the central area of Barcelona, providing evidence of an urban rainfall island over the city [69,70].

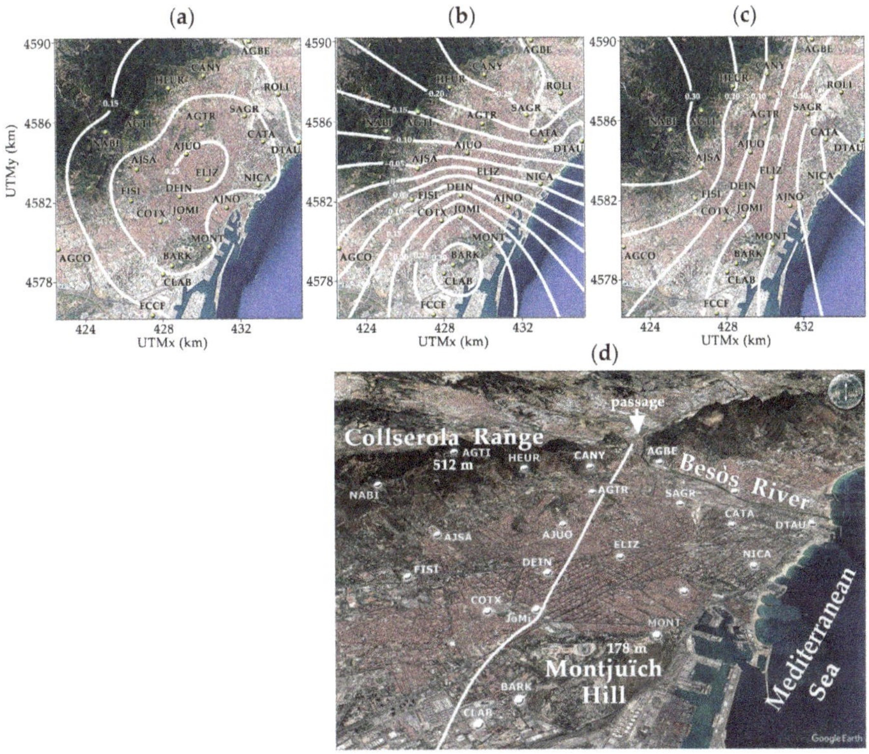

Figure 5. Spatial distribution of the first three principal components (PCA_1 to PCA_3) of rainfall for a duration of 15 min [53]. Contour lines in panels (a–c) represent interpolated PCA scores, highlighting dominant spatial patterns: (a) first component (PCA_1), explaining 30% of the total variance; (b) second component (PCA_2), explaining 11% of the variance; (c) third component (PCA_3), accounting for 9% of the variance. Panel (d) shows the main orographic and geographic features of the city, along with the zero-contour line of PCA_3. This panel uses vertical exaggeration to improve the visualization of relief and to highlight the Collserola corridor. The figures showing satellite views of Barcelona have been adapted by the authors from Google Earth imagery © Google.

For the second principal component (PCA_2), which accounts for 11% of the variance (Figure 5b), we observe a spatial pattern characterized by nearly east–west parallel contours near the city center, reaching both the coast and the mountains of the Collserola range almost perpendicularly, and gradually closing toward the north and south. The central region marks a transition zone with near-zero values, while the scores increase positively toward the south, reaching a maximum near the port (0.30), just behind Montjuïc Hill, suggesting that this small elevation (178 m) may influence the spatial pattern. In contrast, scores decrease negatively toward the north, and the initially east–west-oriented isohyets begin to rotate progressively as they approach the Collserola range in the northwest, aligning perpendicularly to the mountain slope. This reorientation reflects the influence of the topography in modulating rainfall distribution. This pattern suggests two contrasting rainfall regimes. The increasing positive scores toward the southeast indicate that the rain gauges in this region exhibit a stronger correlation with this component, possibly linked to coastal meteorological influences such as sea breezes and convective activity. Conversely, the growing negative scores toward the north, along the Collserola range, suggest that rainfall in this region follows an opposing pattern, likely shaped by orographic effects associated with the mountainous terrain. The transition zone in the central part of the city represents a boundary between these two spatial patterns, highlighting the role of geographical and topographical features in shaping rainfall distribution over Barcelona.

For the third principal component (PCA_3), which accounts for 9% of the variance, the spatial pattern is dominated by almost north–south contours, highlighting the influence of the mountainous terrain to the west and the coastal area to the east (Figure 5c). The zero-score line crosses the city from south to north, aligning with the gap or passage created by the Besòs River as it cuts through the Collserola range (Figure 5d). The scores increase positively toward the west, reaching a maximum of 0.30 over the Collserola range, while decreasing negatively toward the east, reaching −0.20 near the coast. This pattern suggests that rainfall in these areas exhibits two contrasting behaviors, with orographic effects shaping the western sector and coastal influences playing a role in the eastern sector. The following principal components (fourth, fifth, and sixth), which explain an additional 6%, 5%, and 4% of the total variance, respectively, were also examined [53]. However, they do not introduce fundamentally new spatial structures. For instance, the fourth component displays concentric patterns similar to the first, while the fifth resembles the third in its west–east contrast, though with slight differences. Due to their limited additional explanatory power and redundancy with previous components, we chose not to include them in the main analysis. This redundancy is also reflected in the number of components required to explain most of the spatial variance at each timescale. For instance, while up to 12 components are needed to reach approximately 80% explained variance at a 10 min resolution, this number decreases with an increasing duration: seven components for 1 h, four for 4 h, and only two for daily accumulations. This progressive reduction aligns with the observed tendency for rainfall structures to become smoother and more spatially coherent as the accumulation duration increases.

Next, for each rainfall duration, we compute a weighted sum of the first few principal components for each pluviometer k, until their cumulative variance accounts for approximately 80% of the total variance. This yields a composite score S_k per station (Equation (8)), where $PC_{i,k}$ is the score of the i-th principal component at pluviometer k, w_i is the weight (fraction of variance explained by component i), and N is the number of components needed to reach 80% of explained variance [53,58].

$$S_k = \sum_{i=1}^{N} w_i \, PC_{i,k} \tag{8}$$

This weighted combination of the first principal components provides a summary signal S_k for each station, capturing the most relevant spatial variation in rainfall intensity while accounting for the proportion of variance explained by each component. Importantly, because the S_k values result from recombining standardized component scores weighted by their contribution to total variance, they partially reconstruct the dominant structure of the original data. As a result, stations with overall similar rainfall behavior across durations tend to yield similar S_k values. This explains the presence of large, relatively uniform zones in the interpolated maps, where only a few distinct areas emerge. Figure 6 displays the interpolated scores of S_k for durations of 10 min and 1, 4 and 24 h, allowing a visualization of the dominant spatial patterns that emerge when multiple principal components are considered together, as well as how these patterns evolve with the rainfall duration. For durations shorter than 30 min (represented in Figure 6 by the 10 min duration), the contour lines of the interpolated S_k values form enclosed zones, distinguishing between the center–south and northern areas of the city. By the 1 h duration, these enclosed areas disappear, giving way to two broad regions: one more influenced by the mountainous orography, and the other closer to the coast. At the 4 h duration, the separation between coastal and orographic influences persists, though the boundary becomes more curved and a localized closed contour appears in the northern sector of the port area. At 24 h, a central enclosed zone emerges, encompassing most rain gauges while only excluding the stations at the periphery of the network.

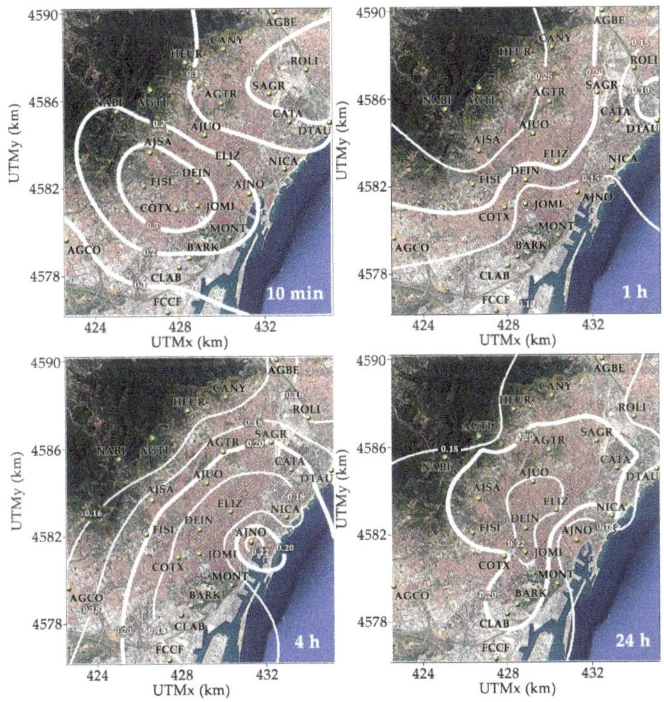

Figure 6. Spatial distribution of the summary signal S_k, obtained as a weighted combination of the first principal components explaining up to 80% of the total variance, for different rainfall durations (10 min, 1 h, 4 h, and 24 h).

To aid interpretation, auxiliary contour lines have been added in Figure 6 for the 1, 4, and 24 h durations. The range of these added levels becomes progressively narrower as the duration increases, reflecting the greater spatial smoothness of the interpolated S_k fields with longer rainfall durations.

The analysis of spatial patterns across different durations reveals how the scale of observation influences rainfall variability. For the shortest duration (10 min), the patterns are more localized, and the network's density and distribution allow for a better resolution of small-scale features, characteristic of the local scale. These small-scale features include topography, urban features, and surface properties, which strongly influence rainfall variability. As we move to longer durations, such as 1 and 4 h, the patterns begin to show broader structures and smooth out compared to shorter durations, but still retain some degree of locality. Both durations (1 and 4 h) represent the small mesoscale and mesoscale range [71], where atmospheric processes like local winds and convective activity play a stronger role, while topographic influences remain significant, especially in areas with notable terrain features. Larger-scale meteorological systems also begin to have a more pronounced influence as the scale increases. For the 24 h duration, the scale increases further, approaching the synoptic scale, where the influence of small, local features becomes less apparent. Here, rainfall patterns are more dominated by broader, large-scale weather systems, and the smaller differences between areas in the city are harder to distinguish due to the averaging effect of the longer time frame.

Having explored how the spatial patterns of rainfall variability evolve across different timescales through PCA, we now turn to examining their relationship with conventional rainfall metrics and indices. For instance, the scaling parameter β has been calculated for each rain gauge in the network, a parameter that has been related to rainfall irregularity as well as to mean annual precipitation [59–61,72]. To explore these characteristics across the city, we interpolated the β values from each rain gauge in the network. Figure 7 presents the resulting β field, illustrating the spatial variation in this parameter over the study area, along with the interpolated field of average annual total precipitation (PRCPTOT). The comparison between both panels does not appear to reveal a strong similarity between the spatial patterns of β and PRCPTOT. Instead, the β field exhibits a structure of concentric bands around the center, resembling the typical urban heat island pattern.

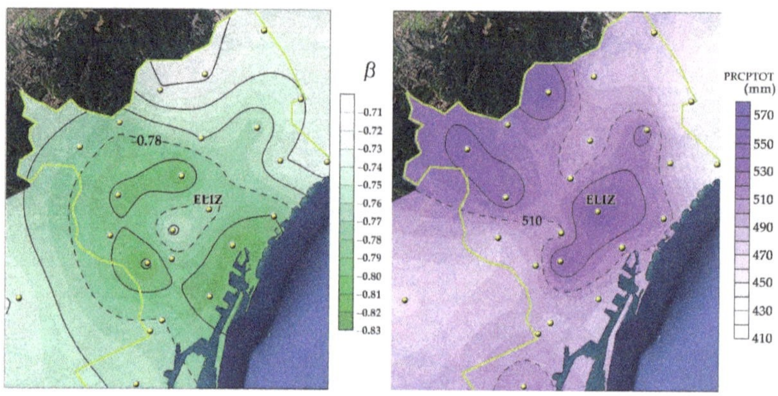

Figure 7. On the left, the interpolated β field, illustrating the spatial variation in rainfall irregularity across the study area. On the right, the interpolated field of average annual total precipitation (PRCPTOT), providing a comparative view of long-term rainfall distribution.

A previous study [59] found that for the Observatori Fabra of Barcelona, located at 411 m above sea level and very close to the AGTI gauge, the β exponent value was -0.78, using both a Hellmann totalizing rain gauge and a Jardí rain intensity recorder, operating in unison between 1927 and 1992. In the present study, small rainfall differences across various urban areas are detected through simple scale analysis. The β values obtained for the different rain gauges range from -0.72 to -0.83 (Figure 7). Except for a small central area (ELIZ and DEIN), the spatial distribution of β shows values below -0.80 in the center of the city, increasing towards the outskirts, with values of -0.73 and -0.74 in the northeast near the Besòs River basin, and -0.74 in AGCO, to the southwest, just outside the municipal area of Barcelona (outlined in green in Figure 7).

Conversely, the spatial pattern of PRCPTOT begins to echo features observed in the PCA-derived fields at intermediate durations (1–4 h), where a transition between a mountainous-influenced western zone and a coastal eastern sector becomes discernible (Figure 6). This separation starts to emerge subtly at 1 h, becomes more noticeable at 4 h, and is more clearly defined at 24 h (Figure 6), where higher values begin to cluster around the city center, particularly at stations such as ELIZ and DEIN, as also seen in

PRCPTOT (Figure 7). Although this separation is not as clearly maintained at longer durations (e.g., 24 h), some of the spatial structure remains consistent with the broader organization revealed by the PCA. A visible transition area crosses the city with values around the mean of 510 mm or less, dividing it into two regions with maxima (above 530 mm), possibly influenced by different meteorological drivers. In the case of long-term rainfall distribution, one of these influences appears to be orography, as the maxima in the northwest are located in the mountains. The other maximum, located to the east near the city center (ELIZ), does not appear to be solely driven by coastal influences, suggesting that additional meteorological factors are involved. However, notice that since PRCPTOT was computed for annual periods, its resolution is not optimal for detecting the finer-scale spatial structures revealed by PCA. To address this, other indices and tailored metrics adapted to shorter timescales have been calculated. The maritime influence becomes more evident when shifting to shorter timescales, for example, when representing the Rx1day index, which captures the maximum daily rainfall at each gauge. This influence becomes even more pronounced at smaller scales, as seen in the tailored indices Rx4h, Rx1h, and Rx10min, which record the maximum rainfall for each duration (4 and 1 h and 10 min, respectively). At the 10 min scale, a clear difference emerges between the northern and southern parts of the city, highlighting the coastal influence, as previously suggested by the PCA (Figure 8).

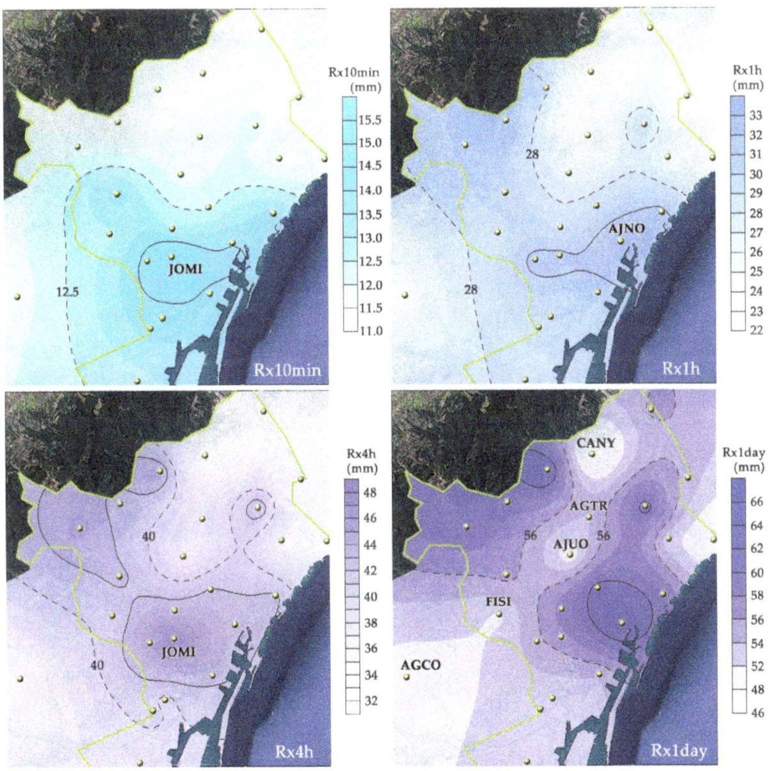

Figure 8. Interpolated fields of indices Rx10min, Rx1h, Rx4h, and Rx1day.

3.2. *Tracking the Rainfall Events Recorded by the Urban Network of Barcelona (1994–2019)*

A total of 2678 rainfall events lasting more than 25 min were detected by at least 10 rain gauges in the urban network of Barcelona. A total of 33% of the initially analyzed rainfall events occur in autumn, a proportion that increases to 57% for the selection of 133 intense events. In contrast, summer accounts for just 14% of the events in the initial dataset. This proportion also increases in the more intense selection, rising to 28%, indicating that although less frequent, summer storms are often among the most intense. Altogether, approximately 85% of the most intense rainfall episodes are concentrated in autumn and summer. All these events were analyzed using tracking techniques based on both storm centroids and first maximum peaks as reference points. The estimations of speed and direction of movement obtained from these two approaches were very similar. Although individual storm estimates may differ due to variations in the behavior of centroids and first maximum peaks throughout the storm lifecycle, both methods lead to consistent and comparable conclusions. Table 2 shows the seasonal distribution of total and intense rainfall events, including their respective percentages.

Table 2. Seasonal distribution of total and intense rainfall events.

	DJF	MAM	JJA	SON
	Winter	**Spring**	**Summer**	**Autumn**
Total events	670 (25%)	741 (28%)	382 (14%)	885 (33%)
Intense events	7 (5%)	13 (10%)	37 (28%)	76 (57%)

In terms of the significance ratio *SR* (Equation (2), [51]), 84% of the storms presented a good fit ($SR \leq 0.85$) when using their centroids, and more than 64% did so when using the first maxima. Building on these findings, an analysis of the rainfall events with a good fit (2245 using the centroid method and 1718 using the first maximum peak method) revealed that the mean motion speed was 30 km/h for events tracked using their centroids, and 24 km/h for those monitored using their first maxima. More than 90% of the speeds fell within the range of 5 to 60 km/h. Moreover, around 65–70% of events moved within a directional range spanning from southeast (SE, towards the coast) to northwest (NW, towards the mountains) through the east (E), northeast (NE), and north (N), while very few moved toward the southwest (SW) (see Figure 9).

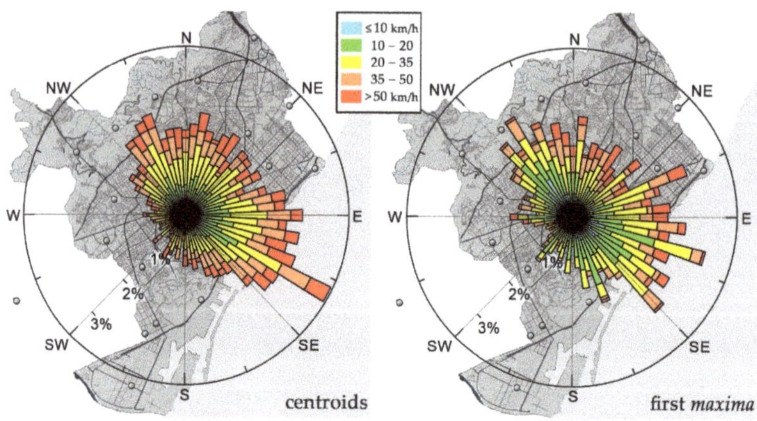

Figure 9. Diagrams with the motion direction and speeds of the well-fitted rainfall events using the centroid method (**left**) and those using the first maximum peak method (**right**).

Figure 10 shows the storm direction and speed diagrams corresponding to the selection of 133 events that reached intensities defined by the SMC [56] as heavy (20 mm in 30 min) and torrential (40 mm in 30 min) at one or more rain gauges. While the full dataset initially displayed a broad range of directions spanning from the northwest (NW) to southeast (SE), the subset of intense events revealed a slightly more balanced distribution. When monitored using the centroid method, this refined selection identified two distinct predominant movements: one toward the east–northeast (ENE), and a more marked one toward the east–southeast (ESE) and southeast (SE), all of them in the direction of the coast. Even so, other quadrants showed relatively higher proportions compared to the full set, indicating a slightly broader diversity in storm trajectories within the subset. In the case of the events tracked using the first maximum, the predominant directions are those toward the coast, with the exception of a subset of events moving toward the north–northwest (NNW), that is, perpendicular to the Collserola range. However, a notable 45% of the intense events recorded by the urban network moved toward the sea, a result consistently observed using both tracking methods.

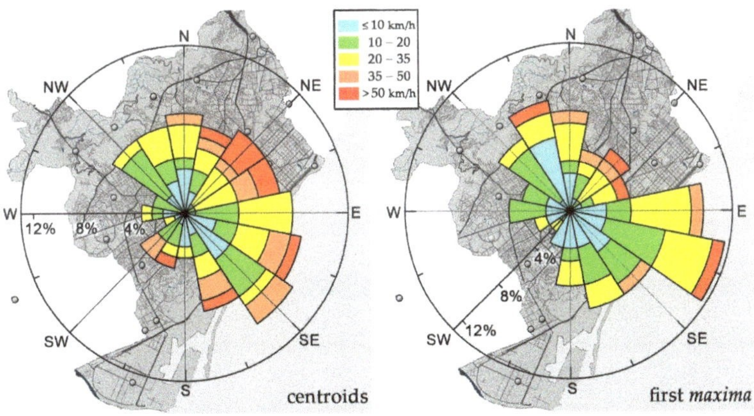

Figure 10. Diagrams with the motion direction and speeds of the well-fitted rainfall events showing heavy and torrential intensities at one or more rain gauges, using the centroid method (**left**, 116 episodes) and those using the first maximum peak method (**right**, 103 events).

Regarding the speeds of the highly intense events, the analysis based on centroid tracking reveals a marked variation across quadrants: in the first quadrant (N to E), the average speed is 32 km/h; in the second (E to S), 24 km/h; in the third (S to W), 25 km/h; and in the fourth (W to N), 15 km/h. The corresponding median speeds for the centroid method are 24, 18, 16, and 14 km/h for quadrants one to four, respectively, reflecting the directional contrasts observed. These median values suggest that the mean speeds, particularly in the first quadrant, are influenced by a few high-velocity outliers. Indeed, more than 11% of the intense events exhibit high velocities (with a mean speed of 32 km/h) clustered in the NNE–NE–ENE range, roughly parallel to the coast. In contrast, the predominant directions moving roughly perpendicular toward the sea, clustered in the ESE–SE–SSE range, have comparatively lower average speeds around 16 km/h.

In comparison, the first maximum method yields generally lower velocities. The median speeds for intense events by quadrant are 21, 13, 10, and 11 km/h for quadrants one through four, respectively. Corresponding mean speeds are 24 km/h in the first quadrant (N to E), 16 km/h in the second (E to S), and 14 km/h in both the third (S to W) and fourth (W to N) quadrants, indicating not only a reduction in overall speeds but also a different distribution pattern across quadrants.

4. Discussion

Accurate analysis of urban precipitation relies on high-quality observational data. Despite the inherent challenges of measuring rainfall in complex urban environments, the dense network used in this study has been carefully maintained for over three decades. Although not all gauges fully meet WMO siting standards, the network was designed with attention to accuracy and has been consistently managed by qualified municipal entities (originally CLABSA, now BCASA) in collaboration with the Barcelona City Council, ensuring reliable data for urban-scale analysis.

Using this robust dataset, the PCA technique applied to the urban network records from 1994 to 2019, along with the resulting interpolated values across a grid covering the study area, has proven highly useful for exploring the spatial distribution of the dominant rainfall patterns in Barcelona. The concentric structure of the first principal component values around the city center revealed a spatial pattern akin to the urban heat island effect (UHI) [11,68], providing the first evidence of an urban rainfall island (URI) over the city. Similarly, the spatial pattern of the scaling parameter β, as shown in the Results section, supports this interpretation, further reinforcing the idea of a spatially concentrated urban rainfall effect, with higher β values observed around the urban fringe and gradually decreasing toward the periphery (while showing a slight local dip near the city core), a pattern consistent with the spatial signature expected under an urban rainfall island scenario. The UHI over Barcelona can be represented by the spatial distribution of the annual average nighttime temperature (ANT) calculated by [68] using MODIS satellite data and GIS analysis. It exhibited a concentric band structure around the city center, coinciding with the location of the ELIZ rain gauge in the network, where the highest ANT is 15.4 °C. The temperature gradually decreases toward the periphery, passing through an average ANT value of 13.8 °C, and reaching a lowest ANT of 11.4 °C in the area of the city deeply embedded in the Collserola range (above the NABI gauge). While a comprehensive assessment of the URI would require reference stations beyond the city limits, the observed internal variations, showing decreasing gradients from the urban core to the periphery, suggest that urban-induced factors, such as enhanced convection due to heat and aerosols, may be influencing precipitation distribution within the city. Indeed, Barcelona has been identified among the large southern European cities with notable urban precipitation anomalies [69], reinforcing this interpretation. Although our dataset lacks external observations beyond the municipal boundaries, the internal spatial gradients support the presence of a URI signal. However, this interpretation is limited by the current extent of our gauge network, especially near the city's edges. For instance, the southwestern periphery, near the mouth of the Llobregat River, may seem to show a rainfall deficit, when in fact this may be due to insufficient observational coverage. As shown by [73], a positive rainfall anomaly was recorded in that area, particularly in autumn, but our network does not detect it, as the nearest station (AGCO, located in the neighboring city of Cornellà) remains several kilometers away from the Llobregat Delta and outside our analysis area.

The second component reveals a north–south contrast, driven by a spatial pattern of nearly east–west-oriented contours, with positive scores increasing toward the south and negative ones toward the north. Small-scale topography (e.g., Montjuïc Hill, 178 m) appears to modulate rainfall patterns on the city's southern fringe. This is consistent with numerous studies and models [49,74,75], which highlight the significant role of small-scale topographic obstacles in storm initiation, as well as the impact of both sea and topography on storm system movement [49,75]. Meanwhile, the Collserola range appears to modulate the rainfall pattern in the northwest, as suggested by the bending of the contours near the mountains. The observed transition zone aligns with the city center and marks a boundary between two rainfall regimes: one to the north, increasingly influenced by topography, and another to the south, likely shaped by maritime influences (e.g., sea breezes and convective dynamics). While the second component suggests a north–south contrast possibly modulated by Montjuïc Hill on the city's southern fringe, its interpretation remains more uncertain. For example, the presence of high component values near Montjuïc could be influenced by localized effects or potentially by edge effects of the network, although the spatial context provides some support for a real signal. Similarly, the negative values observed in the northern area are not easily attributable to the Collserola range. This pattern may instead reflect a spatial shift in rainfall intensity linked to urban factors such as storm duration or UHI-induced convection, as shorter-duration events tend to concentrate more toward the urban core. Nevertheless, this interpretation should be considered tentative and subject to the spatial limitations of the observational network. The third component reveals a distinct spatial structure characterized by nearly north–south-oriented contours, clearly separating the mountainous terrain to the west from the coastal plain to the east. The zero-score line is oriented slightly more northward than the general SW–NE orientation of the coastline aligning with the Besòs River corridor, a natural passage through the Collserola range where the orographic barrier briefly weakens.

The spatial patterns derived from the summed PCA components across different rainfall durations reveal how dominant modes of variability evolve with time. For the shortest duration (e.g., 10 min), the interpolated scores form distinct enclosed zones that separate the center–south from the northern part of the city, an organization that echoes the contrast observed in the second principal component at 15 min, where orographic and maritime influences begin to diverge. As the duration increases to 1 and 4 h, the spatial patterns evolve from localized features to broader structures. At 1 h, the inland–coastal division becomes more apparent, following a relatively straight trajectory, as was found by [3]. By 4 h, the separation acquires a more curved configuration, and a secondary closed structure appears near the northern edge of the port area. By 24 h, a single large, enclosed zone emerges at the center of the network, reflecting the increasing dominance of synoptic-scale processes that override finer-scale urban and topographic variability.

This gradual shift in spatial structure mirrors what is seen in the interpolated fields of climatic indices such as PRCPTOT, Rx1day, and others. However, it is important to note that such indices are typically designed to capture long-term climate variability and change [62], and are often computed over seasonal or annual periods. As such, they are less sensitive to the fine spatial structures that emerge at sub-daily scales. In fact, it is only when these indices are computed at shorter durations, such as 24 h or less, that the patterns hinted at by the PCA become more discernible. For instance, in the Rx1day field, a clear division, roughly following the path of several rain gauges (CANY, AGTR, AJUO, FISI, AGCO), starts to emerge between the part of the city more influenced by the mountains (Collserola) and the sector closer to the coast. As we further reduce the duration to 4 and 1 h, this topographic–maritime contrast becomes more defined, although the influence of the coast progressively gains strength. At 10 min, the spatial organization once again reflects a strong north–south separation, very much in line with the structure captured by the summed PCA components for the same short duration. This parallelism across independent methods underscores the robustness of the observed spatial divisions, and highlights the importance of analyzing rainfall at sub-hourly scales when aiming to resolve the interaction between urban structure, topography, and coastal dynamics.

In order to contextualize the storm speeds observed in Barcelona, the ranges obtained using our tracking methods were compared with values reported in other European studies [50,63,66,67,75]. The speed ranges obtained using tracking methods over the city of Barcelona are consistent with results reported by other authors [37,55], with 90% of the well-fitted storms ($SR \leq 0.85$) falling within a speed range of 5 km/h to 60 km/h. Specifically, using the centroid method, we found an average speed of 30 km/h, while the first maximum peak method yielded an average of 24 km/h. For the selection of highly intense events, the averaged velocities were 27 km/h and 23 km/h, respectively. For comparison, the estimated average speed for 400 storms in Lund (Sweden) was 37 km/h [63], while in England, average speeds of 41 km/h for 110 storms in Cardington and 42 km/h for 133 storms in Winchcombe were reported by [66]. Using data from 107 storms recorded by a network of 20 rain gauges distributed over a 600 km^2 area in northwest England, an average storm speed of 34.5 km/h was estimated by [67]. Based on a 10-gauge rain network in Palermo (Italy), it was found by [50] that 60% of the monitored storms had speeds below 30 km/h, with approximately 30% of the events falling within the 20–30 km/h range.

However, although these comparisons provide useful context, direct contrasts should consider differences in latitude, topography, and prevailing meteorological conditions. Latitude may influence synoptic-scale wind patterns and storm propagation, but local geographic factors, such as the presence of coastal influences and orographic features, also play a crucial role in modulating storm dynamics. Consequently, variations in storm speeds across different regions likely result from a combination of these factors rather than latitude alone.

Maintaining a more local focus, an earlier work [37], based on the initial years of measurements from the Barcelona rainfall network and a selection of 32 storms, identified a predominant eastward propagation direction, with speeds estimated between 3 and 12 m/s (10.8 to 43.2 km/h). Subsequently, and also using data from the rainfall network, ref. [55] analyzed a more limited set of 15 storms and identified predominant propagation directions along the SW–NE and NW–SE axes (parallel and perpendicular to the coastline, respectively), with speeds ranging between 3 and 25 km/h. While limited in scope, both works provided early insights into storm behavior in the region. The results from [37] partially align with the eastward component observed in our current analysis, whereas [55] shows a closer correspondence with the NW–SE propagation paths we found. Based on radar observations, some mesoscale systems with convective cells have been documented to traverse the entire Catalan littoral (approximately 580 km) in under three hours [75,76], which would imply very high speeds. In contrast, other mesoscale systems can remain almost stationary over specific areas, with reported speeds of less than 8.6 km/h [45,75]. More locally, a study conducted in Salou, a coastal town located about 100 km southwest of Barcelona, reported storm propagation speeds ranging between 9 and 61 km/h based on radar measurements and hourly/daily rainfall records during the 2014–2018 period [75]. These findings align well with the speed range observed in our analysis over the city of Barcelona.

The observed difference in average storm speeds obtained by the centroid and first maximum methods can be attributed to their fundamentally different definitions and sensitivities. The first maximum peak method tracks the movement of the localized maximum rainfall intensity, which is spatially and temporally constrained to the convective core of the storm, thereby reflecting primarily the propagation speed of this core. Conversely, the time centroid method computes a weighted average timing over the entire rainfall hyetogram, integrating both convective and stratiform precipitation components, thus representing the overall motion of the storm system. This integration allows the centroid to reflect contributions from diffuse precipitation areas and trailing features that may move more freely or faster than the convective core, leading to higher average speeds estimated by the centroid method.

Regarding the general propagation directions for all analyzed events, two dominant trends emerged consistently with both tracking methods: movement toward the sea and toward the northeast. While centroid-based tracking suggests a slightly more pronounced seaward propagation, this difference mainly reflects the centroid's sensitivity to storm growth dynamics fueled by marine moisture inflow, as previously discussed. The first maximum peak method, focused on the convective core displacement, tends to show somewhat less pronounced movement toward the sea, due to its structural constraints. Focusing specifically on intense events, the predominant motion directions become more structured. Notably, about 45% of these high-intensity storms moved toward the sea, a result consistently observed with both centroid and peak tracking methods. However, only the centroid-based analysis revealed a second pronounced clustering, with a substantial proportion of fast-moving intense storms concentrated in the NNE–NE–ENE directions, nearly parallel to the coastline. This cluster accounted for over 11% of the intense centroid-tracked events, with an average speed of 32 km/h, and it aligns with the gap created by the Besòs River in the Collserola mountain range. Although a direct channeling effect is unlikely given the relatively low elevation of the ridge (~500 m) compared to convective altitudes, the topography and river corridor may still modulate surface-level flows, potentially facilitating inland breeze penetration during sea-breeze regimes. In general, the average speed of events moving in the first quadrant (N to E) was significantly higher than in the others: 32 km/h for the centroid method (with all other quadrants below 25 km/h) and 24 km/h for the first maximum method (others below 16 km/h). Nonetheless, this preferential NE orientation is also observed at broader spatial scales, suggesting a link to mesoscale circulation patterns rather than local orographic forcing alone. For instance, when studying mesoscale convective systems (MCSs) across Catalunya between 2012 and 2016, [76] found that most systems moved predominantly within a directional range from west to east or roughly from WSW to NE. The range of directions reported in that study closely matches those identified in our full dataset, spanning from NNE to SSE, with practically no MCSs moving from east to west.

5. Conclusions

Principal component analysis (PCA) has proven useful in identifying dominant spatial patterns of rainfall variability over the urban area of Barcelona, based on high-resolution data from a dense rain gauge network. The first principal component suggests the presence of a pattern that may be consistent with an urban rainfall island (URI), whose spatial configuration resembles that of the urban heat island (UHI) over the city. This urban-related signal is also reflected in the spatial distribution of the scaling parameter β, calculated from the records of each rain gauge. Topographic influences, particularly from the Collserola mountain range and smaller features like Montjuïc Hill, play a key role in shaping rainfall variability across the city. The second and third principal components highlighted distinct contrasts between inland and coastal areas.

The spatial organization of dominant rainfall modes, as identified through the PCA, varies notably with event duration. At shorter durations (e.g., 10 min), a clear north–south contrast is observed across the city. As the duration increases to 1 h and 4 h, the spatial structure shifts to reveal a more distinct inland mountains versus coastal differentiation. At the 24 h timescale, rainfall patterns appear largely homogeneous across the urban area.

These evolving spatial structures are broadly consistent with those observed in climatic indices such as Rx1day, where a similar inland–coastal differentiation is present, with the maritime influence progressively becoming more prominent when computed at a sub-daily resolution, especially for Rx10min.

Storm tracking analyses revealed a wide range of propagation speeds over the city of Barcelona, with average values and variability consistent with those reported in other European studies. The time centroid method consistently produced higher average propagation velocities compared to the first maximum peak method, reflecting its sensitivity to the integrated motion of the entire precipitation system rather than just the convective core. Additionally, the centroid approach yielded a larger proportion of well-fitted storm tracks, demonstrating its robustness and reliability for capturing storm dynamics at the urban scale. Despite these methodological differences, both tracking techniques revealed spatial propagation patterns broadly consistent with the dominant rainfall variability modes identified by the PCA, effectively capturing the key hydrometeorological features shaping rainfall over Barcelona. Among the most intense events, two predominant storm movement directions were identified for both methods: a dominant flow toward the sea (45% of the intense events), and a faster cluster oriented roughly toward the northeast (NNE–NE–ENE), almost parallel to the coast, which was much more pronounced when using the centroid method. This directional clustering aligns with the gap created by the Besòs River in the Collserola mountain range, although a direct channeling influence is unlikely given the relatively low elevation of the Collserola ridge (~500 m) compared to typical convective altitudes. Moreover, this NE direction is also commonly observed at broader spatial scales, reflecting mesoscale circulation patterns. However, the topography of Collserola and the low-lying corridor formed by the river may still play a role in modulating surface-level flows, potentially favoring inland breeze penetration during sea-breeze regimes.

These findings deepen our understanding of rainfall dynamics in complex urban and coastal environments and underscore the value of high-resolution, long-term observational networks in hydrometeorological research. The analysis of intense rainfall events highlights the capacity of dense urban rain gauge networks to effectively capture storm motion and variability at the city scale.

Supplementary Materials: The following supporting information can be downloaded at: https://www.mdpi.com/article/10.3390/hydrology12070178/s1, Table S1: Date, fitting parameters, speed (v) and direction (θ) of the storms monitored using the time centroid. Table S2: Date, fitting parameters, speed (v) and direction (θ) of the storms monitored using the first maximum.

Author Contributions: Conceptualization, R.R.-S. and M.d.C.C.-C.; methodology, R.R.-S. and X.N.; software, R.R.-S. and X.N.; validation R.R.-S. and M.d.C.C.-C.; formal analysis, M.d.C.C.-C., R.R.-S., and X.N.; investigation, R.R.-S. and M.d.C.C.-C.; resources, M.d.C.C.-C.; data curation, X.N.; writing—original draft preparation, M.d.C.C.-C.; writing—review and editing, M.d.C.C.-C. and R.R.-S.; visualization, M.d.C.C.-C. and R.R.-S.; supervision, M.d.C.C.-C. All authors have read and agreed to the published version of the manuscript.

Funding: This research received no external funding.

Data Availability Statement: The raw data supporting the conclusions of this article will be made available by the authors on request. In addition, selected results are provided in two Supplementary Tables included with the article. The urban rainfall data used in this study can be obtained upon request to info.bcasa@bcn.cat (Departament d'Explotació de BCASA) [77].

Acknowledgments: The authors gratefully acknowledge Barcelona Cicle de l'Aigua, SA (BCASA) for providing the high-resolution urban rainfall data used in this study. We also extend our sincere thanks to Tomeu Rigo, Anna Rius, and the Servei Meteorològic de Catalunya (SMC) for kindly sharing the radar imagery.

Conflicts of Interest: The authors declare no conflicts of interest.

References

1. Martín-Vide, J. Spatial Distribution of a Daily Precipitation Concentration Index in Peninsular Spain. *Int. J. Climatol.* **2004**, *24*, 959–971. [CrossRef]
2. Casas, M.C.; Rodríguez, R.; Redaño, À. Analysis of extreme rainfall in Barcelona using a microscale rain gauge network. *Meteorol. Appl.* **2010**, *17*, 117–123. [CrossRef]
3. Rodríguez, R.; Navarro, X.; Casas, M.C.; Redaño, À. Rainfall spatial organization and areal reduction factors in the metropolitan area of Barcelona (Spain). *Theor. Appl. Climatol.* **2013**, *114*, 1–8. [CrossRef]
4. Tramblay, Y.; El Adlouni, S.; Servat, E. Trends and variability in extreme precipitation indices over Maghreb countries. *Nat. Hazards Earth Syst. Sci.* **2013**, *13*, 3235–3248. [CrossRef]
5. Pérez-Zanón, N.; Casas-Castillo, M.C.; Rodríguez-Solà, R.; Peña, J.C.; Rius, A.; Solé, J.G.; Redaño, À. Analysis of extreme rainfall in the Ebre Observatory (Spain). *Theor. Appl. Climatol.* **2016**, *124*, 935–944. [CrossRef]
6. Ribes, A.; Thao, S.; Vautard, R.; Dubuisson, B.; Somot, S.; Colin, J.; Planton, S.; Soubeyroux, J.M. Observed increase in extreme daily rainfall in the French Mediterranean. *Clim. Dyn.* **2019**, *52*, 1095–1114. [CrossRef]
7. Lana, X.; Rodríguez-Solà, R.; Martínez, M.D.; Casas-Castillo, M.C.; Serra, C.; Burgueño, A. Characterization of standardized heavy rainfall profiles for Barcelona city: Clustering, rain amounts and intensity peaks. *Theor. Appl. Climatol* **2020**, *142*, 255–268. [CrossRef]
8. González-Hidalgo, J.C.; Beguería, S.; Peña-Angulo, D.; Trullenque, V. Catalogue and analysis of extraordinary precipitation events in the Spanish mainland, 1916–2022. *Int. J. Climatol.* **2025**, *45*, e8785. [CrossRef]
9. Naylor, J.; Sexton, A. The Relationship between Severe Weather Warnings, Storm Reports, and Storm Cell Frequency in and around Several Large Metropolitan Areas. *Weather Forecast.* **2018**, *33*, 1339–1358. [CrossRef]
10. Liu, J.; Niyogi, D. Meta-analysis of urbanization impact on rainfall modification. *Sci. Rep.* **2019**, *9*, 7301. [CrossRef]
11. Martín-Vide, J.; Moreno-García, M.C. Probability values for the intensity of urban heat island (Spain). *Atmos. Res.* **2020**, *240*, 104877. [CrossRef]
12. Torelló-Sentelles, H.; Marra, F.; Koukoula, M.; Villarini, G.; Peleg, N. Intensification and changing spatial extent of heavy rainfall in urban areas. *Earth's Future* **2024**, *12*, e2024EF004505. [CrossRef]
13. Shepherd, J.M.; Stallins, J.A.; Jin, M.L.; Mote, T.L. Urbanization: Impacts on Clouds, Precipitation, and Lightning. In *Urban Ecosystem Ecology*; Aitkenhead-Peterson, J., Volder, A., Eds.; Agronomy Monographs; ASA, CSSA, SSSA Books: Madison, WI, USA, 2010. [CrossRef]
14. IPCC. *Climate Change 2021: The Physical Science Basis. Contribution of Working Group I to the Sixth Assessment Report of the Intergovernmental Panel on Climate Change*; Masson-Delmotte, V., Zhai, P., Pirani, A., Connors, S.L., Péan, C., Berger, S., Caud, N., Chen, Y., Goldfarb, L., Gomis, M.I., et al., Eds.; Cambridge University Press: Cambridge, UK; New York, NY, USA, 2021; 2391p. [CrossRef]
15. Myhre, G.; Alterskjær, K.; Stjern, C.W.; Hodnebrog, Ø.; Marelle, L.; Samset, B.H.; Sillmann, J.; Schaller, N.; Fischer, E.; Schulz, M.; et al. Frequency of extreme precipitation increases extensively with event rareness under global warming. *Sci. Rep.* **2019**, *9*, 16063. [CrossRef] [PubMed]
16. Drobinski, P.; Ducrocq, V.; Alpert, P.; Anagnostou, E.; Béranger, K.; Borga, M.; Braud, I.; Chanzy, A.; Davolio, S.; Delrieul, G.; et al. HyMeX: A 10-Year Multidisciplinary Program on the Mediterranean Water Cycle. *Bull. Am. Meteorol. Soc.* **2014**, *95*, 1063–1082. [CrossRef]
17. Ínsua-Costa, D.; Senande-Rivera, M.; Llasat, M.C.; Míguez-Macho, G. A global perspective on western Mediterranean precipitation extremes. *npj Clim. Atmos. Sci.* **2022**, *5*, 9. [CrossRef]
18. Reale, M.; Lionello, P. Synoptic climatology of winter intense precipitation events along the Mediterranean coasts. *Nat. Hazards Earth Syst. Sci.* **2013**, *13*, 1707–1722. [CrossRef]
19. de Lima, M.I.P.; Santo, F.E.; Ramos, A.M.; Trigo, R.M. Trends and correlations in annual extreme precipitation indices for mainland Portugal, 1941–2007. *Theor. Appl. Climatol.* **2015**, *119*, 55–75. [CrossRef]
20. Sunyer, M.A.; Hundecha, Y.; Lawrence, D.; Madsen, H.; Willems, P.; Martinkova, M.; Vormoor, K.; Bürger, G.; Hanel, M.; Kriaučiūnienė, J.; et al. Inter-comparison of statistical downscaling methods for projection of extreme precipitation in Europe. *Hydrol. Earth Syst. Sci.* **2015**, *19*, 1827–1847. [CrossRef]
21. Serrano-Notivoli, R.; Beguería, S.; Saz, M.Á.; de Luis, M. Recent trends reveal decreasing intensity of daily precipitation in Spain. *Int. J. Climatol.* **2018**, *38*, 4211–4224. [CrossRef]

22. Zittis, G.; Almazroui, M.; Alpert, P.; Ciais, P.; Cramer, W.; Dahdal, Y.; Fnais, M.; Francis, D.; Hadjinicolaou, P.; Howari, F.; et al. Climate change and weather extremes in the Eastern Mediterranean and Middle East. *Rev. Geophys.* **2022**, *60*, e2021RG000762. [CrossRef]
23. Vicente-Serrano, S.M.; Tramblay, Y.; Reig, F.; González-Hidalgo, J.C.; Beguería, S.; Brunetti, M.; Kalin, K.C.; Patalen, L.; Kržič, A.; Lionello, P.; et al. High temporal variability not trend dominates Mediterranean precipitation. *Nature* **2025**, *639*, 658–666. [CrossRef] [PubMed]
24. Lionello, P.; Scarascia, L. The relation of climate extremes with global warming in the Mediterranean region and its north versus south contrast. *Reg. Environ. Change* **2020**, *20*, 31. [CrossRef]
25. Vautard, R.; Kadygrov, N.; Iles, C.; Boberg, F.; Buonomo, E.; Bülow, K.; Coppola, E.; Corre, L.; van Meijgaard, E.; Nogherotto, R.; et al. Assessment of the European Climate Projections as Simulated by the Large EURO-CORDEX Regional and Global Climate Model Ensemble. *J. Geophys. Res. D Atmos.* **2021**, *126*, e2019JD032356. [CrossRef]
26. Rodríguez, R.; Navarro, X.; Casas, M.C.; Ribalaygua, J.; Russo, B.; Pouget, L.; Redaño, À. Influence of climate change on IDF curves for the metropolitan area of Barcelona (Spain). *Int. J. Climatol.* **2014**, *34*, 643–654. [CrossRef]
27. Casas-Castillo, M.C.; Rodríguez-Solà, R.; Lana, X.; Serra, C.; Martínez, M.D.; Biere, R.; Arellano, B.; Roca, J. Consecuencias hidrológicas del cambio climático en entornos urbanos. In Proceedings of the XIII International Conference on Virtual City and Territory: Challenges and Paradigms of the Contemporary City, Barcelona, Spain, 2–4 October 2019; p. 8291. [CrossRef]
28. Monjo, R.; Locatelli, L.; Milligan, J.; Torres, L.; Velasco, M.; Gaitán, E.; Pórtoles, J.; Redolat, D.; Russo, B.; Ribalaygua, J. Estimation of future extreme rainfall in Barcelona (Spain) under monofractal hypothesis. *Int. J. Climatol.* **2023**, *43*, 4047–4068. [CrossRef]
29. Huang, X.; Hao, L.; Sun, G.; Yang, Z.-L.; Li, W.; Chen, D. Urbanization aggravates effects of global warming on local atmospheric drying. *Geophys. Res. Lett.* **2022**, *49*, e2021GL095709. [CrossRef]
30. Huang, J.; Fatichi, S.; Mascaro, G.; Manoli, G.; Peleg, N. Intensification of sub-daily rainfall extremes in a low-rise urban area. *Urban Clim.* **2022**, *42*, 101124. [CrossRef]
31. Huang, S.; Gan, Y.; Zhang, X.; Chen, N.; Wang, C.; Gu, X.; Ma, J.; Niyogi, D. Urbanization amplified asymmetrical changes of rainfall and exacerbated drought: Analysis over five urban agglomerations in the Yangtze River Basin, China. *Earth's Future* **2023**, *11*, e2022EF003117. [CrossRef]
32. Llasat, M.C.; Marcos, R.; Turco, M.; Gilabert, J.; Llasat-Botija, M. Trends in Flash Flood Events versus Convective Precipitation in the Mediterranean Region: The Case of Catalonia. *J. Hydrol.* **2016**, *541*, 24–37. [CrossRef]
33. Doswell, C., III; Ramis, C.; Romero, R.; Alonso, S. A Diagnostic Study of Three Heavy Precipitation Episodes in the Western Mediterranean Region. *Weather Forecast.* **1998**, *13*, 102–124. [CrossRef]
34. Alpert, P.; Radian, R.; Halfon, N.; Levin, Z. Urban Rainfall Anomaly under Intensive Development, 1949–2018, Case of Tel-Aviv, Israel. *Atmosphere* **2019**, *10*, 163. [CrossRef]
35. Wilson, J.; Crook, N.; Mueller, C.; Sun, J.; Dixon, M. Nowcasting thunderstorms: A status report. *Bull. Am. Meteorol. Soc.* **1998**, *79*, 2079–2099. [CrossRef]
36. Newton, C.W.; Fankhauser, J.C. Movement and propagation of multicellular convective storms. *Pure Appl. Geophys.* **1975**, *113*, 747–764. [CrossRef]
37. Lorente, J.; Redaño, À. Rainfall rate distribution in a local scale: The case of Barcelona city. *Theor. Appl. Climatol.* **1990**, *41*, 23–32. [CrossRef]
38. Zhang, Y.; Miao, S.; Dai, Y.; Bornstein, R. Numerical simulation of urban land surface effects on summer convective rainfall under different UHI intensity in Beijing. *J. Geophys. Res. Atmos.* **2017**, *122*, 7851–7868. [CrossRef]
39. Hong, S.H.; Jin, H.G.; Han, J.Y.; Baik, J.J. Initiation and evolution of urban-induced precipitation under different background wind speeds: Roles of urban breeze circulation and cold pool. *Theor. Appl. Climatol.* **2024**, *155*, 9457–9470. [CrossRef]
40. Li, Y.; Fowler, H.J.; Argüeso, D.; Blenkinsop, S.; Evans, J.P.; Lenderink, G.; Yan, X.; Guerreiro, S.B.; Lewis, E.; Li, X.F. Strong intensification of hourly rainfall extremes by urbanization. *Geophys. Res. Lett.* **2020**, *47*, e2020GL088758. [CrossRef]
41. Yan, H.; Gao, Y.; Wilby, R.; Yu, D.; Wright, N.; Yin, J.; Chen, X.; Chen, J.; Guan, M. Urbanization further intensifies short-duration rainfall extremes in a warmer climate. *Geophys. Res. Lett.* **2024**, *51*, e2024GL108565. [CrossRef]
42. Rigo, T.; Llasat, M.-C. Analysis of mesoscale convective systems in Catalonia using meteorological radar for the period 1996–2000. *Atmos. Res.* **2007**, *83*, 458–472. [CrossRef]
43. Rigo, T.; Farnell, C. Quasi-Linear Convective Systems in Catalonia Detected Through Radar and Lightning Data. *Remote Sens.* **2024**, *16*, 4262. [CrossRef]
44. Arellano Ramos, B.; Roca-Cladera, J. Urban-CLIMPLAN: La Isla de Calor Urbana en la Región Metropolitana de Barcelona. *ACE-Archit. City Environ.* **2021**, *15*, 10381. [CrossRef]
45. Rigo, T.; Llasat, M.C. A methodology for the classification of convective structures using meteorological radar: Application to heavy rainfall events on the Mediterranean coast of the Iberian Peninsula. *Nat. Hazards Earth Syst. Sci.* **2004**, *4*, 59–68. [CrossRef]

46. Berne, A.; Krajewski, W.F. Radar for hydrology: Unfulfilled promise or unrecognized potential? *Adv. Water Resour.* **2013**, *51*, 357–366. [CrossRef]
47. Thorndahl, S.; Einfalt, T.; Willems, P.; Nielsen, J.E.; ten Veldhuis, M.-C.; Arnbjerg-Nielsen, K.; Rasmussen, M.R.; Molnar, P. Weather radar rainfall data in urban hydrology. *Hydrol. Earth Syst. Sci.* **2017**, *21*, 1359–1380. [CrossRef]
48. Ochoa-Rodríguez, S.; Wang, L.-P.; Willems, P.; Onof, C. A review of radar-rain gauge data merging methods and their potential for urban hydrological applications. *Water Resour. Res.* **2019**, *55*, 6356–6391. [CrossRef]
49. Esbrí, L.; Rigo, T.; Llasat, M.C.; Aznar, B. Identifying storm hotspots and the most unsettled areas in Barcelona by analysing significant rainfall episodes from 2013 to 2018. *Water* **2021**, *13*, 1730. [CrossRef]
50. Lo Conti, F.; Noto, L.; Quatrosi, A.; La Loggia, G. Using high resolution raingauge data for storm tracking analysis in the urban area of Palermo, Italy. In Proceedings of the 8th International Workshop on Precipitation in Urban Areas "Rainfall in the Urban Context: Forecasting, Risk and Climate Change", St. Moritz, Switzerland, 10–13 December 2009; pp. 166–171. Available online: https://hdl.handle.net/10447/41562 (accessed on 10 March 2025).
51. Carbone, M.; Garofalo, G.; Tomei, G.; Piro, P. Storm Tracking based on Rain Gauges for Flooding Control in Urban Areas. *Procedia Eng.* **2014**, *70*, 256–265. [CrossRef]
52. Paz, I.; Tchiguirinskaia, I.; Schertzer, D. Rain gauge networks' limitations and the implications to hydrological modelling highlighted with a X-band radar. *J. Hydrol.* **2020**, *583*, 124615. [CrossRef]
53. Navarro, X. Organització Espacial i Temporal de la pluja a l'àrea Metropolitana de Barcelona. Ph.D. Thesis, Universitat Politècnica de Catalunya, Vilanova I La Geltrú, Spain, 13 October 2021. Available online: http://hdl.handle.net/10803/673995 (accessed on 18 March 2025).
54. Olsson, J.; Niemczynowicz, J. Multifractal analysis of daily spatial rainfall distributions. *J. Hydrol.* **1996**, *187*, 29–43. [CrossRef]
55. Enjamio, C.; Vilar, E.; Redaño, À.; Fontán, F.P.; Ndzi, D. Experimental analysis of microscale rain cells and their dynamic evolution. *Radio Sci.* **2005**, *40*, RS3015. [CrossRef]
56. Servei Meteorològic de Catalunya: Divulgacio/La Predicció Meteorològica/Predicció/. Available online: https://www.meteo.cat/wpweb/divulgacio/la-prediccio-meteorologica/prediccio/ (accessed on 12 April 2025).
57. Platikanov, S.; López, J.F.; Martrat, B.; Martín-Vide, J.; Tauler, R. Temporal and Spatial Relationships Between Climatic Indices and Precipitation Zones in Europe, Spain and Catalonia. *Int. J. Climatol.* **2025**, *45*, e8699. [CrossRef]
58. Jolliffe, I. Principal Component Analysis. In *International Encyclopedia of Statistical Science*; Lovric, M., Ed.; Springer: Berlin/Heidelberg, Germany, 2011. [CrossRef]
59. Rodríguez-Solà, R.; Casas-Castillo, M.C.; Navarro, X.; Redaño, À. A study of the scaling properties of rainfall in Spain and its appropriateness to generate intensity-duration-frequency curves from daily records. *Int. J. Climatol.* **2017**, *37*, 770–780. [CrossRef]
60. Casas-Castillo, M.C.; Llabrés-Brustenga, A.; Rius, A.; Rodríguez-Solà, R.; Navarro, X. A single scaling parameter as a first approximation to describe the rainfall pattern of a place: Application on Catalonia. *Acta Geophys.* **2018**, *66*, 415–424. [CrossRef]
61. Casas-Castillo, M.C.; Llabrés-Brustenga, A.; Rodríguez-Solà, R.; Rius, A.; Redaño, À. Scaling properties of rainfall as a basis for intensity-duration-frequency relationships and their spatial distribution in Catalunya, NE Spain. *Climate* **2025**, *13*, 37. [CrossRef]
62. Karl, T.R.; Nicholls, N.; Ghazi, A. (Eds.) CLIVAR/GCOS/WMO workshop on indices and indicators for climate extremes: Workshop summary, Weather and Climate Extremes. In *Weather and Climate Extremes*; Springer: Dordrecht, The Netherlands, 1999; Volume 42, pp. 3–7. [CrossRef]
63. Niemczynowicz, J. Storm tracking using rain gauge data. *J. Hydrol.* **1987**, *93*, 135–152. [CrossRef]
64. Diskin, M.H. The Speed of Two Moving Rainfall Events in Lund. *Hydrol. Res.* **1990**, *21*, 153–164. [CrossRef]
65. Zawadski, I.I. Statistical properties of precipitation patterns. *J. Appl. Met.* **1973**, *12*, 459–472. [CrossRef]
66. Marshall, R.J. The estimation and distribution of storm movement and storm structure, using a correlation analysis technique and rain gauge data. *J. Hydrol.* **1980**, *48*, 19–39. [CrossRef]
67. Upton, G.J.G. A correlation–regression method for tracking rainstorms using rain-gauge data. *J. Hydrol.* **2002**, *261*, 60–73. [CrossRef]
68. Serra, C.; Lana, X.; Martínez, M.D.; Roca, R.; Arellano, B.; Biere, R.; Moix, M.; Burgueño, A. Air temperature in Barcelona metropolitan region from MODIS satellite and GIS data. *Theor. Appl. Climatol.* **2020**, *139*, 473–492. [CrossRef]
69. Sui, X.; Yang, Z.; Shepherd, M.; Niyogi, D. Global scale assessment of urban precipitation anomalies. *Proc. Natl. Acad. Sci. USA* **2024**, *121*, e2311496121. [CrossRef] [PubMed]
70. Zhao, Y.; Xia, J.; Xu, Z.; Zou, L.; Qiao, Y.; Li, P. Impact of urban expansion on rain island effect in Jinan City, North China. *Remote Sens.* **2021**, *13*, 2989. [CrossRef]
71. Berndtsson, R.; Niemczynowicz, J. Spatial and temporal scales in rainfall analysis—Some aspects and future perspectives. *J. Hydrol.* **1988**, *100*, 293–313. [CrossRef]

72. Casas-Castillo, M.C.; Rodríguez-Solà, R.; Llabrés-Brustenga, A.; García-Marín, A.P.; Estévez, J.; Navarro, X. A Simple Scaling Analysis of Rainfall in Andalusia (Spain) under Different Precipitation Regimes. *Water* **2022**, *14*, 1303. [CrossRef]
73. Mazón, J.; Pino, D. Pluviometric anomaly in the Llobregat Delta. *Tethys Rev. Meteorol. Climatol. Mediterr.* **2009**, *6*, 31–50. [CrossRef]
74. Gascón, E.; Laviola, S.; Merino, A.; Miglietta, M.M. Analysis of a Localized Flash-Flood Event over the Central Mediterranean. *Atmos. Res.* **2016**, *182*, 256–268. [CrossRef]
75. del Moral, A.; Llasat, M.C.; Rigo, T. Connecting Flash Flood Events with Radar-Derived Convective Storm Characteristics on the Northwestern Mediterranean Coast: Knowing the Present for Better Future Scenarios Adaptation. *Atmos. Res.* **2020**, *238*, 104863. [CrossRef]
76. Rigo, T.; Berenguer, M.; Llasat, M.C. An improved analysis of mesoscale convective systems in the western Mediterranean using weather radar. *Atmos. Res.* **2019**, *227*, 147–156. [CrossRef]
77. Barcelona Cicle de l'Aigua, SA. 9. Sol·Licitud d'informació Pluviomètrica. Available online: https://www.bcasa.cat/CAT/solicitud-informacio.asp (accessed on 14 April 2025).

hydrology

MDPI

Article

Uncertainty in Kinetic Energy Models for Rainfall Erosivity Estimation in Semi-Arid Regions

José Bandeira Brasil [1], Ana Célia Maia Meireles [1], Carlos Wagner Oliveira [1], Sirleide Maria de Menezes [1], Francisco Dirceu Duarte Arraes [2] and Maria Simas Guerreiro [3,*]

[1] Postgraduate Program in Sustainable Regional Development—PRODER, Federal University of Cariri, Juazeiro do Norte 63048-080, Brazil; josebandeira@alu.ufc.br (J.B.B.); ana.meireles@ufca.edu.br (A.C.M.M.); carlos.oliveira@ufca.edu.br (C.W.O.); sirleide.menezes@ufrpe.br (S.M.d.M.)

[2] Federal Institute of Education, Science and Technology of Sertão Pernambucano (IFSertãoPE)-Campus Salgueiro, Salgueiro 56000-000, Brazil; dirceu.arraes@ifsertao-pe.edu.br

[3] RISE-Health, Faculty of Science and Technology, Fernando Pessoa University, Praça 9 de Abril 349, 4249-004 Porto, Portugal

* Correspondence: mariajoao@ufp.edu.pt

Abstract

The Brazilian semi-arid Northeast plays a critical role in regional hydrology, where rainfall is marked by pronounced temporal variability and short duration, presenting significant challenges for understanding and managing hydrological and erosive processes. This study aims to evaluate the performance of empirical models for estimating rainfall kinetic energy (KE) and erosivity index (EI_{30}) in this region, for all events and erosive events, using high-resolution rainfall data collected at the Federal University of Cariri (UFCA), Ceará. A total of 283 natural rainfall events were analyzed, with KE and EI_{30} values calculated using multiple models: Wischmeier and Smith, USDA, Van Dijk, a temporal variation-based model (KE_VT), and a regional model developed for Brazil's semi-arid zone, which served as the reference. The results show a predominance of small rainfall events (<5.2 mm), though maximum EI_{30} values exceeded 1300 MJ ha^{-1} mm h^{-1}, highlighting the potential for extreme erosive events. Comparative analysis revealed that all international models significantly underestimated KE and EI_{30} values compared to the regional reference, with the KE_VT model showing the closest approximation (13% underestimation), for all events and erosive events. Statistical assessments using the Wilcoxon test, Nash–Sutcliffe efficiency, and Willmott concordance index confirmed the superior performance of the KE_VT, for all events and erosive events. These findings underscore the importance of considering intra-event rainfall variability and regional calibration when modeling erosivity in semi-arid climates, contributing to more effective soil conservation and hydrological planning.

Keywords: USLE; erosivity index; soil erosion; Wischmeier and Smith; rainfall intensity

1. Introduction

The semi-arid regions of Brazil's Northeast are primarily defined by a distinctive rainfall regime, marked by scarcity, irregular distribution, and a pronounced concentration of precipitation within a short period—typically three to four months—during which sudden and brief rain events occur [1,2]. Rainfall serves as the principal source of soil moisture, making its accurate monitoring and characterization at the event scale crucial, particularly for understanding soil erosion processes and informing hydrological models [3–5].

Understanding rainfall characteristics and their relationship with the dynamics of kinetic energy distribution is essential for assessing rainfall's erosive potential [6–9]. Research on rainfall and its erosivity is vital for implementing effective soil conservation and land management strategies in semi-arid environments [10]. Erosivity, which refers to the ability of rainfall to cause soil erosion, is a key parameter in the Universal Soil Loss Equation [11] and related modeling approaches.

Research in semi-arid environments remains limited, largely due to the high costs associated with acquiring specialized equipment needed to measure detailed rainfall characteristics. In many areas of northeastern Brazil, essential rainfall data—such as event duration and intensity—are often unavailable. Rainfall erosivity is quantified through specific indicators, and its accuracy depends on factors such as the number of rain gauges, the temporal resolution of data recording, and the choice of kinetic energy equation used [12]. Among the various approaches, the most robust kinetic energy models estimate rainfall erosivity by incorporating rainfall intensity, offering a more precise representation of the relationship between rainfall and soil erosion than other variables [13].

Despite the challenges associated with rainfall monitoring, empirical relationships continue to be widely used to estimate kinetic energy and rainfall erosivity [13–15], most commonly based on the foundational model proposed by Wischmeier and Smith [11]. This model employs the average intensity of each rainfall event as a key input; however, rainfall intensity typically varies throughout the duration of an event [6,7,16]. Other widely adopted empirical models—such as those by [17] and the USDA Agricultural Research Service [18], which is incorporated into the RUSLE2 equation—also rely on average intensity values. In contrast, a modified version of the Wischmeier and Smith model that incorporates the temporal variability of rainfall intensity at fixed intervals during an event has shown promise in more accurately capturing the dynamics of rainfall energy and erosivity [10,16].

A significant advancement in this field was achieved by [10] through a study conducted in Iguatu, CE, a semi-arid region in northeastern Brazil. The study involved detailed monitoring of rainfall characteristics, including distribution, intensity, duration, frequency, droplet diameter, and velocity. Based on these measurements, the authors developed a region-specific empirical model tailored to the semi-arid conditions of Brazil. This model was calibrated and validated to improve the accuracy of kinetic energy and rainfall erosivity estimates, thereby minimizing potential errors. It is important to emphasize that the various models discussed in the literature serve as valuable tools for soil and water conservation, supporting informed decision making in land management practices [8,9,15,19,20].

Given this context, obtaining accurate measurements of rainfall kinetic energy is essential for reliably estimating its erosivity potential. Equally important is the evaluation of different empirical models to determine their suitability for application in semi-arid regions. Accordingly, the objective of this study is to analyze rainfall characteristics and compare the uncertainties associated with various kinetic energy and erosivity estimation models, for all events and erosive events, in relation to the reference model specifically developed for Brazil's semi-arid region. By quantifying and comparing these uncertainties, we aim to inform the selection and refinement of models for use in regional and global assessments of soil erosion risk, particularly in the context of changing climate and rainfall patterns.

2. Materials and Methods

2.1. Study Area

The study was conducted at the Center for Agricultural Sciences and Biodiversity (CCAB), located on the campus of the Federal University of Cariri (UFCA) in the municipality of Crato. This area lies within the Salgado sub-basin, part of the Cariri region in

Ceará, northeastern Brazil (Figure 1). The central geographic coordinates of the study site are 7°14′ S and 39°22′ W, with an elevation of 425 m above sea level.

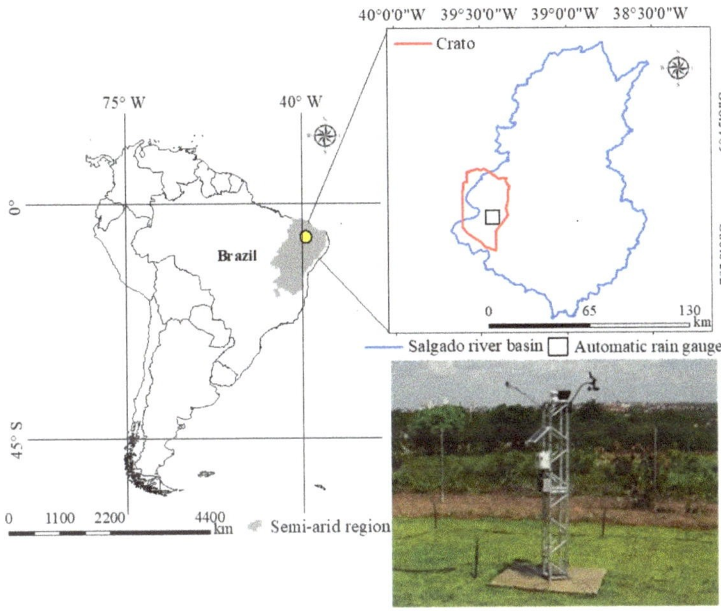

Figure 1. Location of the study area in the municipality of Crato, Ceará, Northeast Brazil.

The study site experiences seasonal climatic variability, with average annual temperatures ranging between 24 °C and 27 °C and clearly defined rainy and dry periods [21]. According to historical data from the Ceará Foundation for Meteorology and Water Resources (FUNCEME), covering the period from 1974 to 2023, the region receives an average annual rainfall of 1116.3 mm. Of this total, approximately 84.4% occurs during the rainy season, which extends from December to April (Figure 2).

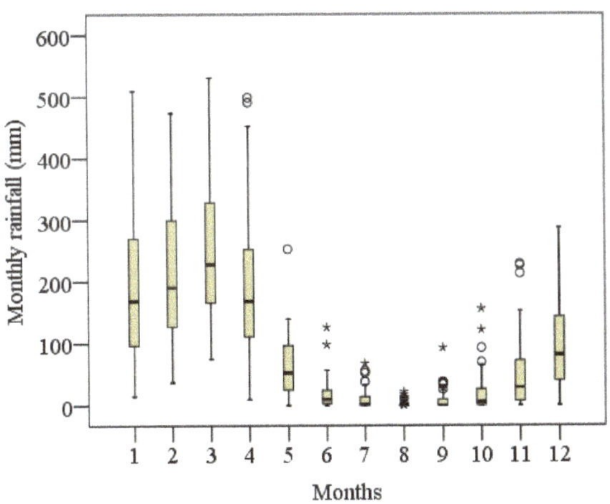

Figure 2. Temporal variability of the historical series of monthly precipitation in the municipality of Crato, Ceará.

The dry season spans from June to November, accounting for only 9% of the total annual rainfall, which underscores the high temporal variability of precipitation characteristic of the state of Ceará [1,5]. The municipality of Crato is located at the base of the Araripe Plateau, in the southernmost part of Ceará, within the Cariri micro-region [21]. Due to its location at the foot of the plateau, the area experiences relatively mild temperatures during the winter months, while summer temperatures tend to be considerably higher.

2.2. Rainfall Data Collection

Rainfall data were collected at the automatic weather station (EMCCAB) at the CCAB/UFCA, equipped with an automatic rain gauge and data logger (Figure 1). The data can be accessed at https://sites.ufca.edu.br/agrotempo/ (accessed 7 September 2024).

The dataset spans from September 2020, when the station began operations, to June 2024. The station utilizes an S-RGF M002 tipping bucket pluviograph with a collection area of 214 cm^2. It has a measurement range of 127 mm h^{-1}, with up to 4000 tips per recording interval. Its accuracy is ±4% for rainfall between 0.2 and 50 mm h^{-1}, and ±5% for rainfall between 50 and 100 mm h^{-1}. Each tip corresponds to a rainfall of 0.2 mm, with data acquisition occurring at 5 min intervals.

Using the rainfall data collected from the automatic rain gauge, the following characteristics were recorded: total rainfall (mm), rainfall duration (h), average rainfall intensity (mm h^{-1}), and the maximum rainfall intensity over 30 min (I_{30}, mm h^{-1}). Individual rain events were classified based on a minimum inter-event time (MIET) of 6 h, as proposed by [2]. This criterion resulted in the identification of 283 distinct natural rainfall events during the study period, of which 69 were classified as erosive events (rainfall greater than 12.7 mm).

2.3. Kinetic Energy Models-KE and Erosivity Index-EI_{30}

The equation proposed by [10], specifically developed for the semi-arid region of Brazil, was obtained, calibrated, and validated for the municipality of Iguatu, CE (Equation (1)). The equation was derived by adjusting the original parameters of the Wischmeier and Smith model based on field measurements and average rainfall intensity obtained from disdrometer data. This model was adopted as the reference for estimating kinetic energy (KE) and erosivity index (EI_{30}) and was used for comparison with globally recognized models from the literature [13,15], for all events and erosive events.

$$\text{KE\_Brasil} = 0.1578 + 0.1049 * \text{Log I} * R \tag{1}$$

where:

- KE is the kinetic energy of rainfall (MJ ha^{-1}),
- I is the average rainfall intensity (mm h^{-1}),
- R is the rainfall height (mm).

The widely used Wischmeier and Smith (WS) model [11], modified by [22], was employed to estimate kinetic energy (Equation (2)).

$$\text{KE\_WS} = 0.119 + 0.0873 * \text{Log I} * \text{R} \tag{2}$$

Additionally, the kinetic energy considering the temporal variation in rainfall intensity (KE_VT) is calculated for each time interval—30 min intra-event—using the Wischmeier and Smith [11] equation, further modified by [22]. The formulas for this calculation are presented in Equations (3) and (4).

$$KE_r = 0.119 + 0.0873 * \text{Log } I_r \quad (3)$$

$$KE\_VT = \left(\sum_{r=1}^{k} KE_r * R_r \right) \quad (4)$$

where:

- KE_r is the kinetic energy of rainfall (MJ ha^{-1} mm^{-1});
- I_r is the rainfall intensity during the adopted time interval (mm h^{-1}).

KE_VT represents the sum of the kinetic energy for each rainfall interval, calculated for each adopted time period (MJ ha^{-1}). This value is multiplied by R_r, which is the rainfall height during the r-th time period of a rainfall event divided into k intervals (R mm).

Kinetic energy (KE_USDA) was estimated using the USDA model [18], which is incorporated into the RUSLE2 equation, as shown in Equation (5):

$$KE\_USDA = 0.29 * (1 - 0.72 * EXP(-0.082 * I)) * R \quad (5)$$

Similarly, kinetic energy (KE_Van Dijk) was estimated using the [17] model, as presented in Equation (6):

$$KE\_Van\ Dijk = 0.283 * (1 - 0.52 * EXP(-0.042 * I)) * R \quad (6)$$

The rainfall erosivity index (EI_{30}) for all events and erosive events was calculated using Equation (7) for all the KE models analyzed:

$$EI_{30} = KE * I_{30} \quad (7)$$

where:

- EI_{30} is the rainfall erosivity index (MJ ha^{-1} mm h^{-1}),
- I_{30} is the maximum rainfall intensity over 30 min during the event (mm h^{-1}).

2.4. Statistical Analysis

Descriptive statistics, linear regression, and boxplots were employed to analyze the rainfall characteristics, as well as the distribution of kinetic energy and erosivity data derived from the models in the literature. The non-parametric Wilcoxon test was used to compare the medians of the KE and EI_{30} data at a 99% confidence level.

The performance of the KE and EI_{30} models from the literature was evaluated through graphical analysis using a 1:1 ratio plot and by statistical indicators, including the coefficient of determination (R^2), mean error (ME), and root mean square error (RMSE). These indicators measure the dispersion between the observed and estimated values, with values closer to zero indicating a better fit.

Further assessment of model performance was carried out using the Nash–Sutcliffe efficiency coefficient [23], which is considered satisfactory when greater than 0.75. The Willmott concordance index (d) [24] was also used, where a value of 1 indicates perfect agreement between observed and predicted values. Additionally, the model confidence index (C) was calculated as the product of the correlation coefficient (r) and the Willmott index (d), with optimal model performance indicated by a C value above 0.85 [25]. For further details on the model performance evaluation, refer to the research by [26,27].

Statistical analyses were performed for all events and erosive events, using the Statistical Package for the Social Sciences (SPSS) version 16.0, MINITAB version 18, and Microsoft Excel.

3. Results and Discussion

3.1. Rainfall Event Characteristics

A total of 283 rainfall events were analyzed (Table 1). The rainfall height (R) ranged from 0.40 mm to 107.80 mm (Table 1), demonstrating the high variability in the temporal distribution of rainfall during these events [2]. A significant frequency of small events was recorded, with 50% of events having a total rainfall of less than 5.20 mm (Q2 = 50%) (Table 1). The predominance of small rainfall events is a typical feature of semi-arid regions [4,5] and plays a crucial role in understanding hydrological processes at the basin scale [3]. These events influence processes such as the redistribution of raindrops under vegetation, the calculation of kinetic energy, and the assessment of the rainfall erosivity index [5,10].

Table 1. Descriptive statistics of rainfall characteristics at EMCCAB (Crato-CE).

Variables	N° of Events	Mean	StDev	Minimum	Q1	Q2 Median	Q3	Maximum
R	283	9.92	13.29	0.40	2.00	5.20	12.20	107.80
D		3.91	4.03	0.08	1.08	2.75	5.17	23.75
I		4.38	6.89	0.21	0.96	2.27	4.89	81.60
I_{30}		9.24	10.96	0.40	2.00	4.40	12.00	53.60
KE_Brasil		2.22	3.43	0.05	0.32	0.92	2.61	29.99
EI_{30}_Brasil		50.91	136.58	0.02	0.63	4.09	32.84	1355.68

R—Rainfall height (mm); D—Rainfall duration (h); I—Average rainfall intensity (mm h^{-1}); I_{30}—Maximum rainfall intensity in 30 min (mm h^{-1}); KE—Rainfall kinetic energy (MJ ha^{-1}); EI_{30}—Rainfall erosivity index (MJ ha^{-1} mm h^{-1}); percentiles—Q1-25%, Q2-50% and Q3-75%; StDev—Standard deviation).

According to the descriptive statistics (Table 1), rainfall duration, average intensity, maximum 30 min intensity, kinetic energy, and the erosivity index all exhibited substantial variability, consistent with the variation observed in rainfall height. The kinetic energy of rainfall (KE) reached a maximum of 29.99 MJ ha^{-1}, which aligns with the upper range of KE values reported for high-quality natural rainfall events [28], as shown in Table 1.

For the rainfall erosivity index (EI_{30}), 75% of the events had values below 32.84 MJ ha^{-1} mm h^{-1} and were associated with rainfall amounts under 12.20 mm, highlighting the high frequency of small rainfall events (Table 1). In Brazil's transitional zone between the Cerrado and Amazon biomes, ref. [8] observed that lower total rainfall in events are associated with reduced rainfall erosivity, which in turn are associated with lower soil erosion risks. These low-intensity events and slower erosion processes, benefit agricultural productivity by minimizing sediment loss [9].

Despite the predominance of low-intensity events, maximum EI_{30} values reached 1355.68 MJ ha^{-1} mm h^{-1}, coinciding with the highest recorded rainfall totals (Table 1), indicating the potential for extreme erosive events even in semi-arid regions. Such high EI_{30} values are not uncommon in northeastern Brazil; for instance, ref. [14] reported values as high as 2806 MJ ha^{-1} mm h^{-1} in the Aiuaba, CE experimental basin. Similarly, the Iguatu, CE experimental basin recorded a maximum EI_{30} of 2868.9 MJ ha^{-1} mm h^{-1} [29]. These elevated values, when combined with reduced vegetation cover—which provides critical protection against raindrop impact—can significantly increase erosion rates. According to [10], vegetation canopy can reduce rainfall kinetic energy (KE) and erosivity (EI_{30}) by

30% and 39%, respectively, with the greatest reduction observed during the period of peak foliage density in the semi-arid region.

High-magnitude rainfall events are key inputs for understanding the variability of the EI_{30} index, as they significantly influence hydrological and erosive processes. Consequently, these events are critical for developing effective soil and water conservation strategies, particularly in the context of forest adaptation to climate change [6]. Ref. [9] emphasize that changes in extreme precipitation can alter rainfall erosivity, thereby impacting soil erosion dynamics. This understanding is essential for informing the implementation of targeted soil and water conservation practices [15].

3.2. Kinetic Energy Models-KE and Erosivity Index-EI_{30}

During the study period, all natural rainfall events were used to compare the performance of the kinetic energy and erosivity index estimation models available in the literature: The model developed by [10] was adopted as the reference (KE_Brasil) and compared with the models for temporal variation in intensity (KE_VT), kinetic energy by [11] (KE_WS), kinetic energy by the [18] model (KE_USDA), and kinetic energy by the [17] model (KE_Van Dijk), as shown in Figure 3.

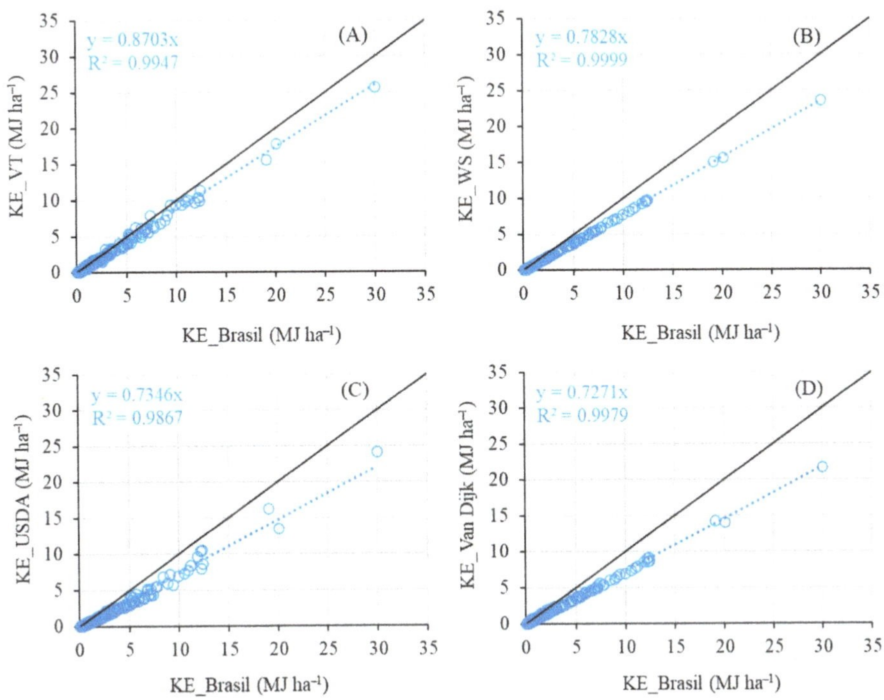

Figure 3. Estimation of kinetic energy (KE) by the Brasil model versus temporal variation in intensity-VT (**A**), Wischmeier and Smith-WS (**B**), USDA (**C**), Van Dijk (**D**).

Among the models evaluated, the kinetic energy model that incorporates temporal variation in rainfall intensity (KE_VT) showed the smallest underestimation relative to the KE_Brasil reference model, with a discrepancy of 13% (Figure 3A). The globally adopted Wischmeier and Smith model (KE_WS) underestimated KE by 22% (Figure 3B), followed by the USDA model with a 27% underestimation (Figure 3C). The Van Dijk model exhibited the greatest deviation, underestimating KE by 28% (Figure 3D). These results demonstrate that while all models tend to underestimate KE compared to the Brasil reference, the degree of underestimation varies. Moreover, statistical analysis using the Wilcoxon test confirmed that all models differ significantly from the KE_Brasil model at the 1% significance level (p-value = 0.000) (Figure 4A).

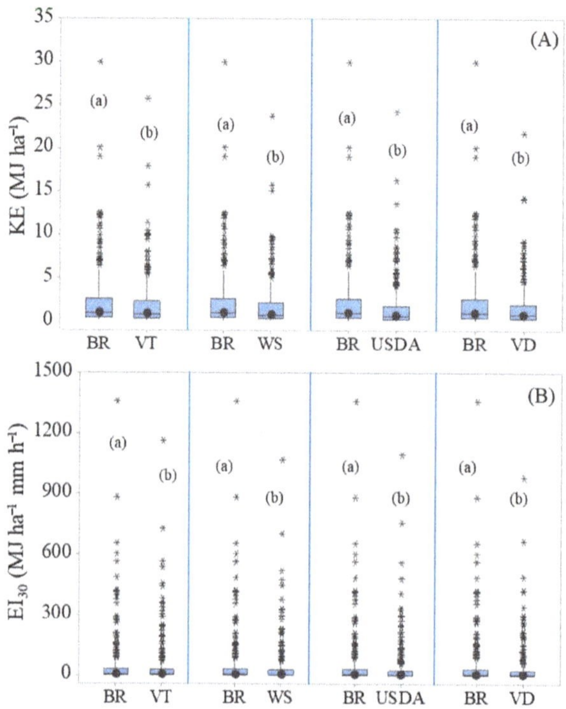

Figure 4. Statistical analysis of kinetic energy-KE (**A**) and estimation of rainfall erosivity potential-EI_{30} (**B**) by the different models. BR-Brasil model; VT-Temporal variation in intensity model; WS-Wischmeier and Smith model; USDA model; VD-Van Dijk model. * Different letters (a, b) represent a statistically different median at the 1% level using the Wilcoxon test.

The higher levels of underestimation observed in the KE_WS, KE_USDA, and KE_Van Dijk models can be attributed to their reliance on average rainfall intensity as the input variable. This approach fails to capture the intra-event variability of rainfall intensity, which is more accurately represented by the KE_VT model. In reality, rainfall intensity fluctuates significantly during an event rather than remaining constant [6,7,16]. Additionally, these empirical models were originally developed under climatic conditions distinct from those of the Brazilian semi-arid region, which further contributes to their reduced accuracy when applied in this context. Consistent with the present findings, ref. [10] also identified the KE_VT model as the most suitable for estimating rainfall kinetic energy in the Iguatu-CE Experimental Basin.

To evaluate model performance (Table 2), all kinetic energy (KE) and rainfall erosivity index (EI_{30}) models were compared against the reference KE_Brasil and EI_{30}_Brasil models. Statistical analyses indicated that all tested models demonstrated strong performance, with Nash–Sutcliffe efficiency coefficients (Nash) exceeding 0.8, indicating high predictive accuracy. Similarly, the Willmott index of agreement (d) values were greater than 0.9, reflecting excellent concordance with the reference model. Furthermore, the model confidence index (C) also exceeded 0.9 for all models, confirming their high reliability when compared to the Brasil-specific reference.

Table 2. Statistical analysis of the methods for estimating kinetic energy-KE and estimating rainfall erosivity potential-EI_{30} in relation to the Brasil model.

	Methods	Indexes				Error	
		Nash	d	C	Performance	RMSE	ME
KE	Brasil × VT	0.97	0.99	0.99	optimal	0.59	0.29
	Brasil × WS	0.93	0.98	0.98	optimal	0.89	0.50
	Brasil × USDA	0.89	0.96	0.96	optimal	1.14	0.68
	Brasil × Van Dijk	0.89	0.96	0.96	optimal	1.12	0.56
EI_{30}	Brasil × VT	0.98	0.99	0.99	optimal	20.46	5.85
	Brasil × WS	0.95	0.98	0.98	optimal	31.14	11.08
	Brasil × USDA	0.94	0.98	0.98	optimal	34.31	13.67
	Brasil × Van Dijk	0.91	0.97	0.97	optimal	39.88	13.81

KE, RMSE, and ME–(MJ ha^{-1}); EI_{30}, RMSE, and ME–(MJ ha^{-1} mm h^{-1}); Brasil model; VT–Temporal variation in intensity model; WS–Wischmeier and Smith model; USDA model; VD–Van Dijk model. Nash > 0.75-Very good; Willmott concordance index (d) > 0.90-Excellent; model confidence index (C) > 0.85-optimal.

The mean error (ME) and root mean square error (RMSE), which assess the dispersion of measured versus estimated values, revealed that the KE_VT and EI_{30}_VT models exhibited the lowest values, demonstrating superior performance in representing both KE and EI_{30} data. As shown in Figure 3A, the KE_VT model was the least prone to underestimating the KE_Brasil data, making it the most accurate for this dataset. In contrast, the Wischmeier and Smith (WS) and USDA models yielded intermediate results, while the Van Dijk model exhibited the poorest performance in estimating KE and EI_{30}, suggesting its limited suitability for semi-arid environments like the one studied. These findings underscore the necessity of adapting models to local climatic conditions for more accurate hydrological predictions (Table 2).

Since the EI_{30} is directly dependent on the KE values obtained by the different models (Figure 3), it was observed that all the models similarly underestimated the EI_{30} data (Figure 5). The model incorporating temporal variation in intensity (EI_{30}_VT) exhibited the smallest underestimation of the EI_{30}_Brasil data, with a 13% difference (Figure 5A). In comparison, the model widely used in soil erosion studies (EI_{30}_WS) underestimated the EI_{30}_Brasil data by 22% (Figure 5B), while the USDA model showed a 23% underestimation (Figure 5C). The EI_{30}_Van Dijk model, however, had the greatest underestimation, at 28% (Figure 5D). All models were statistically different from EI_{30}_Brasil at the 1% significance level (p-value = 0.000) as determined by the Wilcoxon test (Figure 4B).

According to [6,7], South America, including the semi-arid region of Brazil, experiences some of the highest annual rainfall erosivity values. The results presented in Figures 3 and 5 highlight the importance of carefully selecting the appropriate model for estimating kinetic energy and the rainfall erosivity index [10,13,15]. With the ongoing trend in global warming, soil erosion is expected to be significantly affected by changes in rainfall erosivity, and it is of the utmost importance to choose an adequate erosivity quantification model. Extreme rainfall events, in particular, have the potential to aggravate soil erosion, thereby diminishing soil quality, fertility, and the overall health of ecosystems [9].

Given this context, selecting an inappropriate model to estimate EI_{30} can result in misleading conclusions, making it impossible to accurately calculate soil loss at the scale of individual rainfall events [6,7,10,13,15,16], especially in the arid and semi-arid regions of the world. The refinement of rainfall erosivity models has important implications for the soil erodibility factor (K) in the USLE, RUSLE, and RUSLE2 frameworks. To maintain consistency and accuracy in erosion predictions, it's important to recalibrate K when applying updated erosivity methods [30].

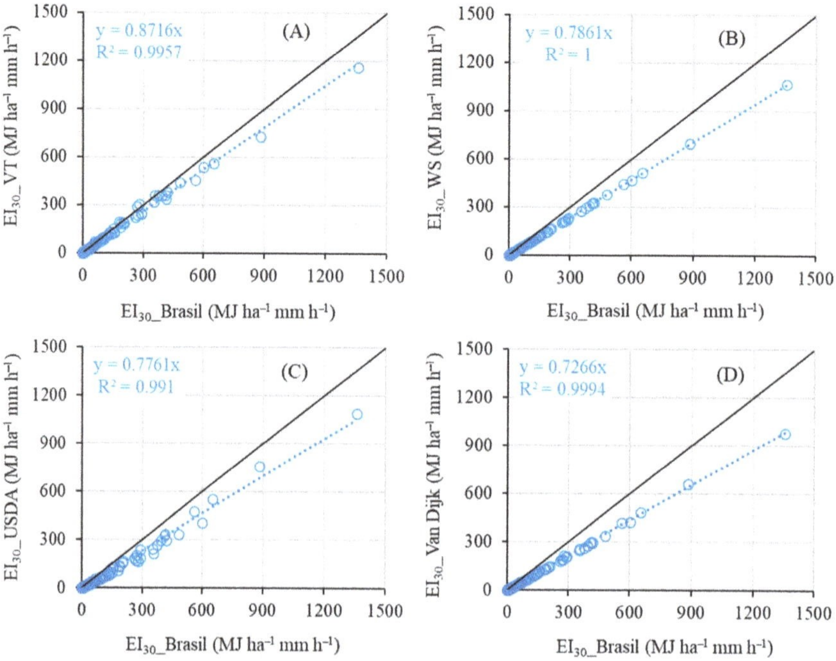

Figure 5. Estimation of rainfall erosivity potential (EI_{30}) by the Brasil model versus temporal variation in intensity-VT (**A**), Wischmeier and Smith_WS (**B**), USDA (**C**), Van Dijk (**D**).

As seen in Figures 3 and 5, all the models show discrepancies, particularly for higher values of KE and EI_{30}, corresponding to more intense rainfall events. However, for lower rainfall intensities, all models perform well when compared to the reference model. This behavior can be attributed to the smaller raindrops, which, even at low speeds, cause minimal erosion, regardless of the rainfall intensity [5,10,13].

Based on the findings of this study, it is evident that optimizing and refining rainfall erosivity estimation models is crucial, and the integration of remote sensing data can be valuable, especially when predicting changes in extreme rainfall erosivity both regionally and globally [9]. Future research should focus on understanding how the inappropriate selection of rainfall erosivity quantification models can impact soil erosion, particularly in the semi-arid regions of northeastern Brazil and other arid and semi-arid regions of the world.

3.3. Erosive Events

The separation of erosion events using a precipitation threshold of 12.7 mm, as defined by Wischmeier and Smith [11], yielded statistically similar results to analyses that included all events (see Supplementary Material). Although this threshold has been widely adopted across diverse regions and climatic contexts [13], alternative criteria have also been employed. For example, Jiang et al. [19] used a 30 mm threshold in their study of erosive rainfall in the karst landscapes of Guizhou Province, China. Such variation underscores the complexity of defining erosive rainfall, as soil erosion is influenced by more than just precipitation amount. Key characteristics such as rainfall intensity, duration, and total depth significantly contribute to erosive processes [13], and these factors are further modulated by local conditions, including climate, topography, data resolution, and methodological choices [20].

Given that our study is based on a relatively recent monitoring effort spanning four years, we opted to include all recorded natural rainfall events in our analysis. This decision is supported by the finding that their inclusion does not significantly influence the uncertainties of the KE and EI30 models when compared to the regional model. Within this dataset, 75% of the events exhibit precipitation totals below 12 mm—a pattern typical of semi-arid environments, where small but frequent rainfall events are common. Although these events fall below conventional erosivity thresholds, they remain important for understanding local hydrological dynamics, particularly in terms of soil moisture replenishment, runoff initiation, and interactions with surface processes.

Future research should incorporate longer-term datasets and critically assess the applicability of commonly used erosive precipitation thresholds, such as the 12.7 mm criterion proposed by Wischmeier and Smith [11]. Given the variability of rainfall characteristics and the influence of local environmental conditions, especially in semi-arid regions, such thresholds may not fully capture the complexity of erosive processes. A more nuanced evaluation could improve the accuracy of erosivity modeling and enhance our understanding of soil erosion dynamics under diverse climatic contexts.

4. Conclusions

The monitoring of 283 rainfall events revealed that 75% of the events had an EI_{30} value of less than 32.84 MJ ha^{-1} mm h^{-1}, with rainfall amounts under 12.20 mm, highlighting the high frequency of small rainfall events and significant temporal variability.

Among the kinetic energy (KE) models, the model incorporating temporal variation (KE_VT) exhibited the least underestimation of KE and EI_{30} data (13%), for all events and erosive events. This was followed by the Wischmeier and Smith (WS) and USDA models, while the Van Dijk model showed the greatest underestimation (28%), for all events and erosive events. Statistically, all models differed significantly at the 1% level (p-value = 0.000) when compared to the reference Brasil model, for all events and erosive events.

The statistical efficiency criteria indicate that while all models performed well, the KE_VT model provided the best representation of the KE and EI_{30} data, for all events and erosive events. Conversely, the Van Dijk model was the least effective for estimating these values, for all events and erosive events.

All the models tested demonstrated some degree of inaccuracy, particularly in estimating higher values of kinetic energy (KE) and rainfall erosivity index (EI_{30}) associated with more intense rainfall events. However, for lower rainfall intensities, all models performed relatively well when compared to the reference model. These findings underscore the need for a more rigorous selection process when choosing models for estimating KE and EI_{30} in studies involving hydrological processes and related modeling.

This study highlights the importance of region-specific calibration and the potential consequences of model misapplication in semi-arid environments, offering valuable insights for future research and model development in rainfall erosivity estimation.

Future research should incorporate long-term datasets and critically evaluate the applicability of erosive precipitation thresholds, especially considering their variability and implications for erosivity modeling in semi-arid regions.

Supplementary Materials: The following supporting information can be downloaded at: https://www.mdpi.com/article/10.3390/hydrology12070181/s1, Figure S1: KE_Brasil model comparison with other models; Figure S2: Statistics of KE and EI30 (B) by the different models; Figure S3: EI30_Brasil model comparison with other models; Table S1: Statistics of KE and EI30 for all models.

Author Contributions: Conceptualization, J.B.B., M.S.G. and A.C.M.M.; methodology, J.B.B., A.C.M.M. and F.D.D.A.; formal analysis, J.B.B., M.S.G., A.C.M.M., F.D.D.A. and S.M.d.M.; investigation, J.B.B. and F.D.D.A.; resources, A.C.M.M. and C.W.O.; data curation, A.C.M.M. and C.W.O.; writing—original draft preparation, J.B.B., M.S.G., A.C.M.M. and F.D.D.A.; writing—review and editing, J.B.B., M.S.G. and A.C.M.M.; supervision, A.C.M.M. and C.W.O.; project administration, A.C.M.M. and C.W.O.; funding acquisition, A.C.M.M. and C.W.O. All authors have read and agreed to the published version of the manuscript.

Funding: This work was supported by Coordination for the Improvement of Higher Education Personnel-Brazil (CAPES) [grant number 88887.807824/2023-00].

Data Availability Statement: The raw data supporting the conclusions of this article will be made available by the authors on request.

Acknowledgments: The authors would like to express their gratitude to the Center for Agricultural Sciences and Biodiversity (CCAB) at the Federal University of Cariri (UFCA) for their logistical support throughout this research. Special thanks are also extended to the Coordination for the Improvement of Higher Education Personnel-Brazil (CAPES) for providing financial assistance to the fellows involved in this study.

Conflicts of Interest: The authors declare that they have no conflicts of interest.

References

1. Campos, D.A.; Andrade, E.M.; Castanho, A.D.; Feitosa, R.C.; Palácio, H.Q. Biomass dynamics in a fragment of Brazilian tropical forest (Caatinga) over consecutive dry years. *Appl. Sci.* **2020**, *10*, 7813. [CrossRef]
2. Brasil, J.B.; Guerreiro, M.S.; Andrade, E.M.; Palácio, H.A.Q.; Medeiros, P.H.A.; Ribeiro Filho, J.C. Minimum Rainfall Inter-Event Time to Separate Rainfall Events in a Low Latitude Semi-Arid Environment. *Sustainability* **2022**, *14*, 1721. [CrossRef]
3. Soares, N.S.; Costa, C.A.G.; Carneiro de Lima, J.B.; Francke, T.; de Araújo, J.C. Method for identification of hydrological seasons in the semi-arid Caatinga biome, Brazil. *Hydrol. Sci. J.* **2024**, *69*, 309–320. [CrossRef]
4. Guerreiro, M.S.; Andrade, E.M.; Sousa, M.M.M.; Brasil, J.B.; Palácio, H.A.Q. Contribution of Non-Rainfall Water Input to Surface Soil Moisture in a Tropical Dry Forest. *Hydrology* **2022**, *9*, 102. [CrossRef]
5. Brasil, J.B.; Andrade, E.M.; Palácio, H.A.Q.; Fernández-Raga, M.; Ribeiro Filho, J.C.; Medeiros, P.H.A.; Guerreiro, M.S. Canopy Effects on Rainfall Partition and Throughfall Drop Size Distribution in a Tropical Dry Forest. *Atmosphere* **2022**, *13*, 1126. [CrossRef]
6. Panagos, P.; Borrelli, P.; Meusburger, K.; Yu, B.; Klik, A.; Jae Lim, K.; Yang, J.E.; Ni, J.; Miao, C.; Chattopadhyay, N.; et al. Global rainfall erosivity assessment based on high-temporal resolution rainfall records. *Sci. Rep.* **2017**, *7*, 4175. [CrossRef]
7. Panagos, P.; Borrelli, P.; Matthews, F.; Liakos, L.; Bezak, N.; Diodato, N.; Ballabio, C. Global rainfall erosivity projections for 2050 and 2070. *J. Hydrol.* **2022**, *610*, 127865. [CrossRef]
8. Santos, W.P.; Avanzi, J.C.; Viola, M.R.; Chou, S.C.; Acuña-Guzman, S.F.; Pontes, L.M.; Curi, N. Projections of rainfall erosivity in climate change scenarios for the largest watershed within Brazilian territory. *Catena* **2022**, *213*, 106225. [CrossRef]
9. Zhao, Y.; Zhu, D.; Wu, Z.; Cao, Z. Extreme rainfall erosivity: Research advances and future perspectives. *Sci. Total Environ.* **2024**, *917*, 170425. [CrossRef]
10. Brasil, J.B.; Andrade, E.M.; Guerreiro, M.S.; Palácio, H.A.Q.; Ribeiro Filho, J.C.; Fernández-Raga, M.; Medeiros, P.H.A. Measurement and modelling of kinetic energy and erosivity of rainfall and throughfall in a tropical semiarid region. *J. Hydrol.* **2024**, *644*, 132088. [CrossRef]
11. Wischmeier, W.H.; Smith, D.D. *Predicting Rainfall Erosion Losses: A Guide to Conservation Planning (No. 537)*; Department of Agriculture, Science and Education Administration: Washington, DC, USA, 1978.

12. Brychta, J.; Podhrázská, J.; Šťastná, M. Review of methods of spatio-temporal evaluation of rainfall erosivity and their correct application. *Catena* **2022**, *217*, 106454. [CrossRef]
13. Wang, L.; Li, Y.; Gan, Y.; Zhao, L.; Qin, W.; Ding, L. Rainfall erosivity index for monitoring global soil erosion. *Catena* **2024**, *234*, 107593. [CrossRef]
14. Medeiros, P.H.A.; de Araújo, J.C. Temporal variability of rainfall in a semiarid environment in Brazil and its effect on sediment transport processes. *J. Soils Sediments* **2014**, *14*, 1216–1223. [CrossRef]
15. Teixeira, D.B.S.; Cecílio, R.A.; Moreira, M.C.; Pires, G.F.; Fernandes Filho, E.I. Recent advancements in rainfall erosivity assessment in Brazil: A review. *Catena* **2022**, *219*, 106572. [CrossRef]
16. Yamaura, Y.; Yamada, Y.; Matsuura, T.; Tamai, K.; Taki, H.; Sato, T.; Hashimoto, S.; Murakami, W.; Toda, K.; Saito, H.; et al. Modeling impacts of broad-scale plantation forestry on ecosystem services in the past 60 years and for the future. *Ecosyst. Serv.* **2021**, *49*, 101271. [CrossRef]
17. Van Dijk, A.I.J.M.; Bruijnzeel, L.A.; Rosewell, C.J. Rainfall intensity–kinetic energy relationships: A critical literature appraisal. *J. Hydrol.* **2002**, *261*, 1–23. [CrossRef]
18. USDA-Agricultural Research Service. Science Documentation Revised Universal Soil Loss Equation Version 2. 2013. Available online: https://www.ars.usda.gov/ (accessed on 2 September 2024).
19. Jiang, Y.; Gao, J.; Yang, L.; Wu, S.; Dai, E. The interactive effects of elevation, precipitation and lithology on karst rainfall and runoff erosivity. *Catena* **2021**, *207*, 105588. [CrossRef]
20. Nearing, M.A.; Yin, S.Q.; Borrelli, P.; Polyakov, V.O. Rainfall erosivity: An historical review. *Catena* **2017**, *157*, 357–362. [CrossRef]
21. Sampaio, M.R.L.; Barboza, E.N.; Crisostomo, N.C.; Bezerra Neto, F.d.C.; Silva, D.H.d.; Felizardo, R.d.O.; Pinto, N.A. Estudo comportamental da precipitação pluviométrica no município de Crato-Ceará, no período entre 1974–2009. *Rev. Bras. De Gestão Ambient.* **2019**, *13*, 27–33. Available online: https://www.gvaa.com.br/revista/index.php/RBGA/article/view/7729 (accessed on 2 September 2024).
22. Foster, G.R.; McCool, D.K.; Renard, K.G.; Moldenhauer, W.C. Conversion of the universal soil loss equation to SI metric units. *J. Soil Water Conserv.* **1981**, *36*, 355–359. [CrossRef]
23. Nash, J.E.; Sutcliffe, J.V. River flow forecasting through conceptual models part I—A discussion of principles. *J. Hydrol.* **1970**, *10*, 282–290. [CrossRef]
24. Willmott, C.J. On the validation of models. *Phys. Geogr.* **1981**, *2*, 184–194. [CrossRef]
25. Camargo, A.D.; Sentelhas, P.C. Avaliação do desempenho de diferentes métodos de estimativa da evapotranspiração potencial no Estado de São Paulo, Brasil. *Rev. Bras. De Agrometeorol.* **1997**, *5*, 89–97.
26. da Costa, S.A.T.; de Souza, L.S.B.; de Assunção Montenegro, A.A.; de Souza, C.A.A.; de Morais, J.E.F.; de Carvalho Lopes, D.; Neto, A.J.S.; de Souza Santos, J.P.A.; Jardim, A.M.D.R.F.; Alves, C.P.; et al. Calibration and validation of the AquaCrop model for forage cactus production systems under different management interventions in the semi-arid region of Brazil. *Phys. Chem. Earth Parts A/B/C* **2024**, *136*, 103716. [CrossRef]
27. Carvalho, T.; Brosinsky, A.; Foerster, S.; Teixeira, A.; Medeiros, P. Reservoir sediment characterisation by diffuse reflectance spectroscopy in a semiarid region to support sediment reuse for soil fertilization. *J. Soils Sediments* **2022**, *22*, 2557–2577. [CrossRef]
28. Nanko, K.; Tanaka, N.; Leuchner, M.; Levia, D.F. Throughfall Erosivity in Relation to Drop Size and Crown Position: A Case Study from a Teak Plantation in Thailand. In *Forest-Water Interactions*; Levia, D.F., Carlyle-Moses, D.E., Iida, S., Michalzik, B., Nanko, K., Tischer, A., Eds.; Ecological Studies; Springer: Cham, Switzerland, 2020; Volume 240. [CrossRef]
29. Santos, J.C.N.; Andrade, E.M.; Medeiros, P.H.A.; Palácio, H.A.Q.; Araújo, J.R. Sediment delivery ratio in a small semi-arid watershed under conditions of low connectivity1. *Rev. Ciência Agronômica* **2017**, *48*, 49–58. [CrossRef]
30. Gupta, S.; Borrelli, P.; Panagos, P.; Alewell, C. An advanced global soil erodibility (K) assessment including the effects of saturated hydraulic conductivity. *Sci. Total Environ.* **2024**, *908*, 168249. [CrossRef]

hydrology

MDPI

Article

Evolution of Rainfall Characteristics in Catalonia, Spain, Using a Moving-Window Approach (1950–2022)

Carina Serra [1,*], María del Carmen Casas-Castillo [2], Raül Rodríguez-Solà [1] and Cristina Periago [3]

1 Department of Physics, Barcelona, School of Industrial Engineering (ETSEIB), Universitat Politècnica de Catalunya · BarcelonaTech, Diagonal 647, 08028 Barcelona, Spain; raul.rodriguez@upc.edu
2 Department of Physics, The School of Industrial, Aerospace and Audiovisual Engineering of Terrassa (ESEIAAT), Universitat Politècnica de Catalunya · BarcelonaTech, Colom 1, 08222 Terrassa, Spain; m.carmen.casas@upc.edu
3 Department of Physics, Barcelona East School of Engineering (EEBE), Universitat Politècnica de Catalunya · BarcelonaTech, Eduard Maristany 16, 08019 Barcelona, Spain; cristina.periago@upc.edu
* Correspondence: carina.serra@upc.edu

Abstract

A comprehensive analysis of the evolution of rainfall characteristics in Catalonia, NE Spain, was conducted using monthly data from 72 rain gauges over the period 1950–2022. A moving-window approach was applied at annual, seasonal, and monthly scales, calculating mean values, coefficients of variation (*CV*), and trends across 43 overlapping 31-year periods. To assess trends in these moving statistics, a modified Mann–Kendall test was applied to both the 31-year means and *CV*s. Results revealed a significant 10% decrease in annual rainfall, with summer showing the most pronounced decline, as nearly 90% of stations exhibited negative trends, while the *CV* showed negative trends in coastal areas and mostly positive trends inland. At the monthly scale, February, March, June, August, and December exhibited negative trends at more than 50% of stations, with rainfall reductions ranging from 20% to 30%. Additionally, the temporal evolution of Mann–Kendall trend coefficients within each 31-year moving window displayed a fourth-degree polynomial pattern, with a periodicity of 30–35 years at annual and seasonal scales, and for some months. Finally, at the annual scale and in two centennial series, the 80-year oscillations found were inversely correlated with the large-scale climate indices North Atlantic Oscillation (NAO) and Atlantic Multidecadal Oscillation (AMO).

Keywords: rainfall oscillations; moving windows; trends; Mann–Kendall test; coefficient of variation; Catalonia; Spain

1. Introduction

Analyzing precipitation trends is essential for understanding climate change. According to the findings presented the IPCC Cross-Chapter Paper 4 on the Mediterranean Region (CCP4.1, [1]), the observed trends in annual precipitation in the Mediterranean are significant only in some areas and some periods, and they are stationary in the long term throughout the region. Several authors have reported a decrease in average precipitation since the 1960s [2–4]. However, a decreasing trend in annual rainfall amounts was found [5] only in 18% of the Spanish territory for the period 1951–2019. A similar result was reported in [6], which analyzed annual rainfall trends in Spain for the period 1901–2010. Focusing on the Spanish eastern region and for the period 1951–2000, an overall reduction of 12% in annual rainfall across 90% of the studied areas was found by [7]. Regarding total precipitation in Spain for the period 1971–2022, a trend toward a drier climate cannot be confirmed with a significant level of statistical accuracy [8]. In addition, [9] concluded that there is a lack of statistically significant trends in long-term precipitation records in Southwestern Europe, while [10] reported that, although trends in individual Spanish regions may not be statistically significant when analyzed separately, their aggregation reveals a significant decreasing trend in total precipitation across Southwestern Europe, including France, Spain, and Portugal. However, this trend contrasts with other parts of Europe, which exhibit stable or increasing precipitation patterns [11,12].

On the other hand, precipitation extremes have increased in some northern Mediterranean areas and are projected to continue rising, potentially leading to a higher incidence of flash floods, with no change in the southern part [1]. An increasing trend in extreme rainfall events in Mediterranean coastal areas of the Iberian Peninsula was reported by [13,14]. This aligns with studies in other areas, such as those conducted along the Mediterranean coast of France [15–18] and Morocco [19–21]. Nevertheless, a recent study by [22] found no evidence of an increase in the frequency of the monthly RX1 day (the highest daily precipitation amount within a given month) exceeding 100 and 200 mm in the Spanish mainland from 1916 to 2022, which aligns with other works [23,24]. In Catalonia, an increase in rainfall irregularity, based on the temporal evolution of its scaling behavior [25,26], has been reported. Additionally, studies have observed a rise in the time concentration and intensity of convective-type events [27], as well as their contribution, particularly in areas where this type of precipitation is dominant [28]. These shifts in precipitation patterns, linked to an increase in variability, lead to significant implications for water management, agriculture, and urban planning [29–31]. The spatial distribution of maximum daily precipitation in Catalonia for several established return periods was found by [32], providing rainfall intensity-frequency values for sub-daily durations. In addition, drought has also been studied in this area. For instance, [33] analyzed patterns of monthly rainfall shortages and excesses, while [34,35] examined trends and statistical distributions of dry spells across the region.

The mentioned studies are some examples of the extensive list of articles that have contributed to a general understanding of precipitation patterns in this region, although they often focused on fixed time periods without accounting for variability across different time scales. The use of moving windows (MW) offers a dynamic and continuous perspective, allowing for the detection of more subtle and temporal changes in the time series, and may help explain the differences in conclusions across studies. In a recent paper, [36] examined the spatial and temporal variations in seasonal precipitation patterns for the period 1916–2015, applying moving windows of 30 and 60 years. In addition, [37] applied MW analysis to the long series with more than two centuries of registers in Barcelona. In [38], the secular series of the Fabra Observatory were analyzed and compared for three reference periods of 30 years. Although the aforementioned articles employed moving windows, their objectives and findings differ from those of the present work.

The present study aims to detect multidecadal oscillations in rainfall amount and coefficient of variation in Catalonia and their trends when 31-year moving windows (MW31) are applied. This 31-year period has been chosen because it closely aligns with the 30-year periods recommended by the World Meteorological Organization (WMO, [39]) for calculating climatological averages. Analyzing the key statistical features of these periods provides further insight into their underlying dynamics. This approach enables the assessment of how trends may vary when the study periods shift, revealing multidecadal periodicities and providing another point of view of the evolution of precipitation, contributing to the development of more effective long-term water resource management and territorial planning strategies.

The paper is organized as follows. Section 2 presents the selected database, the study region, and the methodology used. Section 3 illustrates the results obtained, and Section 4 discusses the results. Finally, Section 5 summarizes the main conclusions.

2. Materials and Methods

2.1. Database

A complete monthly rainfall database corresponding to 72 rain gauges has been obtained from Servei Meteorològic de Catalunya (SMC, [40]) for the period 1950–2022. Most of the rain gauges (70) are located in Catalonia, while the other two are in Ransol (Andorra) and Perpinyà (France), labeled as 20 and 33 in Figure 1, respectively. All these series have been subjected to rigorous quality control and time series homogenization [41,42]. Figure 1 shows the main orographic features of Catalonia, the location of the selected stations, and the annual average rainfall computed from the stations used in this study.

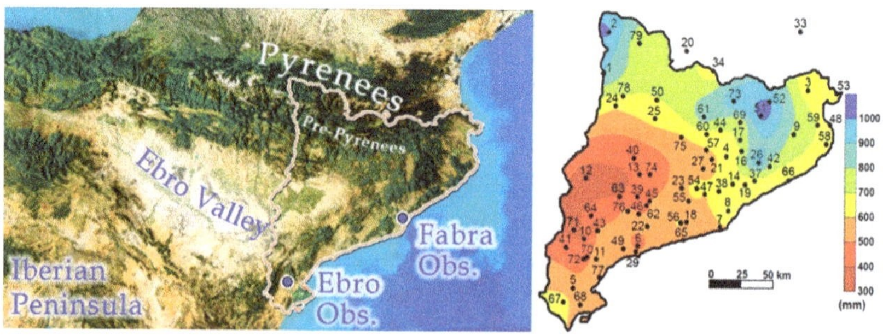

Figure 1. Left: Location of Catalonia within the Iberian Peninsula, including key topographic features and sites mentioned in the text. **Right**: Locations of rain gauges with their corresponding identification numbers and annual average rainfall.

The code number of each station, shown in Figure 1, is also included in Table 1 along with the corresponding name, coordinates, and altitude. Furthermore, for comparison with temperature behavior, 23 thermometric series from Catalonia, obtained from the SMC, have also been used; these are highlighted in bold in Table 1.

Table 1. Name, code, coordinates UTM X and UTM Y (31T), and altitude (m.a.s.l.) of rain gauges. The names of the gauges with both thermometric and rainfall records are shown in bold.

Code	Name	UTM X (km)	UTM Y (km)	Altitude (m.a.s.l.)	Code	Name	UTM X (km)	UTM Y (km)	Altitude (m.a.s.l.)
1	**El Pont de Suert**	314.390	4696.649	823	44	Prats de Lluçanès	419.397	4651.026	701
2	**Vielha**	319.403	4729.904	975	45	Rocafort	356.457	4593.22	560
3	**Figueres**	496.369	4684.011	31	46	Montblanc	346.606	4582.81	340
4	**Moià**	424.938	4629.639	735	47	Esparraguera	405.436	4599.078	187
5	**Ebro Observatory**	288.703	4522.022	49	48	Torroella	516.672	4655.775	2
6	Reus Airport	345.585	4556.658	73	49	Montbrió	333.136	4554.598	132
7	**Barcelona Airport**	419.544	4572.631	3	50	Organyà	362.27	4675.133	566
8	**Fabra Observatory**	426.879	4585.783	412	51	La Vall d'en Bas	455.339	4663.113	510
9	**Girona**	484.094	4648.026	72	52	Castellfollit	462.603	4674.424	296

Table 1. *Cont.*

Code	Name	UTM X (km)	UTM Y (km)	Altitude (m.a.s.l.)	Code	Name	UTM X (km)	UTM Y (km)	Altitude (m.a.s.l.)
10	**Flix**	298.285	4562.288	53	53	Cadaqués	522.557	4681.476	24
11	**Tivissa**	309.325	4546.042	313	54	El Bruc	398.259	4603.715	489
12	**Lleida**	299.694	4611.309	192	55	St. Quintí de Mediona	390.919	4593.443	325
13	**Tàrrega**	347.109	4614.634	380	56	Castellví	384.319	4575.694	198
14	**Caldes**	430.803	4607.309	176	57	Balsareny	406.601	4635.019	327
16	**Malla–Torrellebreta**	438.458	4634.027	570	58	Palafrugell	513.318	4640.267	64
17	Vic	437.077	4643.040	499	59	Jafre	505.127	4655.771	15
18	**Vilafranca**	389.362	4576.479	176	60	Puig-Reig	406.818	4647.462	455
19	Granollers	441.629	4606.726	202	61	Berga	404.203	4661.652	704
20	**Ransol**	388.463	4715.122	1645	62	Sarral	353.598	4589.634	467
21	**Sallent**	411.665	4627.476	278	63	Els Omellons	329.778	4596.644	385
22	**Valls**	354.383	4572.742	230	64	La Granadella	305.008	4580.475	505
23	**Igualada**	384.760	4603.934	333	65	Cubelles	386.684	4561.776	17
24	**Tremp**	325.636	4670.183	473	66	Malgrat	479.729	4610.62	2
25	**Oliana**	360.703	4659.884	490	67	Mas de Barberans	280.386	4510.733	240
26	**El Turó de l'Home**	453.364	4624.999	1668	68	Amposta	295.436	4508.513	3
27	Manresa	403.719	4619.630	291	69	Torelló	437.113	4657.362	508
29	Vila-seca	344.301	4552.210	53	70	Benissanet	301.215	4547.698	32
33	**Perpinyà**	489.425	4731.437	44	71	Riba-roja	289.19	4569.44	76
34	Puigcerdà	412.395	4699.199	1214	72	Miravet	298.143	4546.014	43
37	Llinars	449.806	4610.209	198	73	Campdevànol	431.279	4674.834	738
38	Terrassa	418.373	4601.603	290	74	Cervera	356.469	4614.555	554
39	Rocallaura	345.288	4596.651	645	75	Riner	384.044	4644.945	830
40	Agramunt	341.843	4627.918	337	76	Vimbodí	336.933	4585.049	446
41	Vilalba	282.449	4555.323	442	77	L'Ametlla de Mar	312.141	4531.061	93
42	Breda	463.111	4621.813	169	78	La Pobla de Segur	332.041	4678.363	508
43	Cabacers	310.237	4568.904	357	79	Esterri	346.296	4721.003	957

In addition, two of these rain gauges—Fabra Observatory (Barcelona, 412 m a.s.l.) and Ebro Observatory (Roquetes, Tarragona, 49 m a.s.l.)—have data series extended to longer historical periods to support further analysis and interpretation of the results. The Fabra Observatory series spans from 1914 to 2024 and was obtained from the Reial Acadèmia de Ciències i Arts de Barcelona [43]. The Ebro Observatory record, covering the period from 1905 to 2023, was provided by the Ramon Llull University (URL) and the Spanish National Research Council (CSIC) through the Observatori de l'Ebre [44]. The 31-year running averages of these two centennial stations will be correlated with the corresponding values for the North Atlantic Oscillation index (NAO) and the Atlantic Multidecadal Oscillation index (AMO) obtained from the official website of the National Oceanic and Atmospheric Administration (NOAA, [45]).

2.2. Methodology

2.2.1. Main Characteristics of Each 31-Years Window

A 31-year moving-window approach, MW31, with a 1-year shift, has been applied to annual, seasonal and monthly amounts for the 72 rain gauges from 1950 to 2022 in order to analyze possible temporal oscillations and time trends. Then, two characteristics define every one of the resulting 43 moving windows for each station. These characteristics are the average rainfall amount, $\overline{R}$, and the coefficient of variation, CV.

For a first insight into the variation in the main annual characteristics, the comparison between the two reference periods, 1960–1990 and 1990–2020, has been analyzed. For this purpose, a relative difference for the mean annual rainfall, in percentage, has been computed for each rain gauge. These relative differences have also been computed for the CV and summarized in two respective histograms.

In addition, the main characteristics are graphically represented for all 43 windows and 72 stations at annual, seasonal and monthly scales. Seasons were defined as winter (DJF), spring (MAM), summer (JJA) and autumn (SON). The global linear trends of these 43 × 72 points are obtained by means of the ordinary least squares regression method (OLS). The percentage of these global trends is calculated using Equation (1), where the function y represents the linear trend, and 1965 and 2007 are the central years of the first and last MW, respectively.

$$Global\ trend(\%) = \frac{y(\text{lastMW}) - y(\text{firstMW})}{y(\text{firstMW})} \times 100 = \frac{y(2007) - y(1965)}{y(1965)} \times 100. \quad (1)$$

Furthermore, the main values of the principal characteristics at annual, seasonal, and monthly scales, along with some results, will be presented in the following sections.

2.2.2. Trends Within Each 31-Years Window

The linear trend of rainfall amounts, Tr, for each window and station, has been computed. These trends have been studied through the ordinary least squares method, OLS, and their significance assessed by means of the Mann–Kendall (MK) trend test [46]. The significance has been evaluated by comparing the standardized test statistic $Z = \frac{S}{\sqrt{V(s)}}$ with the standard normal variate at the desired significance level. In this article, the significance level is set to 5%, corresponding to a 95% confidence level (p-value = 0.05). In addition, the 85% confidence level (p-value = 0.15) is also considered. Finally, the time evolution of the MK coefficients will be graphically represented and fitted to a fourth-degree polynomial.

2.2.3. Trends of Moving Window Characteristics

The evolution of the moving average values and coefficient of variation along the 43 moving windows and 72 stations has also been studied. These new trends have also been computed by means of the OLS. These computations correspond to the trends of the moving averages, $Tr\overline{R}$, and of the moving coefficient of variation, $TrCV$.

However, for the significance of these trends, when dealing with autocorrelated series, as is often the case with moving window averages, applying the MK test can lead to erroneous conclusions because the test assumes that the data are independent and randomly ordered. Therefore, this study has utilized the modified Mann–Kendall (mMK) test for autocorrelated data [47] to assess the significance of detected trends in the time series processed using the moving-window technique.

The modified variance is computed as Equation (2), where $\frac{n}{n_s^*}$ represents a correction due to the autocorrelation in the data, computed using Equation (3).

$$V^*(S) = V(S)\frac{n}{n_s^*} = \frac{n(n-1)(2n+5)}{18}\frac{n}{n_s^*}. \quad (2)$$

$$\frac{n}{n_s^*} = 1 + \frac{2}{n(n-1)(n-2)}\sum_{j=1}^{n-1}(n-j)(n-j-1)(n-j-2)\rho_s(j). \quad (3)$$

In both Equations (2) and (3), n denotes the number of sample elements; in this article, $n = 43$, which corresponds to the number of moving windows. In Equation (3), $\rho_s(j)$ is the autocorrelation function of the rainfall amount ranks, at time lag j.

Finally, the trends of the 31-year moving averages and the CV will be spatially represented over Catalonia at annual, seasonal and monthly scales. The stations with a confidence level greater than or lower than 95% (p-value = 0.05) are also indicated. Figure 2 summarizes the main steps of this research.

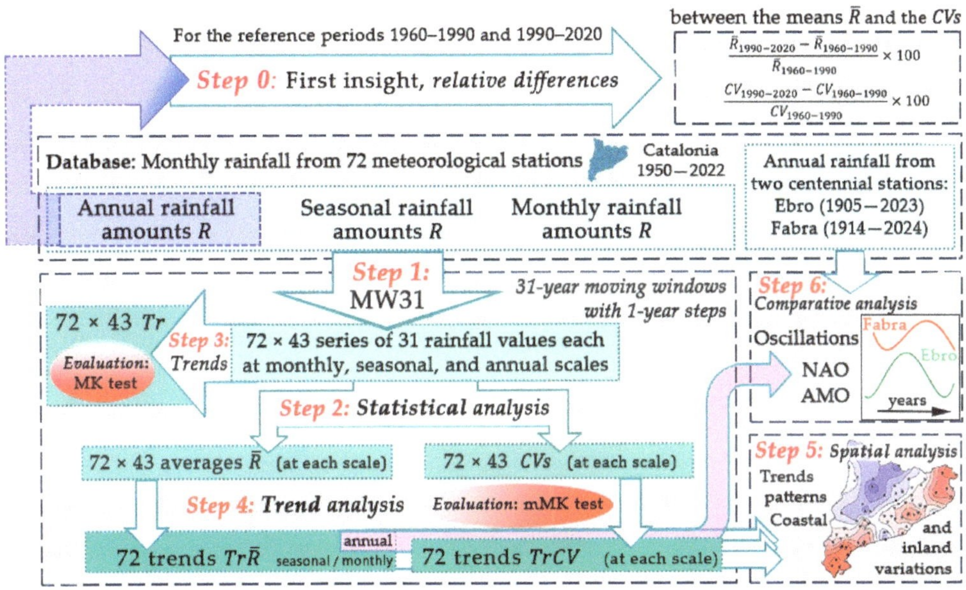

Figure 2. Methodology framework.

2.2.4. Two Centennial Series and Correlation with NAO and AMO

In order to compare the results for the years 1950–2022 with those obtained over longer periods, for two centennial series (Fabra and Ebro Observatories) and at an annual scale, the 31-year running averages will be represented and the oscillations studied. In addition, the correlations with the MW31 NAO and AMO indices will be analyzed. The significance of the correlation between two autocorrelated series is obtained using the effective number of degrees of freedom, N_{eff}, by means of the approximation given in Equation (4) [48,49], where N is the sample size and $\rho_{xx}(j)$ and $\rho_{yy}(j)$ are the autocorrelations of two sampled time series x and y at time lag j, respectively.

$$\frac{1}{N_{eff}} \approx \frac{1}{N} + \frac{2}{N}\sum_{j=1}^{N} \frac{N-j}{N}\rho_{xx}(j)\rho_{yy}(j). \quad (4)$$

3. Results

3.1. Annual Amounts

3.1.1. First Insight

Figure 3 shows three examples of the evolution of the mean annual rainfall amounts and the corresponding coefficient of variation for the 43 moving windows of 31 years. The three selected stations have different types of climate. Viella (station code 2), with high annual amounts, is located on the north face of the Pyrenees, with a NE orientation and

Atlantic influences. Barcelona Airport (code 7) is on the central coast of Catalonia, with typical Mediterranean weather, and Lleida (code 12), with a more continental climate, is located in the Central Basin and experiences lower annual rainfall amounts. Although their climates are quite different, the linear trends of the mean values are negative and statistically significant at the 95% confidence level for the three stations.

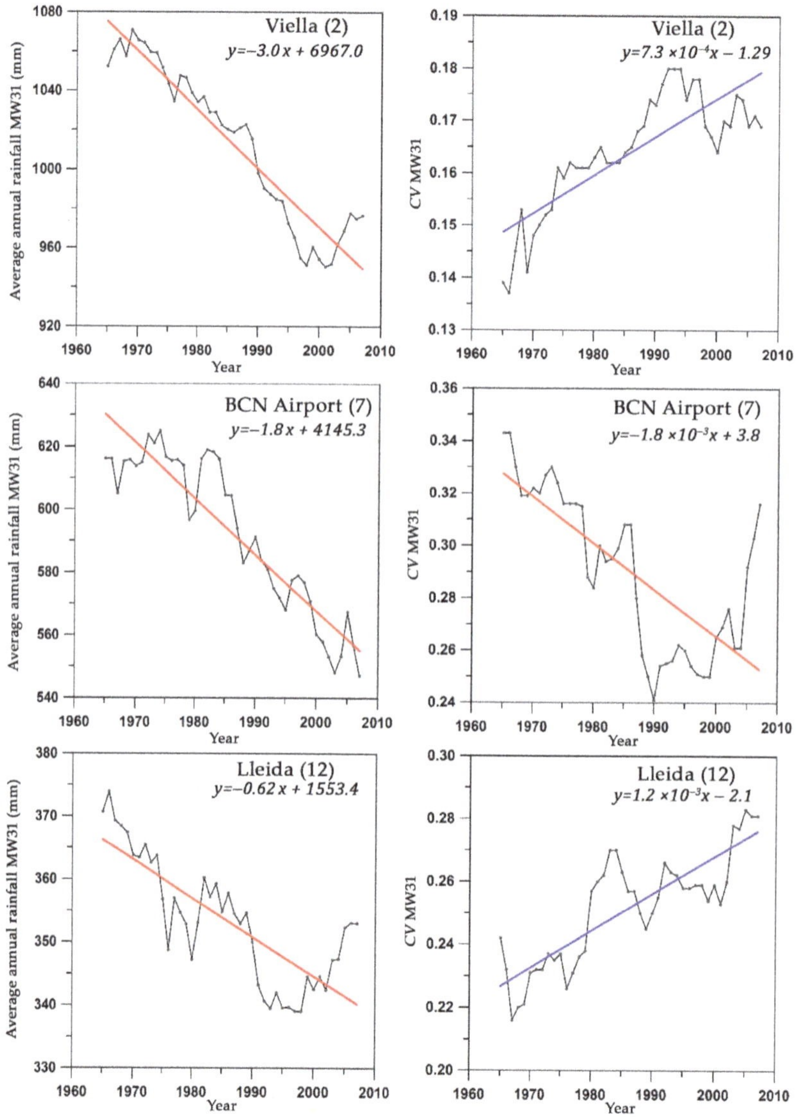

Figure 3. Three examples of the moving annual average and the corresponding coefficient of variation (CV) for MW31 with a 1-year shift. Positive linear trends are shown in blue, and negative trends in red. The x-axis represents the central year of MW31.

For instance, the annual mean in Barcelona Airport has been decreasing, with some oscillations, from an initial value of 616.2 mm for the first window (1950–1980) to 547.1 mm for the last window (1992–2022), which represents a decreasing percentage of 11.2%. On the other hand, for the coefficient of variation, the linear trend is negative in the Barcelona Airport, while in the other two rain gauges it is positive. All three trends are statistically significant at a 95% confidence level.

Figure 4 illustrates the histograms of the relative differences between the two reference periods, 1960–1990 and 1990–2020, for the mean and CV values and each station. For the mean values, the differences are negative for 62 of 72 stations. Furthermore, a diminution between 5% and 10% has been accomplished for 28 locations. However, for CV, the number of stations with positive and negative differences are more variably distributed.

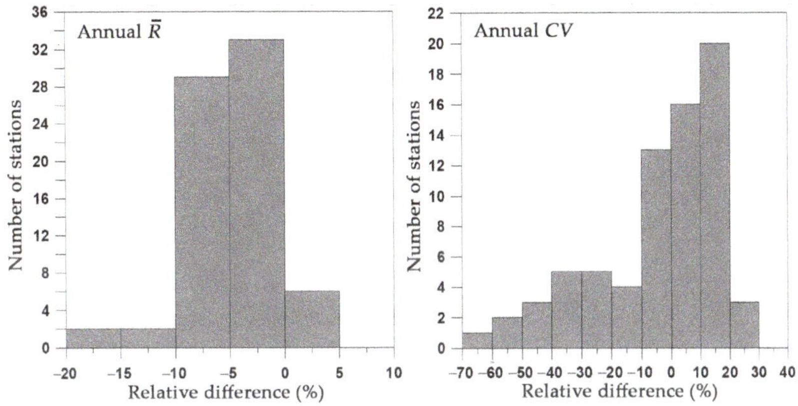

Figure 4. Relative differences between the periods 1960–1990 and 1990–2020 for average values $\overline{R}$ and for CV.

3.1.2. Mean Annual Rainfall and CV for Each Window and Station

The averages of the annual rainfall amounts, $\overline{R}$, for the 72 rain gauges and the 43 moving windows are represented in Figure 5. These values range from 339.0 mm to 1111.8 mm (Table 2). The global trend is negative with −1.5 mm/year, implying a reduction close to 10% for the MW31 mean annual value. For the moving CV, the values of the first window are more dispersed (0.14–0.41) than the CV of the last windows (0.17–0.32), and the mean trend, as in the case of average amounts, is approximately −10%.

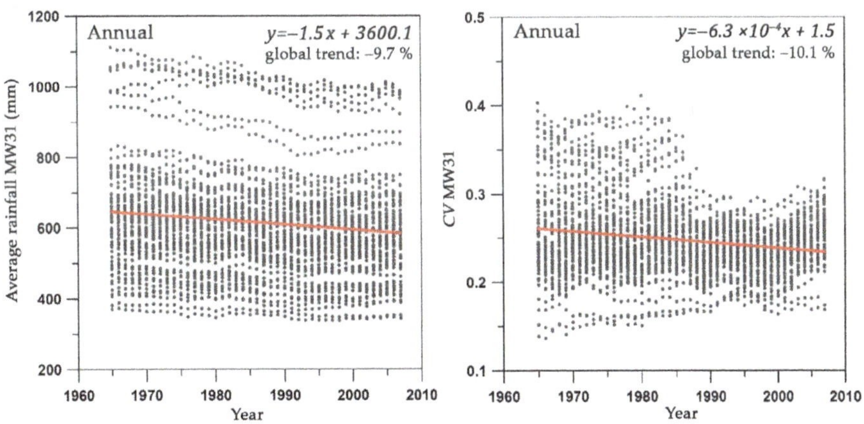

Figure 5. Annual averages and CV for each MW31 and 72 rain gauges. The solid line represents the global linear trend.

Table 2. Minimum (MIN), maximum (MAX), average (AVG) and standard deviation (S) of the average rainfall ($\overline{R}$) and coefficient of variation (CV) for each MW31 and rain gauge.

	$\overline{R}_{MIN}$ (mm)	$\overline{R}_{MAX}$ (mm)	$\overline{R}_{AVG}$ (mm)	$\overline{R}_S$ (mm)	CV_{MIN}	CV_{MAX}	CV_{AVG}	CV_S
Annual	339.0	1111.8	615.7	156.9	0.14	0.41	0.25	0.042
Winter	60.1	289.2	120.5	37.3	0.34	0.81	0.58	0.086
Spring	101.8	320.5	166.0	40.6	0.20	0.81	0.45	0.094
Summer	43.1	364.5	130.2	59.9	0.26	0.84	0.51	0.114
Autumn	101.2	336.0	199.0	39.4	0.30	0.83	0.47	0.091
January	20.5	110.6	40.9	13.2	0.54	1.43	1.05	0.131
February	12.9	77.2	32.1	11.8	0.53	1.41	0.94	0.147
March	21.2	99.6	43.7	12.7	0.48	1.52	0.88	0.158
April	28.1	103.9	55.6	13.6	0.38	1.40	0.70	0.174
May	32.3	133.0	66.7	18.5	0.30	1.24	0.68	0.161
June	17.9	125.8	50.3	20.8	0.34	1.62	0.82	0.235
July	6.4	113.0	30.0	19.3	0.45	1.87	0.97	0.253
August	11.4	133.0	50.0	21.8	0.39	1.30	0.76	0.163
September	34.8	108.5	67.5	14.8	0.40	1.58	0.76	0.194
October	36.0	132.3	74.1	16.9	0.50	1.75	0.83	0.153
November	26.0	119.0	57.5	14.6	0.52	1.86	0.90	0.154
December	20.1	116.6	47.4	14.6	0.46	1.58	0.99	0.129

3.1.3. Trends of Annual Amounts Within Each 31-Years Window

For each window and station, the trends of the annual amounts have also been computed. The MK coefficients corresponding to these MW31 trends and the 72 available stations are represented in Figure 6a, and Table 3 summarizes some values of the trends and MK coefficients at the annual scale and, in addition, for the analyses in the following sections, at seasonal and monthly scales.

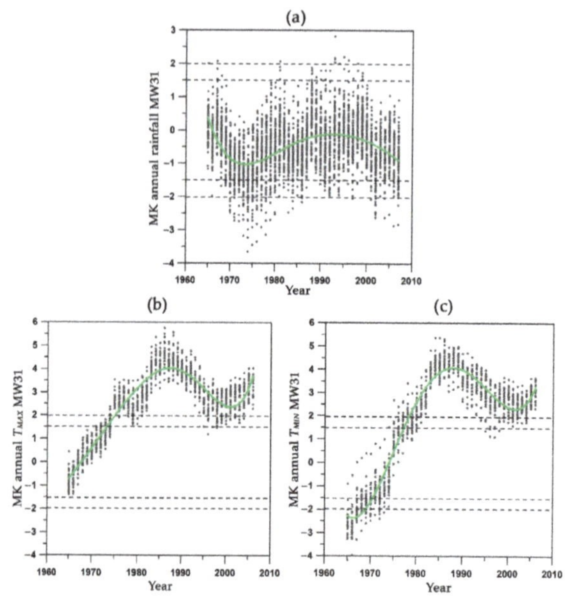

Figure 6. Mann–Kendall trend coefficients of the (**a**) annual rainfall, (**b**) maximum temperature, and (**c**) minimum temperature for each MW31 and available station. Dashed lines are the 85% and 95% confidence levels, and green lines are the polynomial fittings.

Table 3. Minimum (MIN), maximum (MAX), average (AVG) and standard deviation (S) of trends (Tr) and Mann–Kendall test coefficient (MK). for each MW31 and rain gauge.

	Tr_{MIN} (mm/y)	Tr_{MAX} (mm/y)	Tr_{AVG} (mm/y)	Tr_S (mm/y)	MK_{MIN}	MK_{MAX}	MK_{AVG}	MK_S
Annual	−14.8	8.93	−1.72	2.74	−3.59	2.72	−0.51	0.83
Winter	−6.23	7.68	−0.42	1.27	−3.26	2.89	−0.42	0.86
Spring	−7.59	4.99	−0.10	1.68	−3.43	2.48	−0.09	1.13
Summer	−5.82	3.70	−0.72	1.08	−3.32	2.92	−0.53	0.84
Autumn	−7.97	6.31	−0.47	2.21	−3.60	2.97	−0.16	1.05
January	−3.43	4.28	0.14	0.72	−2.52	2.47	0.08	0.70
February	−2.89	2.77	−0.13	0.60	−3.45	2.77	−0.28	1.03
March	−3.06	2.96	−0.18	0.91	−3.64	3.52	−0.18	1.31
April	−2.53	2.95	0.20	0.72	−2.02	3.28	0.39	0.85
May	−3.75	4.19	−0.12	0.81	−3.16	2.67	−0.14	0.88
June	−2.51	1.82	−0.41	0.62	−3.06	2.37	−0.62	0.90
July	−2.62	2.66	−0.07	0.53	−3.28	3.46	−0.04	1.02
August	−3.23	2.08	−0.24	0.65	−2.84	2.69	−0.33	0.88
September	−5.35	4.28	−0.43	1.37	−3.62	4.28	−0.26	1.31
October	−4.50	4.35	−0.11	1.12	−3.16	2.58	0.08	0.89
November	−4.09	3.26	0.07	1.04	−3.33	3.30	0.01	1.00
December	−4.76	3.42	−0.44	1.04	−3.35	2.74	−0.47	1.05

Few stations have values greater than +2.0 or less than −2.0, corresponding to the 95% confidence level (p-value = 0.05), for $N = 31$. On the one hand, only some stations with significant negative trends are detected during the periods centered between 1970 and 1985, with a minimum MK value of −3.59 (99.89% or p-value = 0.00112). The 31-year period with the highest number of significant negative trends, centered in 1974, was 1959–1989, with 26% of stations. However, for the 85% confidence level (p-value = 0.15), 54% of stations show significant negative trends during the same period. For the more recent years, the 31-period centered in 2002 (1987–2017) shows 11% of stations with significant negative trends. On the other hand, very few stations with positive significant trends (coefficients greater than +2.0) are observed around 1995, with a maximum MK value of 2.72 (98.94% or p-value = 0.0106). A comparative study with temperatures in this region, based on the 23 available thermometric gauges (Table 1), is shown in Figure 6b,c for maximum and minimum temperatures.

The difference between temperature and rainfall behavior is evident. For both maximum and minimum temperatures, trends have been significant and positive since the MW31 centered in 1975 for maximum temperature and in 1978 for minimum temperature.

3.1.4. Trends in Characteristics of Annual Moving Windows

Figure 7 represents the relative trends, in percentage, and the significance, obtained with the mMK test, at the 95% confidence level of the moving average amounts, $Tr\overline{R}$, and the moving coefficient of variation, $TrCV$. This percentage has been calculated relative to the value of the first MW31 (1950–1980).

Table 4 summarizes the number of stations with significant positive and negative moving trends for the mean values and the CV at annual, seasonal and monthly scales. For the moving annual average, 63 of 72 stations (88%) have significant negative trends with a relative percentage as low as −20%. The minimum MK value is −5.99 (p-value < 0.000001), which corresponds to a station located in the southern corner of Catalonia.

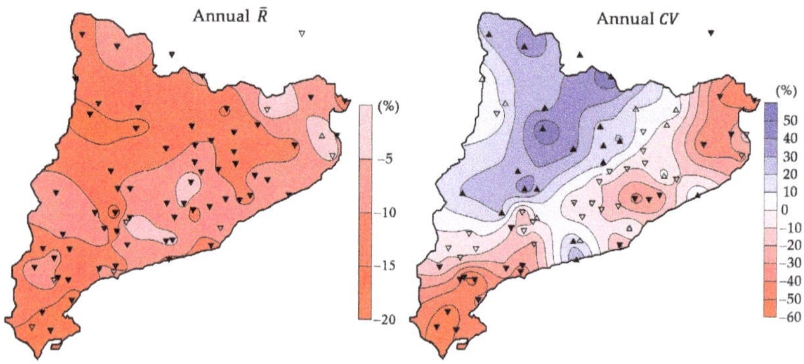

Figure 7. Relative trend (in percentage) for the MW31 annual average and CV. Solid triangles indicate statistically significant trends at the 95% confidence level, while empty triangles represent non-significant trends: ▲ for positive and ▼ for negative trends.

Table 4. Number and percentage of rain gauges with significant positive and negative trends (95% confidence level) for the moving average and CV. Positive trends are shown in blue, and negative trends in red.

	$Tr\bar{R}_{MW31}$				$TrCV_{MW31}$			
	Number/%		Number/%		Number/%		Number/%	
Annual	0	0.0	63	87.5	19	26.4	21	29.2
Winter	0	0.0	45	62.5	57	79.2	3	4.2
Spring	6	8.3	28	38.9	10	13.9	29	40.3
Summer	1	1.4	64	88.9	32	44.4	3	4.2
Autumn	19	26.4	5	6.9	5	6.9	31	43.1
January	12	16.7	4	5.5	15	20.8	17	23.6
February	0	0.0	47	65.3	41	56.9	4	5.6
March	0	0.0	41	56.9	21	29.2	1	1.4
April	24	33.3	2	2.8	0	0.0	47	65.3
May	0	0.0	27	37.5	21	29.2	15	20.8
June	0	0.0	56	77.7	61	84.7	0	0.0
July	6	8.3	19	26.4	11	15.3	13	18.1
August	3	4.2	46	63.9	18	25.0	16	22.2
September	5	6.9	29	40.3	6	8.3	39	54.2
October	34	47.2	7	9.7	4	5.6	38	52.8
November	16	22.2	0	0.0	7	9.7	25	34.7
December	0	0.0	54	75.0	27	37.5	2	2.8

However, for the corresponding CV, some inland regions and the Pyrenees show significant positive trends (19 out of 72), while in coastal areas, trends are mostly negative (21 out of 72), especially in the southern and northern corners.

These results indicate that regions, such as the Pyrenees, with high amounts of annual precipitation and a low irregularity, are experiencing a decrease in the total amounts and an increase in irregularity in the recent years.

3.1.5. Two Centennial Series and Correlation with NAO and AMO

In Figure 8, the mean annual rainfall of two long-term series, Fabra Observatory (1914–2024) and Ebro Observatory (1905–2023), is represented. This figure shows that, while the MW31 annual averages are decreasing during the study period of this article, earlier periods also exhibited positive trends. For instance, at the Ebro Observatory, a positive trend is observed from the central year 1920 (period 1905–1935) to 1960 (period 1945–1975), and at the Fabra Observatory, from 1940 (period 1925–1955) to 1985 (period 1970–2000). The MW31 averages at Fabra Observatory range from 586.8 to 660.4 mm, with a mean value of 620.8 mm.

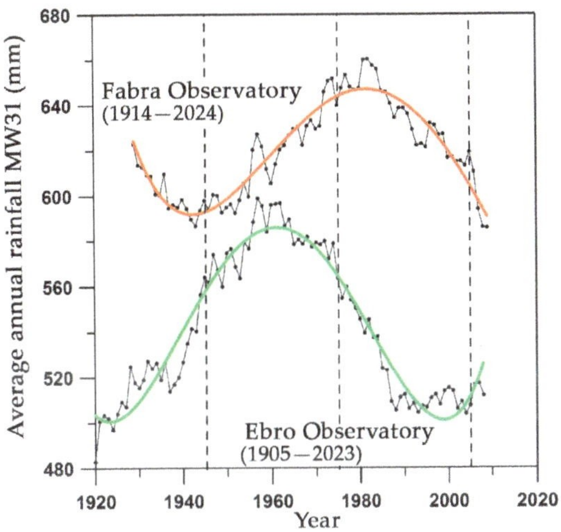

Figure 8. Time evolution of the annual moving average rainfall (MW31) and its polynomial adjustments for two centennial stations: Fabra Observatory (1914–2024) and Ebro Observatory (1905–2023). The x-axis represents the central year of the window. Dashed lines indicate the average rainfall for three reference periods: 1930–1960, 1960–1990, and 1990–2020.

For the Ebro Observatory, the first two reference periods have similar values, both higher than that of the last period. The difference between the minimum (482.9 mm) and the maximum (599.1 mm), for this station, is greater than 100 mm. A periodicity close to 80 years is detected in Figure 8. Although for rainfall amounts, several studies correlate precipitation in Catalonia mainly with Mediterranean indices, such as Mediterranean Oscillation (MO), Western Mediterranean Oscillation (WeMO), or upper-level MO (ULMO) [50–52], for MW31 values these oscillations could be explained by the quasi-periodic multidecadal variability of the North Atlantic Oscillation (NAO) or by the Atlantic Multidecadal Oscillation (AMO). To this end, the MW31 averages of the NAO (NAO_{MW31}) and AMO (AMO_{MW31}) are calculated to compare them with the mean annual averages, $\overline{R}_{MW31}$, of Ebro ($Ebro_{MW31}$) and Fabra ($Fabra_{MW31}$), respectively. While for the NAO a minimum is centered in 1960, a maximum is observed for the Ebro Observatory (Figure 9a), explaining an inverse correlation between the NAO index and rainfall at this station. The coefficient correlation between the reconstructed NAO_{MW31} [53,54] and annual $\overline{R}_{MW31}$ of Ebro, from the central years 1920 to 1986, is equal to −0.79 with a confidence level of 91% (p-value = 0.09), obtained with the effective number, N_{eff}, from Equation (4). The coefficient of correlation between NAO_{MW31} of NOAA [45] and annual $\overline{R}_{MW31}$ of Ebro for the period of central years 1920 to 1996 is equal to −0.63 with a confidence level of 86% (p-value = 0.14).

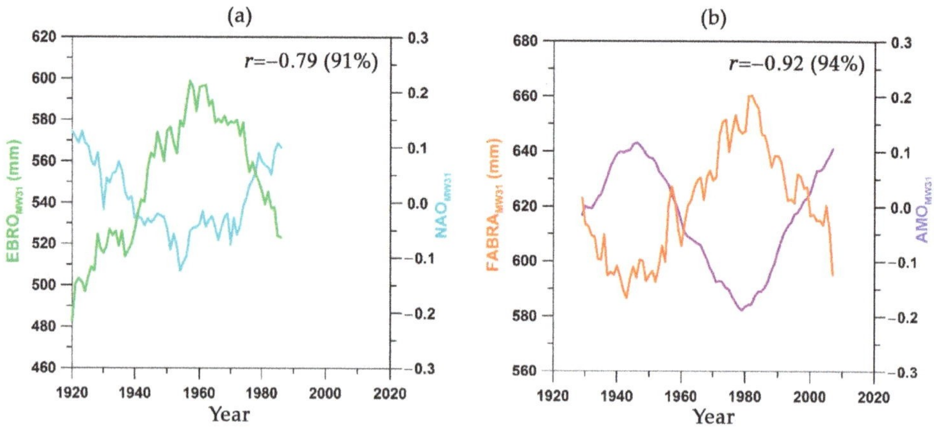

Figure 9. Time evolution of the annual (**a**) Ebro$_{MW31}$ and NAO$_{MW31}$ and (**b**) Fabra$_{MW31}$ and AMO$_{MW31}$.

Figure 9b shows the evolution of the Fabra$_{MW31}$ annual rainfall and the AMO$_{MW31}$ index. The correlation coefficient is equal to −0.92 with a confidence level of 94% obtained by means of Equation (4). The high negative correlation between both magnitudes is noteworthy. Figure 10 shows the correlation between Ebro$_{MW31}$ and Fabra$_{MW31}$ for different lags. The maximum correlation is equal to 0.88 with a 20-year lag.

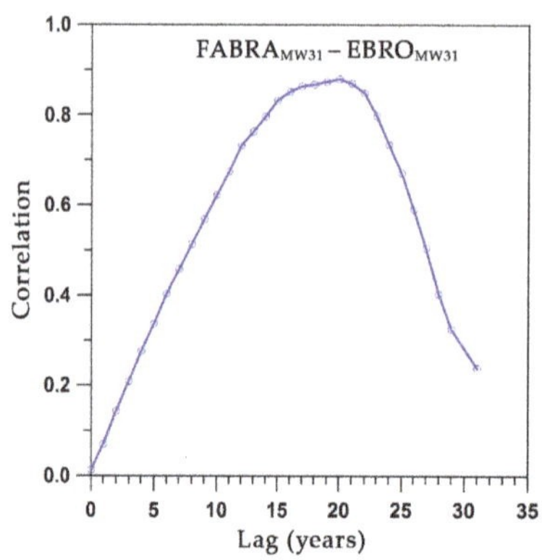

Figure 10. Correlations between mean MW31 rainfall at Ebro and Fabra for different year-lags.

3.2. *Seasonal Amounts*

3.2.1. Mean Seasonal Rainfall and *CV* for Each Window and Station

Figure 11 represents the moving seasonal means for the MW31 with a one-year shift. Trends are globally negative except for autumn, without a clear trend. The summer season presents the highest global negative trend (−0.76 mm/year), equivalent to a 21.5% decrease. For instance, the station with the highest summer moving values is a pre-Pyrenees station (Campdevànol, code 73) with a first mean value of 363.6 mm and a last value of 292.4 mm and with a monotonic negative trend. The greatest seasonal standard deviation of the moving average and moving *CV* values corresponds to summer, with 59.9 mm and 0.114 (Table 2), respectively. This season presents at the same time the lowest (43.1 mm) and the highest moving averages (364.5 mm). The second season with the highest negative average trend is winter (−0.43 mm/year) with a global negative trend of −14.6% with some oscillations. Autumn is the season with the highest mean (199.0 mm) of all the moving averages and stations.

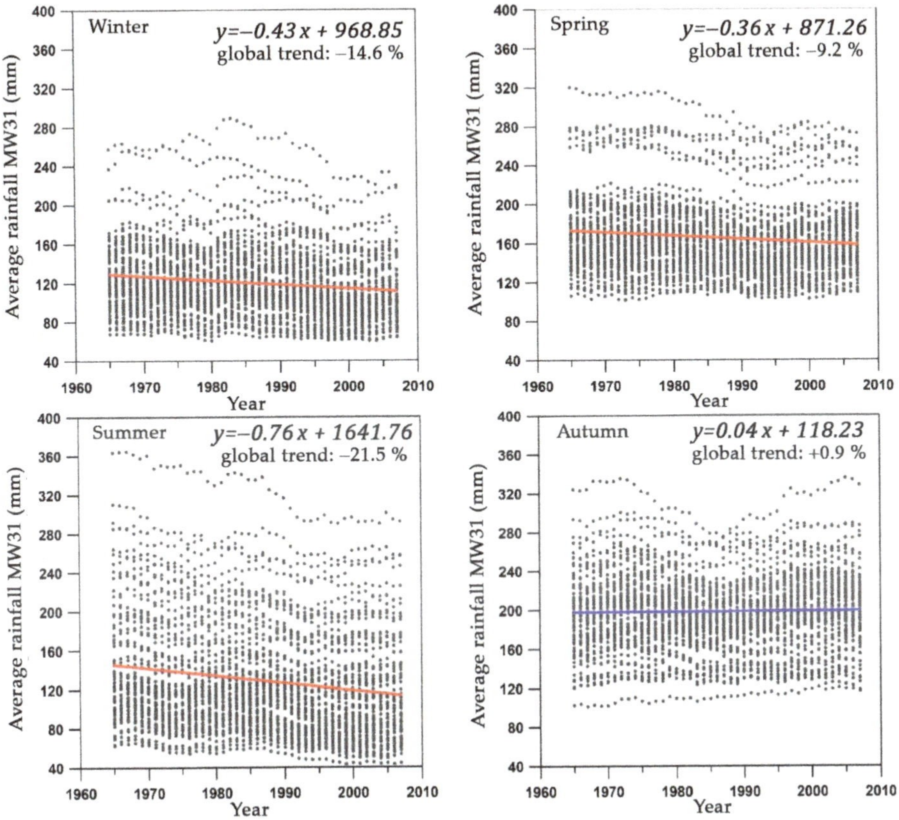

Figure 11. Seasonal mean rainfall for each MW31 and rain gauge. The solid line represents the average linear trend. Positive linear trends are shown in blue, and negative trends in red.

The evolution of the moving seasonal *CV* is represented in Figure 12. While winter and summer have positive trends, indicating an increase in the irregularity since 1950 for these two seasons, spring and autumn show global negative trends. Notably, the *CV* increase in winter is particularly prominent.

3.2.2. Trends in Seasonal Amounts Within Each 31-Year Window

Figure 13 depicts the evolution of the MK coefficients corresponding to the trends of the seasonal amounts within each MW31 and for each rain gauge. For spring, the period of the central years from 1981 to 1991 shows a non-negligible number of rain gauges with significant negative coefficients. Autumn is the season with the greatest minimum and maximum MK coefficient values, −3.60 and 2.97 (Table 3), which correspond to *p*-values of 0.00109 and 0.00571, respectively. A fourth-degree polynomial is adequate to fit the average evolution of the MK coefficients for the four seasons, revealing a phase opposition between spring and autumn. For these two seasons, a clear oscillation with a periodicity of 30–35 years is observed.

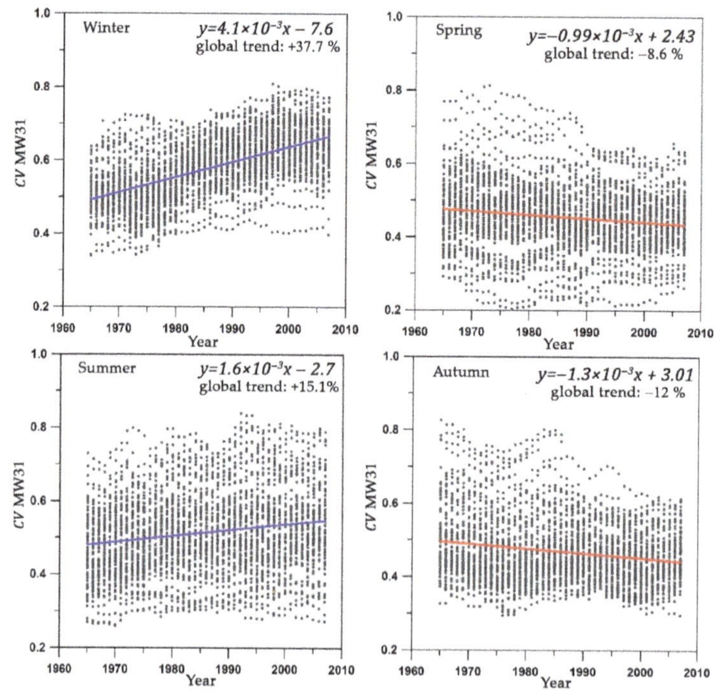

Figure 12. Seasonal CV for each MW31 and rain gauge. The solid line represents the average linear trend. Positive linear trends are shown in blue, and negative trends in red.

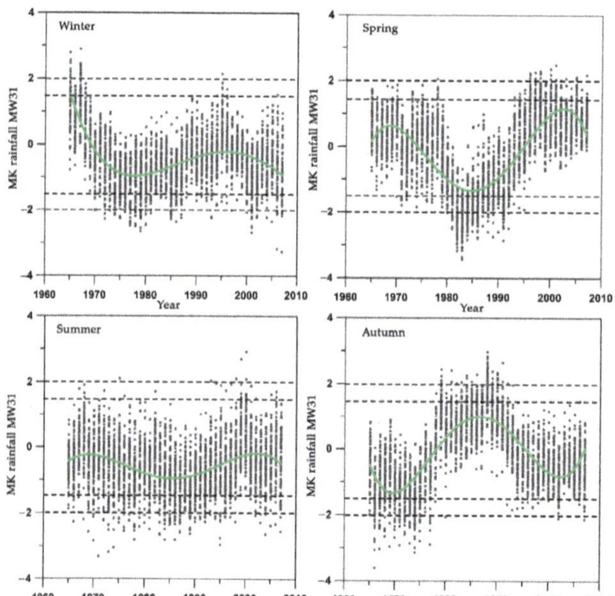

Figure 13. Mann–Kendall coefficients for the seasonal rainfall trends for each MW31 and rain gauge. Dashed lines represent the 85% and 95% confidence levels, and green lines show the fourth-degree polynomial fits.

3.2.3. Trends in the Characteristics of Seasonal Moving Windows

Figure 14 depicts the spatial distribution of the significant and non-significant trends in the moving averages for each rain gauge. For winter and especially summer, trends are negative for a high number of stations: 45 and 64, respectively (Table 4). The red (negative) color is predominant in these two figures. However, for spring and autumn, there are regional differences with positive (blue) and negative (red) trends. For the moving CV (Figure 15) in winter, the positive trends are notable, especially in the inland areas with trends above 90%. In addition, for autumn, these inland regions also present positive trends (over 40%), while coastal regions and the Pyrenees show negative trends (up to −50%).

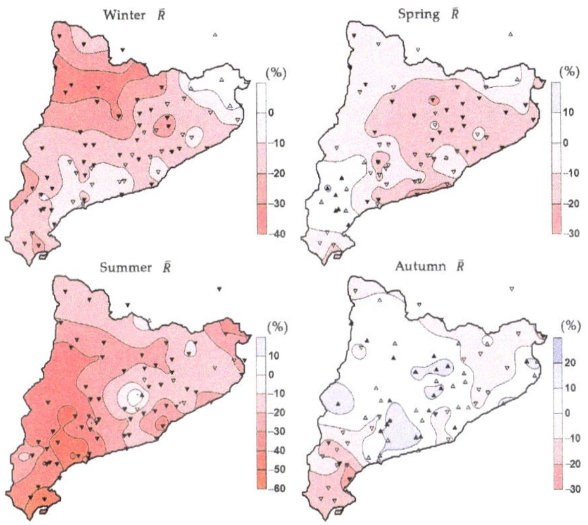

Figure 14. Relative trend (in percentage) for the MW31 seasonal average. Solid triangles indicate statistically significant trends at the 95% confidence level, while empty triangles represent non-significant trends: ▲ for positive and ▼ for negative trends.

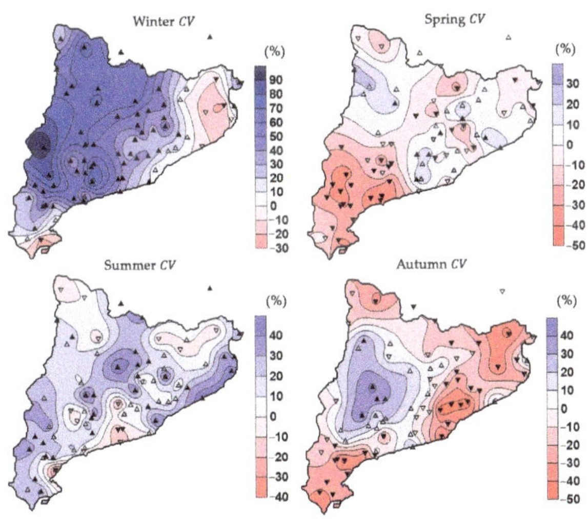

Figure 15. Relative trend (in percentage) for the MW31 seasonal CV. Solid triangles indicate statistically significant trends at the 95% confidence level, while empty triangles indicate non-significant trends: ▲ for positive and ▼ for negative trends.

3.3. Monthly Amounts

3.3.1. Mean Monthly Rainfall and CV for Each Window and Station

Mean monthly values for each MW31 and rain gauge are represented in Figure 16. The solid line represents the global linear trend, considering all 43 × 72 values. The highest global negative trends are observed in February, March, June, August and December. June shows the highest negative trend with −0.42 mm/year, equivalent to a 30% decrease with respect to the first value of the linear fit. However, January, April, October and November show global positive trends.

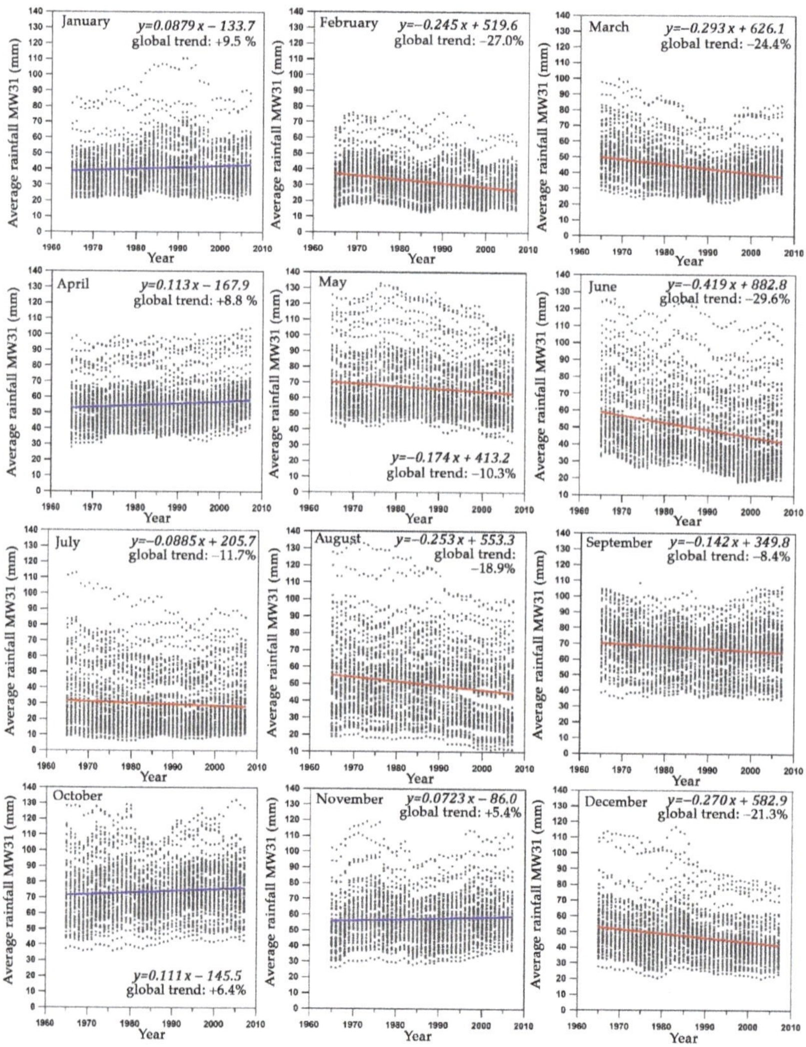

Figure 16. Monthly mean rainfall for each MW31 and rain gauge. Solid line represents the global linear trend. Positive linear trends are shown in blue, and negative trends in red.

The monthly CV values for each MW31 and rain gauge are depicted in Figure 17. Observing the global trends, moving CV shows the highest positive trends in June and February and the most negative in September, April, October and November. June has the highest global positive trend for CV and at the same time the highest negative for the average amounts. Table 2 summarized the main statistics of the MW31. The maximum values of $\overline{R}$ are reached in May, August and October, and the minimum in July. While summer months (June, July and August) show the highest standard deviation of moving averages, February has the lowest. These deviations can also be observed in Figure 16.

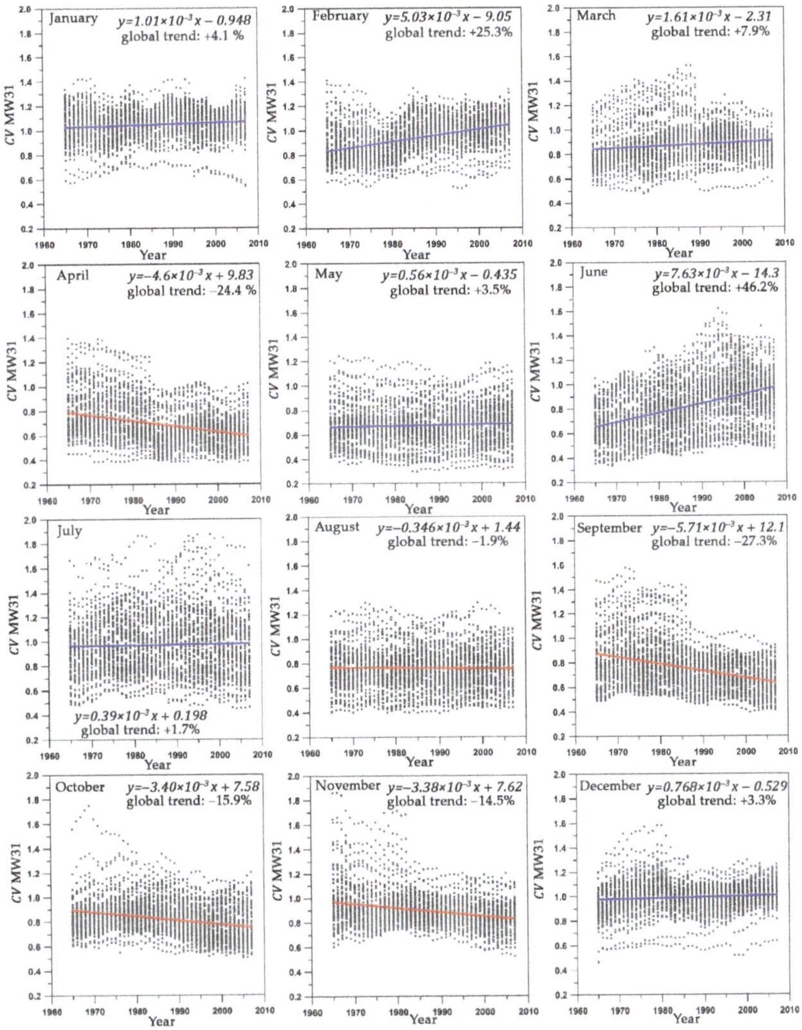

Figure 17. Monthly CV for each MW31 and rain gauge. Solid line represents the global linear trend. Positive linear trends are shown in blue, and negative trends in red.

For CV, the minimum value is reached in May, and the maximum in July and November. In contrast, the highest deviations for CV occur in two summer months (June and July), and the lowest in two winter months (January and December). Four months show the highest positive trends for certain MW31 windows and rain gauges. These months are January, May, September, and October, with values greater than +4.0 mm/year (Table 3). Conversely, the highest negative trends are obtained in autumn months (September, October and November) and in December, all trends with values lower than −4.0 mm/year.

Regarding the average trend considering all 43 × 72 trends, the highest trends are negative and correspond to June, September, and December. The MK coefficients of March and September present high positive and negative values. Other months with high positive MK values are April, July, and November.

3.3.2. Trends of Monthly Amounts Within Each 31-Year Window

Figure 18 depicts the MK coefficients for each MW31 and rain gauge, along with the fourth-degree polynomial fits. Similarities in the evolution of the MK coefficients between September and December can be observed, with a minimum around the central year 1970 and a maximum near 1990. February and March show the opposite behavior, with positive trends during the most recent periods, which aligns with [55]. September has the highest positive MK value, 4.28 (99.98% or p-value = 0.00017, Table 3), which is found in the MW31 centered in 1991. The polynomial fits for January and August show lower oscillations, and January is the month with the fewest significant trends.

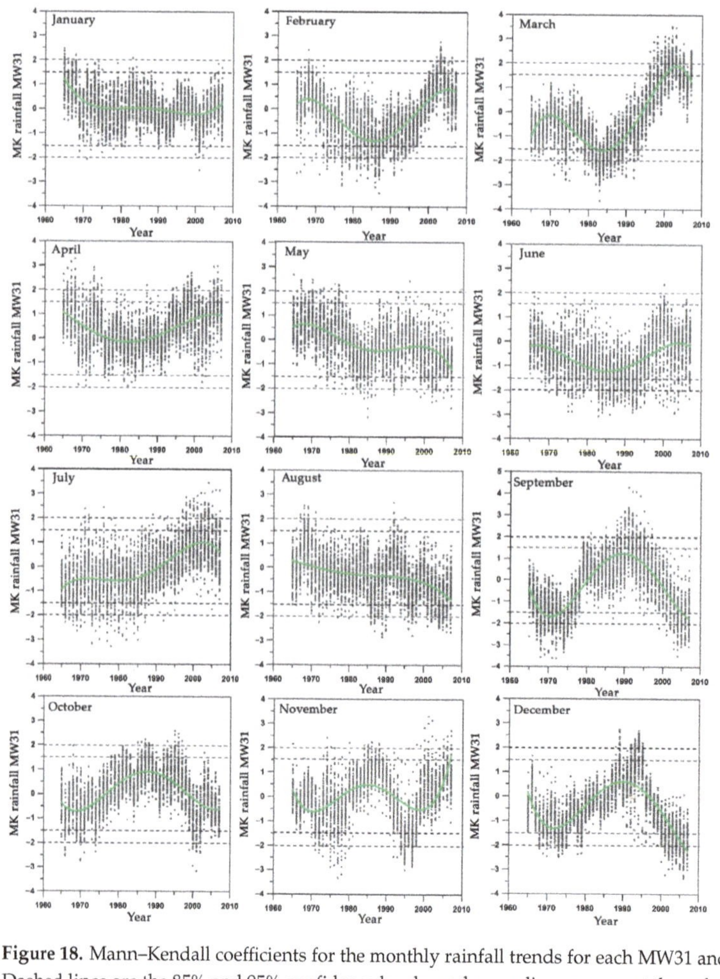

Figure 18. Mann–Kendall coefficients for the monthly rainfall trends for each MW31 and rain gauge. Dashed lines are the 85% and 95% confidence levels, and green lines represent the polynomial fits.

3.3.3. Trends in Characteristics of Monthly Moving Windows

Figures 19 and 20 show the spatial distribution of the MW31 relative trends for the mean monthly rainfall and the corresponding CV, including their statistical significance at the 95% confidence level. Five months (February, March, June, August, and December) show a high number of significant negative (red) trends (Table 4) for the mean MW31 rainfall. June and December have the highest number of rain gauges with negative trends: 56 and 54 out of 72, respectively, and none with positive trends. Three months, May, July and September, also display predominantly negative trends. Only two months (April and October) have a high number of rain gauges with significant positive (blue) trends.

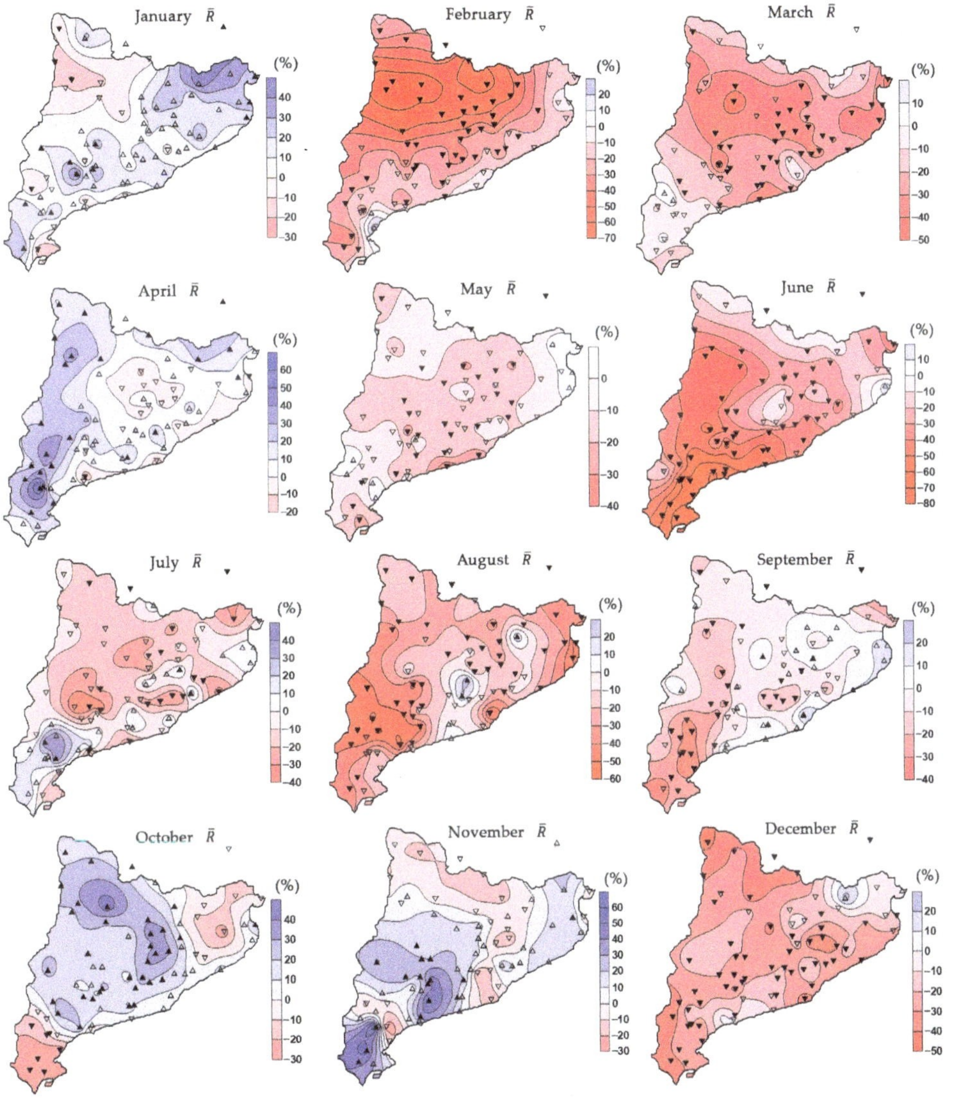

Figure 19. Relative trends, in percentage, for the MW31 monthly averages. Solid triangles indicate statistically significant trends at the 95% confidence level, while empty triangles indicate non-significant trends: ▲ for positive and ▼ for negative trends.

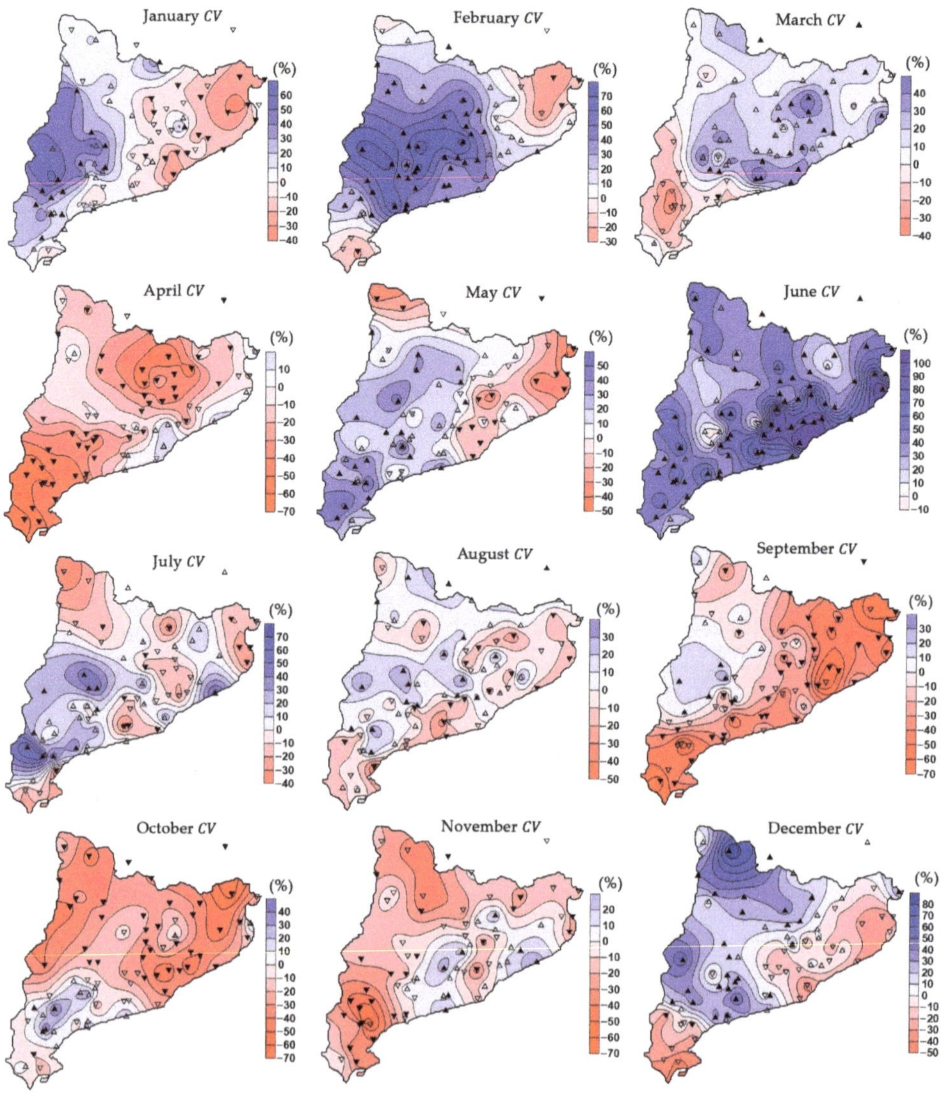

Figure 20. Relative trends, in percentage, for the MW31 monthly *CV* Solid triangles indicate statistically significant trends at the 95% level, while empty triangles indicate non-significant trends: ▲ for positive and ▼ for negative trends.

For *CV*, June is the month with the highest number of positive trends (61 out of 72), and February also shows a large number with 41. On the other hand, significant negative *CV* trends are observed in April, September and October.

4. Discussion

The comparison between the trend results obtained in the present work (Table 4), using moving windows for Catalonia (1950–2022), and those reported in other studies analyzing total precipitation trends in the region (Table 5) shows that the annual results are consistent with those obtained by [7], who found a decrease in precipitation across more than 90% of the Mediterranean Spanish coast for the period 1951–2000.

Table 5. Rainfall trends in six previous analyses [5,7,36,56–58] and in the present work. All columns show the percentages of stations with a significant (positive/negative) trend, except for column [7], which shows the percentage of stations with negative trends and the trend magnitudes in parentheses.

	[56]	[7]	[58]	[57]	[5]	[36]	Present Work
Scale	**Monthly**	**Seasonal**	**Monthly**	**Monthly, Annual**	**Monthly, Seasonal, Annual**	**Seasonal, Annual**	**Monthly, Seasonal, Annual MW31**
Region	Mediterranean Spanish coast	Mediterranean Spanish coast	Spain	Catalonia	Spain	Spain	Catalonia
Confidence Label	95% (+/−)	95%	90% (+/−)	85% (+/−)	95% (+/−)	90% (+/−)	95% (+/−)
Annual		90 (−12.4)		(4.2/26.0)	(0.74/18.00)	(1.2/22.5)	(0.0/87.5)
Winter		64 (−7.3)			(0.00/15.00)	(0.6/7.0)	(0.0/62.5)
Spring		82 (−19.3)			(0.11/9.60)	(0.8/40.6)	(8.3/38.9)
Summer		85 (−22.5)			(0.01/47.71)	(0.4/20.0)	(1.4/88.9)
Autumn		61 (−5.2)			(0.89/0.19)	(3.9/1.8)	(26.4/6.9)
January	(16.3/7.7)		(0.0/6.1)	(1.0/1.0)	(0.0/1.0)		(16.7/5.5)
February	(18.7/36.1)		(2.4/1.9)	(0.0/30.2)	(0.0/1.0)		(0.0/65.3)
March	(2.8/72.4)		(0.0/68.9)	(0.0/32.3)	(0.0/27.5)		(0.0/56.9)
April	(24.8/20.5)		(5.0/3.6)	(19.8/1.0)	(7.5/2.0)		(33.3/2.8)
May	(12.2/6.2)		(0.0/0.6)	(3.1/8.3)	(0.5/1.0)		(0.0/37.5)
June	(4.4/42.6)		(0.1/31.8)	(0.0/52.1)	(0.0/65.0)		(0.0/77.7)
July	(20.8/25.2)		(6.1/6.2)	(5.2/18.8)	(0.0/12.0)		(8.3/26.4)
August	(19.3/9.0)		(1.6/5.1)	(0.0/15.6)	(1.0/2.5)		(4.2/63.9)
September	(13.7/19.6)		(1.3/0.3)	(3.1/7.3)	(7.5/4.0)		(6.9/40.3)
October	(3.0/34.3)		(33.7/1.4)	(1.0/5.2)	(1.5/2.5)		(47.2/9.7)
November	(5.1/6.2)		(3.3/0.0)	(7.3/1.0)	(0.5/1.0)		(22.2/0.0)
December	(2.9/18.6)		(0.0/0.8)	(2.1/31)	(0.0/4.0)		(0.0/75.0)

For the seasonal study, the results are also aligned with those of [56] for the period 1951–2000. In both cases, summer is the season with the highest number of negative trends, 89% and 85%, respectively, with an average decrease of 22% in both studies. For winter, the proportion of negative trends are similar around 64% in both cases, with an overall decrease between 7% and 9%. The largest differences between the two studies are observed in spring and autumn. In spring, although both studies report negative trends, the percentage is higher for the entire Mediterranean coast than for Catalonia, due to the high number of negative trends along the southern coast of Spain. In autumn, while the present study shows a predominance of positive trends (26%), the broader Mediterranean Spanish coast exhibits negative trends in 61% of the territory.

In Catalonia, June is the month with the highest number of negative trends, which also aligns with [5] for the whole of Spain. In addition, December and February have a high number of stations with negative trends in Catalonia and along the Mediterranean Coast; however, these percentages are very low in the analysis of Spain [5,56]. Regarding positive trends in Catalonia, some discrepancies are observed compared to the different results shown in Table 5. For instance, while, for October, a high percentage of positive trends is obtained in the present work and in [58], the remaining studies report very low percentages for this month. With respect to the MW31 *CV* trends, winter and summer show increasing values, mainly due to the positive trends in February and June. On the other hand, spring and autumn exhibit a decrease in irregularity, primarily due to April, September, October and November, while August is the month with the highest stability in MW31 CV values.

Although in [57] trends are computed for monthly rainfall amounts and for the entire studied period, some similarities can be observed. Specifically, the number of stations showing statistically significant trends at the 95% confidence level using the MW31 method is comparable to those identified at the 85% confidence level when trends in monthly rainfall amounts are analyzed. Despite the application of a modified Mann–Kendall trend test, the MW31 approach yields a greater number of significant trends than the analysis based on monthly totals over the full period.

The moving-window approach used in this study offers key advantages. It enables the detection of temporal variability in rainfall trends at different time scales and reveals potential oscillatory behaviors that may remain hidden in fixed-period analyses. By applying the method consistently over overlapping 31-year periods, we capture evolving patterns in both trend direction and variability. However, this approach also presents limitations. First, it assumes local stationarity within each window, which may not fully account for abrupt shifts or non-linear changes within those periods. Second, although precautions were taken to reduce the effects of autocorrelation in the overlapping windows, the greater number of statistically significant trends observed, especially in comparison with studies using full-period analyses (e.g., [57]), may reflect residual autocorrelation effects that artificially inflate significance levels. This possibility highlights the need for careful interpretation of trend significance in moving-window analyses. Nevertheless, the consistency of our findings with previous research, along with the ability to detect multi-decadal oscillations, supports the methodological value of this approach.

Concerning the trends in the amounts within each MW31, oscillations and periodicities of approximately 30–35 years, particularly evident in spring and autumn, are consistent with findings from other studies reporting similar cycles in annual rainfall and the North Atlantic Oscillation (NAO). For instance, in [59] a 33.3-year periodicity in annual rainfall in Croatia was identified, while [60] reported several NAO cycles, including one lasting 34 years. From a monthly perspective, some months show periodicity and oscillations, for example, February, March, September, October, November, and December. At the annual scale, however, lower-frequency periodicities are also observed, suggesting the presence of different underlying patterns. The contrast with temperature behavior indicates the existence of two distinct modes of variability. This aspect represents a promising direction for future research, particularly through the analyses of temperature trend variability using moving windows of different lengths. Additionally, the opposing behavior observed between spring and autumn should be further investigated in future studies.

With respect to the oscillations observed in annual MW31 amounts for the two long series, Fabra and Ebro, similar cycles to those shown in Figure 8 were found in [53] for the NAO and AMO, but with opposite phases and based on 21-year running averages. In addition, the inverse correlation found between the NAO index and MW31 annual rainfall in Ebro as represented in Figure 9a is consistent with the findings of [61]. Although Fabra and Ebro have a similar periodicity close to 80 years, both cycles are shifted by 20 years, confirmed by the cross-correlation analysis (Figure 10), indicating a different behavior between these two gauges, which are 180 km apart. These differences are likely caused by the orography, orientation of the coastline, the more southwestern location of the Ebro Observatory, which results in a greater influence from the Ebro Valley, and the different altitudes: 412 m for Fabra and 49 m for Ebro. It is surprising to observe the coincidence of the 20-year lag between $Ebro_{MW31}$ and $Fabra_{MW31}$ and between NAO_{MW31} and AMO_{MW31}, as found by [53].

In addition to the influences of the NAO and AMO on the MW31 rainfall oscillations, the Gleissberg 80-year solar cycle should also be considered [62,63]. Solar cycles influence the Earth's climate and, consequently, rainfall regimes. While the 11-year period of the solar cycle is well known for its correlation with climatological variability, the 80-year cycle also appears to have a significant impact on rainfall variability. Several studies have identified periodicities close to 80 years in rainfall, floods, lakes and sea level, droughts, tree rings and temperatures [64–66].

Future research should aim to extend these analyses to the widest possible set of long-term datasets available across the Iberian Peninsula, both spatially and across other meteorological variables. In particular, applying this moving-window approach to temperature records in the same region may help reveal potential links between thermal and rainfall variability. Comparing trends in precipitation and temperature over similar time frames could offer a more comprehensive understanding of the climatic dynamics affecting the Mediterranean basin and provide further insights to support long-term water resource planning.

5. Conclusions

This detailed study, conducted using 31-year moving windows of annual, seasonal, and monthly precipitation values from 72 stations in Catalonia over the period 1950–2022, shows that the annual moving average values have decreased in nearly 90% of the stations, with a global reduction close to 10%. Rainfall variability, measured by the coefficient of variation (CV), has decreased along the coast but increased in inland areas and the Pyrenees. On a seasonal scale, summer stands out as the season with the most significant negative trend in MW31 averages, while autumn shows an almost overall neutral trend, although some regions in central Catalonia exhibit a positive trend. The winter CV has significantly increased in about 80% of the territory.

On the monthly scale, June, December and February exhibit the highest percentage of stations with negative trends, and for June and February, they also show the greatest reductions in precipitation, reaching up to 80%. At the same time, these two months show a significant increase in variability.

Additionally, the analysis of precipitation trend values within each MW31 shows Mann–Kendall (MK) test coefficients that, although statistically significant only in certain years and stations, reveal a fourth-degree polynomial fit that indicates whether the trend is becoming increasingly positive or negative. A periodicity of 30–35 years is especially evident in spring and autumn, with the completely opposite behavior between these two seasons being particularly striking. Furthermore, at least six months display similar variability patterns. A comparable oscillation of 33–34 years has also been identified in the NAO index and in annual rainfall in previous articles.

Finally, a periodicity of approximately 80 years has been identified in the 31-year moving averages of two centennial stations, Fabra Observatory (Barcelona) and Ebro Observatory (Tarragona). This periodicity may be attributed to the oscillations of the AMO and NAO indices and the 80-year Gleissberg solar activity cycle. This insight can support future projections and contribute to more effective long-term water resources management. However, the study period of this research (1950–2022) falls within a phase of negative trend, but it may suggest that this decline is approaching its end.

Beyond the scientific implications, these findings may also hold relevance for water resource management. The results presented in this study provide valuable insights into the historical evolution of precipitation patterns in the Mediterranean region. Although the analysis is retrospective, identifying when and where significant shifts in rainfall occurred can help water resource managers understand the timing and spatial extent of past changes. This historical perspective is crucial for evaluating long-term variability, benchmarking current conditions, and informing adaptive water planning strategies, especially in regions with increasing climatic and hydrological pressures.

Author Contributions: Conceptualization, C.S.; methodology, C.S.; software, C.S.; validation, C.S., M.d.C.C.-C., R.R.-S. and C.P.; formal analysis, C.S., M.d.C.C.-C., R.R.-S. and C.P.; investigation, C.S. and R.R.-S.; resources, C.S., M.d.C.C.-C., R.R.-S. and C.P.; data curation, C.S.; writing—original draft preparation, C.S.; writing—review and editing, C.S., M.d.C.C.-C. and R.R.-S.; visualization, C.S., M.d.C.C.-C., R.R.-S. and C.P.; supervision, C.S., M.d.C.C.-C. and R.R.-S. All authors have read and agreed to the published version of the manuscript.

Funding: This research received no external funding.

Data Availability Statement: Rainfall data are available on the websites of the Servei Meteorológic de Catalunya (Generalitat de Catalunya) [40], the Fabra Observatory (RACA, Barcelona) [43], and the Ebro Observatory [44]. The NAO and AMO indices are available on the NOAA website [45].

Acknowledgments: The authors acknowledge the Servei Meteorològic de Catalunya (Generalitat de Catalunya), the Fabra Observatory (RACA, Barcelona), the Ebro Observatory (Ramón Llull University and CSIC, Spanish Government) and NOAA for providing the rainfall data used in this study. The authors thank the academic support of the Qurbis and DF-GeoTech research groups from the Universitat Politècnica de Catalunya (UPC). The authors also thank the editor and the anonymous reviewers for their valuable comments and suggestions to improve the manuscript.

Conflicts of Interest: The authors declare no conflicts of interest.

References

1. IPCC. Cross-Chapter Paper 4: Mediterranean Region. In *Climate Change 2022: Impacts, Adaptation and Vulnerability. Contribution of Working Group II to the Sixth Assessment Report of the Intergovernmental Panel on Climate Change*; Pörtner, H.-O., Roberts, D.C., Tignor, M., Poloczanska, E.S., Mintenbeck, K., Alegría, A., Craig, M., Langsdorf, S., Löschke, S., Möller, V., et al., Eds.; Cambridge University Press: Cambridge, UK, 2022. Available online: https://www.ipcc.ch/report/ar6/wg2/chapter/ccp4/ (accessed on 13 July 2025).
2. Longobardi, A.; Villani, P. Trend analysis of annual and seasonal rainfall time series in the Mediterranean area. *Int. J. Climatol.* **2010**, *30*, 1538–1546. [CrossRef]
3. Gudmundsson, L.; Senevirante, S.I. Anthropogenic climate change affects meteorological drought risk in Europe. *Envirom. Res. Lett.* **2016**, *11*, 044005. [CrossRef]
4. Caloiero, T.; Coscarelli, R.; Ferrari, E.; Mancini, M. Precipitation change in Southern Italy linked to global scale oscillation indexes. *Nat. Hazards Earth Syst. Sci.* **2011**, *11*, 1683–1694. [CrossRef]
5. Senent-Aparicio, J.; López-Ballesteros, A.; Jimeno-Sáez, P.; Pérez-Sánchez, J. Recent precipitation trends in Peninsular Spain and implications for water infrastructure design. *J. Hydrol. Reg.* **2023**, *45*, 101308. [CrossRef]
6. Merino, A.; López, L.; Hermida, L.; Sánchez, J.L.; García-Ortega, E.; Gascón, E.; Fernández-González, S. Identification of drougth phases in a 110-year record from Western Mediterranean basin: Trends, anomalies and periodicity analysis for Iberian Peninsula. *Glob. Planet. Change* **2015**, *113*, 96–108. [CrossRef]
7. De Luis, M.; González-Hidalgo, J.C.; Longares, L.A.; Štepánek, P. Seasonal precipitation trends in the Mediterranean Iberian Peninsula in second half of 20th century. *Int. J. Climatol.* **2009**, *29*, 1312–1323. [CrossRef]
8. Arellano, B.; Zheng, Q.; Roca, J. Analysis of Climate Change Effects on Precipitation and Temperature Trends in Spain. *Land* **2025**, *14*, 85. [CrossRef]
9. Peña-Angulo, D.; Vicente-Serrano, S.M.; Domínguez-Castro, F.; Murphy, C.; Reig, F.; Tramblay, Y.; Trigo, R.M.; Luna, M.Y.; Turco, M.; Noguera, I.; et al. Long-term precipitation in Southwestern Europe reveals no clear trend attributable to anthropogenic forcing. *Environ. Res. Lett.* **2020**, *15*, 9. [CrossRef]
10. Platikanov, S.; López, J.F.; Martrat, B.; Martín-Vide, J.; Tauler, R. Temporal and spatial relationships between climatic indices and precipitation zones in Europe, Spain and Catalonia. *Int. J. Climatol.* **2024**, *45*, e8699. [CrossRef]

11. André, J.; D'Andrea, F.; Drobinski, P.; Muller, C. Regimes of precipitation change over Europe and the Mediterranean. *J. Geophys. Res.* **2024**, *129*, e2023JD040413. [CrossRef]
12. Zittis, G.; Bruggeman, A.; Lelieveld, J. Revisiting future extreme precipitation trends in the Mediterranean. *Weather Clim. Extrem.* **2021**, *34*, 100380. [CrossRef] [PubMed]
13. Miró, J.J.; Estrela, M.J.; Caselles, V.; Gómez, I. Spatial and temporal rainfall changes in the Júcar and Segura basins (1955–2016): Fine-scale trends. *Int. J. Climatol.* **2018**, *38*, 4699–4722. [CrossRef]
14. Miró, J.J.; Lemus-Canovas, M.; Serrano-Notivoli, R.; Olcina-Cantos, J.; Estrela, M.J.; Martín-Vide, J.; Sarricolea, P.; Meseguer-Ruiz, O. A component-based approximation for trend detection of intense rainfall in the Spanish Mediterranean coast. *Weather Clim. Extrem.* **2022**, *38*, 100513. [CrossRef]
15. Ribes, A.; Thao, S.; Vautard, R.; Dubuisson, B.; Somot, S.; Colin, J.; Planton, S.; Soubeyroux, J.M. Observed increase in extreme daily rainfall in the French Mediterranean. *Clim. Dyn.* **2019**, *52*, 1095–1114. [CrossRef]
16. Blanchet, J.; Blanc, A.; Creutin, J.D. Explaining recent trends in extreme precipitation in the Southwestern Alps by changes in atmospheric influences. *Weather Clim. Extrem.* **2021**, *33*, 100356. [CrossRef]
17. Nouaceur, Z.; Murarescu, O.; Muratoreanu, G. Statistical analysis of heavy rains and floods around the French Mediterranean basin over one half a century of observations. *Geosciences* **2022**, *12*, 447. [CrossRef]
18. Blanchet, J.; Creutin, J.D. Instrumental agreement and retrospective analysis of trends in precipitation extremes in the French Mediterranean Region. *Environ. Res. Lett.* **2022**, *17*, 074011. [CrossRef]
19. Khomsi, K.; Mahe, G.; Tramblay, Y.; Sinan, M.; Snoussi, M. Trends in rainfall and temperature extremes in Morocco. *Nat. Hazards Earth Sys. Sci.* **2015**, *16*, 1079–1090. [CrossRef]
20. Ouatiki, H.; Boudhar, A.; Ouhinou, A.; Arioua, A.; Hssaisoune, M.; Bouamri, H.; Benabdelouahab, T. Trend analysis of rainfall and drought over the Oum Er-Rbia River Basin in Morocco during 1970–2010. *Arab. J. Geosci.* **2019**, *12*, 128. [CrossRef]
21. Hadria, R.; Boudhar, A.; Ouatiki, H.; Lebrini, Y.; Elmansouri, L.; Gadouali, F.; Lionboui, H.; Benabdelouahab, T. Combining use of TRMM and ground observations of annual precipitations for meteorological drought trends monitoring in Morocco. *Am. J. Remote Sens.* **2019**, *7*, 25–34. [CrossRef]
22. González-Hidalgo, J.C.; Beguería, S.; Peña-Angulo, D.; Trullenque, V. Catalogue and analysis of extraordinary precipitation events in the Spanish mainland, 1916–2022. *Int. J. Climatol.* **2025**, *45*, e8785. [CrossRef]
23. Merino, A.; Fernández-Vaquero, M.; López, L.; Fernández-González, S.; Hermida, L.; Sánchez, J.L.; García-Ortega, E.; Gascón, E. Large-scale patterns of daily precipitation extremes on the Iberian Peninsula. *Int. J. Climatol.* **2016**, *36*, 3873–3891. [CrossRef]
24. Serrano-Notivoli, R.; Beguería, S.; Saz, M.A.; de Luis, M. Recent trends reveal decreasing intensity of daily precipitation in Spain. *Int. J. Climatol.* **2018**, *38*, 4211–4224. [CrossRef]
25. Casas-Castillo, M.C.; Llabrés-Brustenga, A.; Rius, A.; Rodríguez-Solà, R.; Navarro, X. A single scaling parameter as a first approximation to describe the rainfall pattern of a place: Application on Catalonia. *Acta Geophys.* **2018**, *66*, 415–424. [CrossRef]
26. Casas-Castillo, M.C.; Rodríguez-Solà, R.; Llabrés-Brustenga, A.; García-Marín, A.P.; Estévez, J.; Navarro, X. A simple scaling analysis of rainfall in Andalusia (Spain) under different precipitation regimes. *Water* **2022**, *14*, 1303. [CrossRef]
27. Llasat, M.C.; Marcos, R.; Turco, M.; Gilabert, J.; Llasat-Botija, M. Trends in flash flood events versus convective precipitation in the Mediterranean region: The case of Catalonia. *J. Hydrol.* **2016**, *541*, 24–37. [CrossRef]
28. Llasat, M.C.; del Moral, A.; Cortés, M.; Rigo, T. Convective precipitation trends in the Spanish Mediterranean region. *Atmos. Res.* **2021**, *257*, 105581. [CrossRef]
29. Casas-Castillo, M.C.; Rodríguez-Solà, R.; Lana, X.; Serra, C.; Martínez, M.D.; Biere, R.; Arellano, B.; Roca, J. Consecuencias hidrológicas del cambio climático en entornos urbanos. In *XIII CTV 2019 Proceedings: XIII International Conference on Virtual City and Territory: "Challenges and Paradigms of the Contemporary City", UPC, Barcelona, Spain, 2–4 October 2019*; CPSV: Barcelona, Spain, 2019; p. 8291. [CrossRef]
30. Esbrí, L.; Rigo, T.; Llasat, M.C.; Aznar, B. Identifying storm hotspots and the most unsettled areas in Barcelona by analysing significant rainfall episodes from 2013 to 2018. *Water* **2021**, *13*, 1730. [CrossRef]
31. Casas-Castillo, M.C.; Navarro, X.; Rodríguez-Solà, R. Rainfall organization and storm tracking in urban Barcelona, NE Spain, using a high-resolution rain gauge network. *Hydrology* **2025**, *12*, 178. [CrossRef]
32. Casas-Castillo, M.C.; Llabrés-Brustenga, A.; Rodríguez-Solà, R.; Rius, A.; Redaño, À. Scaling properties of rainfall as a basis for intensity-duration-frequency relationships and their spatial distribution in Catalunya, NE Spain. *Climate* **2025**, *13*, 37. [CrossRef]
33. Lana, X.; Serra, C.; Burgueño, A. Patterns of monthly rainfall shortage and excess in terms of the standardized precipitation index for Catalonia (NE Spain). *Int. J. Climatol.* **2001**, *21*, 1669–1691. [CrossRef]
34. Serra, C.; Burgueño, A.; Martínez, M.D.; Lana, X. Trends in dry spells across Catalonia (NE Spain) during the second half of the 20th century. *Theor. Appl. Climatol.* **2006**, *85*, 165–183. [CrossRef]

35. Lana, X.; Martínez, M.D.; Burgueño, A.; Serra, C.; Martín-Vide, J.; Gómez, L. Distributions of long dry spells in the Iberian Peninsula, years 1951–1990. *Int. J. Climatol.* **2006**, *26*, 1999–2021. [CrossRef]
36. González-Hidalgo, J.C.; Trullenque-Blanco, V.; Beguería, S.; Peña-Angulo, D. Seasonal precipitation changes in the western Mediterranean Basin: The case of the Spanish mainland, 1916–2015. *Int. J. Climatol.* **2024**, *44*, 1800–1815. [CrossRef]
37. Lana, X.; Serra, C.; Casas-Castillo, M.C.; Rodríguez-Solà, R.; Prohom, M. Two Centuries of Monthly Rainfall in Barcelona (NE Spain): Disparity Trends, Correlation of Autumnal Rainfall with the WeMO Index and Its Contribution to Annual Amounts. *Climate* **2024**, *12*, 166. [CrossRef]
38. Martín-Vide, J.; Moreno-García, M.C. Análisis de la serie secular de precipitación anual del Observatorio Fabra (Barcelona) (1914–2020). *Geographicalia* **2021**, *73*, 81–94. [CrossRef]
39. World Meteorological Organization. WMO Climatological Normals. Available online: https://community.wmo.int/en/activity-areas/climate-services/climate-products-and-initiatives/wmo-climatological-normals (accessed on 6 July 2025).
40. Servei Meteorològic de Catalunya (Generalitat de Catalunya). Climatologia, Dades i Productes Climàtics. Sèries Climàtiques des de 1950. Available online: https://www.meteo.cat/wpweb/climatologia/dades-i-productes-climatics/series-climatiques-des-de-1950/ (accessed on 13 July 2025).
41. Llabrés-Brustenga, A.; Rius, A.; Rodríguez-Solà, R.; Casas-Castillo, M.C.; Redaño, À. Quality control process of the daily rainfall series available in Catalonia from 1855 to the present. *Theor. Appl. Climatol.* **2019**, *137*, 2715–2729. [CrossRef]
42. Prohom, M.; Domonkos, P.; Cunillera, J.; Barrera-Escoda, A.; Busto, M.; Herrero-Anaya, M.; Aparicio, A.; Reynés, J. CADTEP: A new daily quality-controlled and homogenized climate database for Catalonia (1950–2021). *Int. J. Climatol.* **2023**, *43*, 4771–4789. [CrossRef]
43. Reial Acadèmia de Ciències i Arts de Barcelona (RACAB). Meteorologia. Dades Meteorològiques i Climàtiques. Available online: https://observatorifabra.cat/meteorologia/dades-metereologiques-i-climatiques/ (accessed on 13 July 2025).
44. Portal de Datos del Observatori de l'Ebre: Dades Climàtiques. Precipitació. Available online: https://www.obsebre.es/es/es-datos-climaticos/en-precipitacion (accessed on 13 July 2025).
45. National Oceanic and Atmospheric Administration, NOAA, U.S. Government. Physical Science Laboratory: 20CR Climate Indices: North Atlantic Oscillation (NAO). Available online: https://psl.noaa.gov/data/20thC_Rean/timeseries/monthly/NAO/ (accessed on 15 January 2025).
46. Sneyers, R. *On the Statistical Analysis of Series of Observations*; Technical Note No. 143, WMO No. 415; World Meteorological Organization: Geneva, Switzerland, 1990; 192p. Available online: https://library.wmo.int/idurl/4/30743 (accessed on 13 July 2025).
47. Hamed, K.H.; Rao, A.R. A Modified Mann-Kendall Trend Test for Autocorrelated Data. *J. Hydrol.* **1998**, *204*, 182–196. [CrossRef]
48. Pyper, B.J.; Peterman, R.M. Comparison of methods to account for autocorrelation in correlation analyses of fish data. *Can. J. Fish. Aquat. Sci.* **1998**, *55*, 2127–2140. [CrossRef]
49. Li, J.; Sun, C.; Jin, F. NAO implicated as a predictor of Northern Hemisphere mean temperature multidecadal variability. *Geophys. Res. Lett.* **2013**, *40*, 5497–5502. [CrossRef]
50. Martín-Vide, J.; López-Bustins, J.A. The Western Mediterranean Oscillation and rainfall in the Iberian Peninsula. *Int. J. Climatol.* **2006**, *26*, 1455–1475. [CrossRef]
51. Ríos-Cornejo, D.; Penas, A.; Álvarez-Esteban, R.; del Río, S. Links between teleconnection patterns and precipitation in Spain. *Atmos. Res.* **2015**, *156*, 14–28. [CrossRef]
52. Martínez-Artigas, J.; Lemus-Cánovas, M.; López-Bustins, J.A. Precipitation in peninsular Spain: Influence of teleconnection indices and spatial regionalisation. *Int. J. Climatol.* **2020**, *41*, E1320–E1335. [CrossRef]
53. Sun, C.; Li, J.; Jin, F. A delayed oscillator model for the quasi-periodic multidecadal variability of the NAO. *Clim. Dym.* **2015**, *45*, 2083–2099. [CrossRef]
54. Luterbacher, J.; Schmutz, C.; Gyalistras, D.; Xoplaki, E.; Wanner, H. Reconstruction of monthly NAO and EU indices back to AD 1675. *Geophys. Res. Lett.* **1999**, *26*, 2745–2748. [CrossRef]
55. Benetó, P.; Khodayar, S. On the need for improved knowledge on the regional-to-local precipitation variability in eastern Spain under climate change. *Atmos. Res.* **2023**, *290*, 106795. [CrossRef]
56. González-Hidalgo, J.C.; Lopez-Bustins, J.A.; Štepánek, P.; Martín-Vide, J.; de Luis, M. Monthly precipitation trends on the Mediterranean fringe of the Iberian Peninsula during the second-half of the twentieth century (1951–2000). *Int. J. Climatol.* **2009**, *29*, 1415–1429. [CrossRef]
57. Lana, X.; Casas-Castillo, M.C.; Rodríguez-Solà, R.; Serra, C.; Martínez, M.D.; Kirchner, R. Rainfall regime trends at annual and monthly scales in Catalonia (NE Spain) and indications of CO_2 emissions effects. *Theor. Appl. Climatol.* **2021**, *146*, 981–996. [CrossRef]

58. González-Hidalgo, J.C.; Brunetti, M.; de Luis, M. A new tool for monthly precipitation analysis in Spain: MOPREDAS database (monthly precipitation trends December 1945–November 2005). *Int. J. Climatol.* **2011**, *31*, 715–731. [CrossRef]
59. Gajić-Čapka, M. Periodicity of annual precipitation in different climate regions of Croatia. *Theor. Appl. Climatol.* **1994**, *49*, 213–226. [CrossRef]
60. Seip, K.L.; Grøn, Ø.; Wang, H. The North Atlantic Oscillations: Cycle Times for the NAO, the AMO and the AMOC. *Climate* **2019**, *7*, 43. [CrossRef]
61. Paredes, D.; Trigo, R.M.; Garcia-Herrera, R.; Trigo, I.F. Understanding precipitation changes in Iberia in early spring: Weather typing and storm-tracking approaches. *J. Hydrometeorol.* **2006**, *7*, 101–113. [CrossRef]
62. Gleissberg, W. The eighty-year solar cycle in auroral frequency numbers. *J. Br. Astron. Assoc.* **1965**, *75*, 227–231. Available online: https://adsabs.harvard.edu/pdf/1967SoPh....2..231G (accessed on 13 July 2025).
63. Peristykh, A.N.; Damon, P.E. Persistence of the Gleissberg 88-year solar cycle over the last 12,000 years: Evidence from cosmogenic isotopes. *J. Geophys. Res.* **2003**, *108*, SSH 1-1–SSH 1-15. [CrossRef]
64. Burroughs, W.J. *Weather Cycles Real or Imaginary*, 2nd ed.; Cambridge University Press: Cambridge, UK, 2003; 330p, ISBN 9780521528221.
65. Hameed, S.; Yeh, W.M.; Li, M.T.; Cess, R.D.; Wang, W.C. An analysis of periodicities in the 1470 to 1974 Beijing precipitation record. *Geophys. Res. Lett.* **1983**, *10*, 436–439. [CrossRef]
66. Yousef, S.M. The solar Wolf-Gleyssberg cycle and its influence on the earth. In Proceedings of the International Conference for Environment Hazard (ICEHM2000), Cairo, Egypt, 9–12 September 2000; pp. 267–293. Available online: https://www.virtualacademia.com/pdf/cli267_293.pdf (accessed on 13 July 2025).

hydrology

MDPI

Article

Method for Establishing Heavy Rainfall Equations Based on Regional Characteristics: Transformation of Maximum Daily Precipitation

Laura Thebit de Almeida [1], Roberto Avelino Cecílio [2,*], Marcel Carvalho Abreu [3] and Ivana Patente Torres [4]

1 Federal Institute of Education, Science and Technology of Northern Minas Gerais, Januária 39480-000, Brazil
2 Department of Forest and Wood Sciences, Federal University of Espírito Santo, Jerônimo Monteiro 29550-000, Brazil
3 National Center for Monitoring and Alerts for Natural Disasters, São José dos Campos 12247-016, Brazil
4 Department of Agricultural Engineering, Federal University of Viçosa, Viçosa 36570-900, Brazil
* Correspondence: roberto.cecilio@ufes.br

Abstract

Modeling heavy rainfall patterns is essential for designing hydraulic structures, planning land use and water resources, and predicting disasters, among others. Usually, heavy rainfall is characterized by curves that relate the intensity, duration, and frequency (IDF), adjusted from the analysis of pluviograms. Alternatively, these equations can be adjusted using disaggregated daily rainfall data, whose reliability is currently questioned due to the use of common coefficients to disaggregate the maximum daily precipitation (h_{day}) into rainfall associated with shorter durations. This study proposes the Transformation of Maximum Daily Precipitation method (TMDP) using the maximum daily precipitation of the station of interest and the curve of heavy rainfall of the nearest location, denoting the local characteristic, to transform the h_{day} associated with a return period into rainfall intensities for shorter durations. The TMDP proved to be slightly superior to the most widely used rainfall disaggregation method in Brazil, particularly in regions with a higher density of data for the IDF equation. The TMDP is a potential tool for regions with low density of rainfall data, although it has limitations in regions where such data are scarce.

Keywords: intensity-duration-frequency; rainfall disaggregation; extreme precipitation; hydrological modeling

1. Introduction

Heavy rainfall is regarded as high rainfall volumes in a short time. These events can have considerably undesirable effects on human activities and the environment, such as losses in agricultural production [1], water erosion [2], floods [3], landslides [4], silting of lowlands and riverbeds [5], damage to hydraulic structures [6], among others.

Models providing reliable estimates of heavy rainfall are essential for hydraulic structure design, land use planning, water resource management, and forecasting environmental disasters [7]. Currently, one of the most common ways of characterizing heavy rainfall in a given location is through intensity, duration, and frequency (IDF) curves [8], represented by Equation (1), which has local parameters, adjusted from data on the most intense rainfalls of different durations, associated with different return periods [9]. Usually, the data needed to adjust the equations comes from pluviographs.

$$I_m = \frac{K\, T^a}{(d+b)^c} \tag{1}$$

in which I_m represents the average maximum rainfall intensity, in mm h^{-1}; T is the rainfall's return period, in years; d is its duration, in minutes; K, a, b, c are parameters statistically adjusted to each location.

An IDF curve, following the standard method, is established from a series of sub-daily rainfall data of maximum rainfall associated with different durations. The largest database of IDF curves for Brazil, a continental country with more than 8.5 million km^2, accounts for only 373 equations established in this way, in the Plúvio 2.1 application. This number represents the density of one equation (rainfall station) for every 22,754 km^2, a number much higher than the maximum limit of one station for every 3000 km^2, suggested by the World Meteorological Organization [10]. In countries with large territorial extensions and demographic, social, economic, and environmental differences, such as Brazil, the distribution of the rainfall network is uneven, aggravating the situation of representativeness of IDF equations in remote regions.

To increase the availability of IDF equations, an alternative is to establish them from daily rainfall data, disaggregating the maximum daily precipitation into maximum rainfalls with sub-daily durations. This method, called "Rainfall Depth-Duration" (RDD), is the most widely used in Brazil in studies to obtain the IDF equation [11]. The vast majority of common applications of RDD in Brazil use generic disaggregation coefficients (*dc*) developed by "Companhia Ambiental do Estado de São Paulo" [12], assumed to be valid for the entire country. However, said disaggregation coefficients were established from the ratio of maximum rainfalls of different durations from short and old rainfall series (between 1957 and 1979), which do not necessarily reflect the specific temporal and regional/local peculiarities of intense rainfall for the whole of Brazil [13–15]. The methods described and traditionally employed are essentially site-specific, lacking a spatial relationship among the generated IDF equations. Their application in areas without rain gauges, therefore, relies on spatialization techniques, which can only be implemented after the development of the equations.

Some efforts have been made to address regional/local peculiarities of heavy rainfall by establishing specific disaggregation coefficients in places with sub-daily extreme rainfall data available to represent a specific coverage area [14,16–19]. However, the procedure is limited by low data availability to determine specific disaggregation coefficients, due to the insufficiency of a monitoring network and/or the time lag and progressive gaps in the series [20,21].

The major challenge to studying alternative methods for obtaining IDF relationships in Brazil is to cover broader areas with disaggregation coefficients in regions without observations of pluviometric data. It is believed, as a premise of this work, that a better way of establishing IDF curves can be achieved by associating the ease of application of the "Rainfall Depth-Duration" with the regional peculiarities of intense rainfall. Local and regional characteristics of extreme rainfall events can be utilized to improve methods for deriving IDF equations, allowing for a more accurate representation of spatial and temporal variability.

Hence, the present work aims to propose a new method, called Transformation of Maximum Daily Precipitation (TMDP), to obtain the curves of intense rainfall, from the transformation of daily rainfall data from the desired location into maximum sub-daily rainfall intensities. The information on intense rainfall is obtained through the product of the local maximum daily precipitation by a transformation coefficient (*tc*), established from data extracted from the IDF curve of the nearest rain gauge station, denoting the regional characteristic.

2. Materials and Methods

2.1. Study Area

Brazil exhibits a diverse rainfall regime (Figure 1a) primarily governed by its vast territorial extension, tropical location, and complex atmospheric circulation patterns. Annual precipitation ranges from over 3000 mm in the Amazon Basin to less than 500 mm in the semi-arid northeast. The majority of the country experiences a marked wet season during the austral summer (approximately November to March), driven by the South American Monsoon System. In contrast, the northeast region is characterized by irregular and scarce rainfall, often influenced by the Intertropical Convergence Zone (ITCZ) and sea surface temperature anomalies in the Atlantic and Pacific oceans. Coastal areas and the southern region present more evenly distributed rainfall throughout the year.

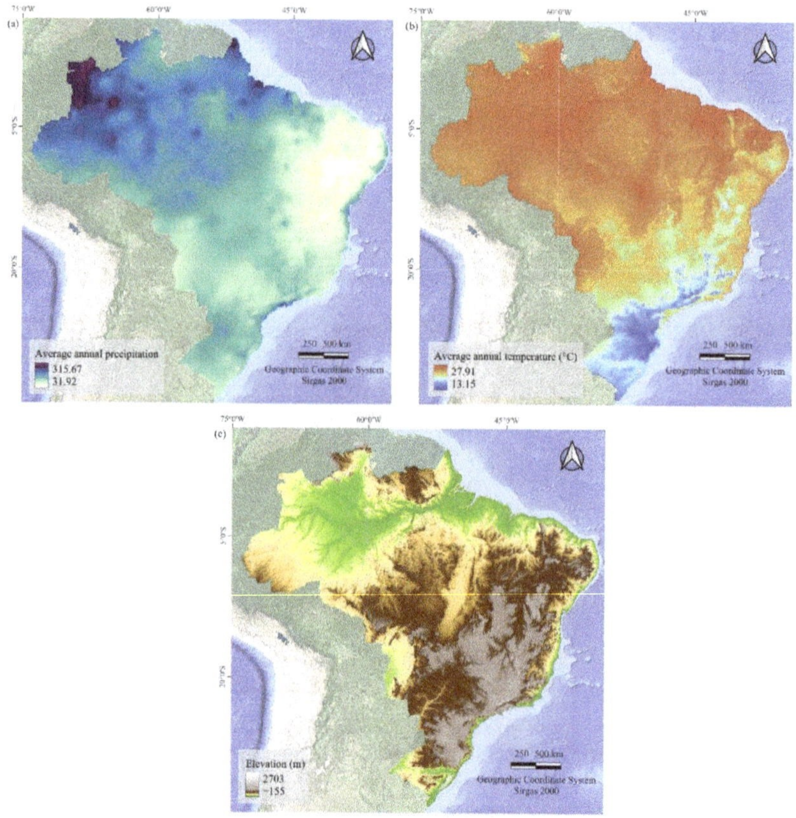

Figure 1. Spatial distribution of (**a**) annual total precipitation, (**b**) annual mean air temperature, and (**c**) digital elevation model (DEM) across Brazil. Precipitation and temperature data from WorldClim versão 2.1 (1970–2000). The DEM data from GMTED2010 (Global Multi-resolution Terrain Elevation Data 2010).

The temperature regime (Figure 1b) is predominantly influenced by its tropical and subtropical latitudinal position, low topographic variation in most regions, and extensive landmass. The country experiences generally high average temperatures throughout the year, with mean annual temperatures ranging from approximately 18 °C in the southern highlands to over 28 °C in the northern lowlands. Seasonal temperature variation is relatively low in equatorial regions but becomes more pronounced in the southern states due to subtropical influences. Although Brazil does not experience severe winters, occasional cold air incursions from higher latitudes can affect the South, leading to temporary drops in temperature, particularly during austral winter (June–August).

Brazil's topography (Figure 1c) is characterized predominantly by extensive plateaus, low mountain ranges, and broad sedimentary basins. The Brazilian Highlands (Planalto Brasileiro) dominate much of the central and eastern territory, with elevations typically between 300 and 900 m above sea level. In contrast, the Amazon Basin in the north features vast lowland plains with gentle relief, playing a crucial role in the country's hydrological and ecological systems. Mountainous regions are relatively limited, with the highest elevations found in the southeastern and southern parts of the country, such as the Mantiqueira and Serra do Mar ranges. Overall, the Brazilian terrain lacks extreme altitudinal variation but contributes significantly to regional climatic and hydrological dynamics.

2.2. Proposed Method

The proposed method was called Transformation of Maximum Daily Precipitation (TMDP), and it uses a transformation coefficient ($tc_{d,T}$) designed to transform the maximum daily precipitation depth (h_{day}) of a specific station into the maximum average rainfall intensity, associated with a specific duration (d) and return period (T). The maximum rainfall intensity associated with a pair of d and T ($\hat{I_{d,T}}$ in mm/h) is obtained by the product between $tc_{d,T}$ and the maximum daily precipitation depth associated with the same T ($h_{day,T}$ in mm), according to Equation (2).

$$\hat{I_{d,T}} = tc_{d,T} \,.\, h_{day,T} \quad (2)$$

The $h_{day,T}$ represents the local magnitude of maximum rainfall events. It is calculated through frequency analysis and estimation of quantiles of interest of the maximum annual daily rainfall for the station of interest, as described in Section 2.2.1.

The $tc_{d,T}$ to be applied are not derived from the station or location of interest. Instead, they pertain to the nearest location for which an IDF equation is available (the support station). Therefore, the $tc_{d,T}$ represent the regional patterns of heavy rainfall. The methodology for establishing the $tc_{d,T}$ is outlined in Section 2.2.2.

Once the transformations of $h_{day,T}$ have been performed for several $\hat{I_{d,T}}$ associated with a series of pairs of d (1440, 720, 360, 240, 180, 120, 60, 50, 40, 30, 20, and 10 min) and T (2, 5 10, 20, 50, and 100 years), it is possible to adjust the IDF equation according to the traditional procedure for establishing these equations [16], estimating the parameters K, a, b and c of Equation (1), using the Gauss–Newton method.

2.2.1. Maximum Daily Precipitation Depth Associated with a Return Period ($h_{day,T}$)

The $h_{day,T}$ is determined from an annual series of maximum daily precipitation (h_{day}) from the rainfall station of interest. Annual extreme sub-daily rainfall time series are rarely not random or independent [13]. In this study, the series were therefore considered independent.

Probability density functions (PDF) are fitted to the data series; the most commonly used are: Gumbel (GUM), Generalized Extreme Events (GEV), 2-parameter Log-normal (LN2), 3-parameter Log-normal (LN3), or Gamma (G) [22]. PDFs are commonly employed for the estimation of the probable maximum precipitation, particularly in association with return periods. This approach enables the characterization of rainfall extremes and their recurrence, providing a statistical basis for hydrological design and risk assessment [13,16,22].

The adequacy of the probability distributions to the observed data, as well as the classification of their performance, was evaluated using goodness-of-fit tests at a 5% significance level. It can also classify the PDFs according to the level of adequacy of each PDF to the data series [13]. The best PDF is chosen through goodness-of-fit (GOF) tests, which verify the adequacy of the PDF to the data series with a significance level of 5%. It can also classify the PDFs according to the level of adequacy of each PDF to the data series. The GOF tests used were: Kolmogorov–Smirnov (KS), Chi-square (X^2), and Anderson–Darling (AD). The classification of PDFs was performed by analyzing the p-value in each GOF test, which represents the probability of obtaining a test statistic equal to or more extreme than that observed in a sample, under the null hypothesis that the PDF is adherent to the data series. The p-values below the significance level ($\alpha = 0.05$) indicate PDFs that are not suitable for the h_{day} data series, and in such cases, these distributions are not employed to estimate the probable maximum precipitation. The highest p-value in each GOF indicates which PDF is most appropriate to the h_{day} series. If the adherence tests regarding the best PDF disagree among themselves, that is, GOF tests resulted in different PDFs with the best performance, the PDF with the best performance in the chi-square test (X^2) was considered. The X^2 test was selected because it presented good performance in the selection of PDFs, with less uncertainty in the estimation of quantiles of interest and good statistical performance in the representation of h_{day} [22].

2.2.2. Coefficients to Transform Daily Rainfall ($tc_{d,T}$)

The transformation coefficients are used to consider the regional characteristics of intense rainfall with sub-daily duration, based on a geographic criterion. Therefore, they must be established with the daily rainfall data and with the IDF equation of the support station. The $tc_{d,T}$ are established through Equation (3), which expresses the relationship between the maximum rainfall intensity calculated by the IDF equation ($I_{IDF\ d,\ T}$) and the $h_{day,T}$ of the support station. Several combinations between d (1440, 720, 360, 240, 180, 120, 60, 50, 40, 30, 20, 10 min) and T (2, 5 10, 20, 50, 100 years) are considered, to establish a specific $tc_{d,T}$ of these combinations.

$$tc_{d,T} = \frac{I_{IDF\ d,T}}{h_{\ day,T}} \tag{3}$$

The $h_{day,T}$ is the quantile of interest calculated as described in Section 2.2.1, using precipitation data for the same period (same years of analysis) for which the IDF equation was originally adjusted.

Assuming a spatial dependence [23] and similar characteristics of heavy rainfall for nearby locations, the established $tc_{d,T}$ are applied to transform the maximum daily precipitation in nearby rainfall stations, thus reflecting the regional behavior of sub-daily heavy rainfall.

2.3. *Method's Application and Assessment*

2.3.1. Establishing $tc_{d,T}$ for Brazil

First, the $tc_{d,T}$ of the TMDP method was established for the 373 locations in Brazil where adjusted IDF equations exist [24–30], as in Section 2.2.2. Figure 2 shows the spatial distribution of these stations in Brazil. The temporal extent of the extreme rainfall series ranges from 5 to 26 years, exhibiting variability in both commencement and termination dates. The majority of these series initiated in the 1980s and terminated towards the late 1990s [11]. The highest density of IDF equations, defined as the high-density group (HD), is in the southeast of the country, except for the state of São Paulo, and in the south of Bahia, with approximately one equation every 3400 km^2. Equations concentrated in the states of São Paulo, Santa Catarina, and Paraná correspond to the region with medium density (MD) of IDF equations distributed by area, with one station every 9000 km^2. The lowest density group (LD), represented by the other stations spread throughout the country, has one station every 150,000 km^2. It is noteworthy that all observed densities are still below those recommended by the WMO.

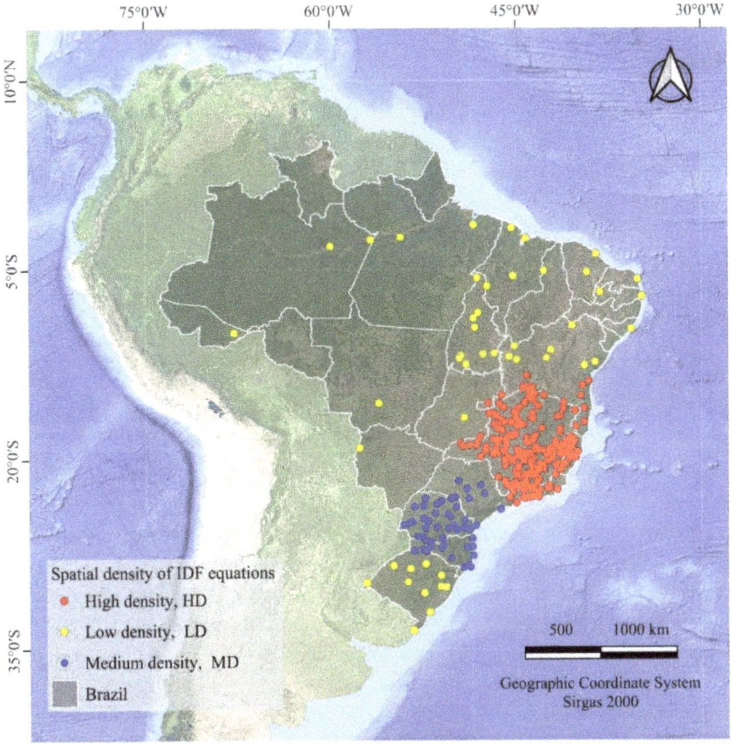

Figure 2. Location of rainfall stations classified according to their density with established IDF equations.

2.3.2. Assessment

The TMDP was applied to the same locations where rainfall data and established IDF equations are available. The rainfall data from the stations were obtained from the Santa Catarina Agricultural Research and Rural Extension Company—Epagri, the Department of Water and Energy (DAEE), and the Hidroweb portal of the National Water and Basic Sanitation Agency [31]. The IDF equations were obtained from publications in peer-reviewed journals. It should be noted that, for this stage, the $tc_{d,T}$ used were those of the station closest to the one of interest, and not from the station itself.

For each station, the $I_{IDF\ d,T}$ were obtained using the parameters K, a, b, and c of Equation (1), for the combinations of durations (d) of 1440, 720, 360, 240, 180, 120, 60, 50, 40, 30, 20, 10 min with the return periods (T) of 2, 5 10, 20, 50, 100 years.

The performance of TMDP was assessed by comparing its results with the Rainfall Depth-Duration method (RDD), due to its wide application in Brazil [11] and the availability of its application in the same stations where new IDF equations were established via the RDD method. Therefore, the two methods that configured the estimated data (E_i) were evaluated: TMDP, using the $tc_{d,T}$ of the nearest support station, and RDD, applied with the disaggregation coefficients (dc) from Table 1, which are the most widely used in Brazil [16].

Table 1. Disaggregation coefficient (dc) for the transformation of maximum daily precipitation [12].

24 h/ 1 d	12 h/ 24 h	10 h/ 24 h	8 h/ 24 h	6 h/ 24 h	4 h/ 24 h	2 h/ 24 h	1 h/ 24 h	50 min/ 1 h	40 min/ 1 h	30 min/ 1 h	20 min/ 30 min	10 min/ 30 min	5 min/ 30 min
1.14	0.85	0.82	0.78	0.72	0.54	0.48	0.42	0.74	0.91	0.81	0.7	0.54	0.34

The reference data was from the IDF equations of each location made from observed sub-daily precipitation data. The local IDF equation (using the parameters K, a, b, and c) was applied for each duration (d) and return period (T), to obtain the average maximum intensity for each d and T ($I_{IDF\ d,T}$). The estimates of $I_{IDF\ d,T}$ at each station of interest, obtained by the alternative methods (TMDP and RDD), were considered for comparison with the reference data, to verify the performance and applicability of these methods.

The equivalence between the estimated and the reference $I_{IDF\ d,T}$ was performed using Student's t-test for the intercept (β_0) and angular coefficient (β_1) parameters of the linear regression ($Y_j = \beta_0 + \beta_1 Y_1 + \varepsilon i$) between the data from the original IDF equations (Y_1) and the maximum rainfall intensities estimated with the application of alternative methods (Y_j): TMDP method ($\hat{I_{d,T}}_{\ TMDP}$) and the RDD method ($\hat{I_{d,T}}_{\ RCDD}$). The Student's t-test was performed to test the null hypotheses (H_0) that $\beta_1 = 1$ and $\beta_0 = 0$. When the regression line has an intercept equal to zero and an angular coefficient equal to 1, it passes through the origin (point: 0, 0), defining a line with 45°, which guarantees an equivalent rate of modification of Y_j with the modification of Y_1. Therefore, equivalence is assumed between the reference data and the estimated one.

In addition to equivalence, fit and error statistics such as root mean square error (*RMSE*, in mm/h), mean percentage error (*MPE*, in percentage), coefficient of determination (R^2), and Kling–Gupta efficiency (*KGE*) were calculated. The *hydroGOF* package [32] was used to obtain the fit statistics and error indices through the *gof* function in the R environment [33]. The *RMSE*, *MPE*, R^2, and *KGE* equations are available in Equations (4), (5), (6), and (7), respectively, in which, n is the number of observations; O_i is the maximum intensity in the i-th reference observation; $\overline{O}$ is the average of the reference values of maximum precipitation intensity; E_i is the maximum intensity in the i-th observation estimated by each alternative method; $\overline{E}$ is the average of the estimated values of maximum precipitation intensity; r is the Pearson correlation coefficient; σ is the standard deviation; μ is the average of the simulation/bias term.

$$RMSE = \sqrt{n^{-1}\sum_{i=1}^{n}(O_i - E_i)^2} \tag{4}$$

$$MPE = \frac{\sum_{i=1}^{n}\frac{(E_i - O_i)}{O_i}100}{n} \tag{5}$$

$$R^2 = \frac{\sum_{i=1}^{n}(O_i - \overline{O})(E_i - \overline{E})}{\left\{\left[\sum_{i=1}^{n}(O_i - \overline{O})^2\right]\left[\sum_{i=1}^{j}(E_i - \overline{E})^2\right]\right\}^{0,5}} \tag{6}$$

$$KGE = 1 - \sqrt{\left\{(r-1)^2 + \left(\frac{\sigma O_i}{\sigma E_i} - 1\right)^2 + \left(\frac{O_i}{E_i} - 1\right)^2\right\}} \tag{7}$$

The performance parameters used aim to analyze the precision and accuracy of the estimates of $I_{IDF\ d,T}$ by the proposed method and the most common alternative method. The R^2 is a measure of precision (repeatability or reproducibility of results) and its values range from 0 to 1, and the closer to the unity, the greater the precision of the estimates. The *KGE* provides an interesting diagnostic decomposition of the Nash-Sutcliffe efficiency, facilitating the analysis of the relative importance of its different components: correlation, bias, and variability within hydrological modeling. The error indices (*RMSE* and *MPE*) indicate the magnitude of the errors on average, with the lowest values of these indices indicating a smaller deviation between observed and estimated data of $I_{IDF\ d,T}$.

3. Results and Discussion

3.1. $tc_{d,T}$ *for Brazil*

From 373 IDF equations available in the literature, the $tc_{d,T}$ was adjusted in 301 stations. In the remaining stations, there was no daily rainfall data series available coinciding with the period used to adjust the IDF equation as well as incomplete and gapped series. Therefore, in 72 stations, the method's performance was not assessed. Figure 3 illustrates an example of the values of one of the 96 $tc_{d,T}$ adjusted for the transformation of maximum daily precipitation (in mm) into maximum rainfall intensity (mm/h), with a duration of 10 min and a return period of 2 years ($tc_{10,2}$).

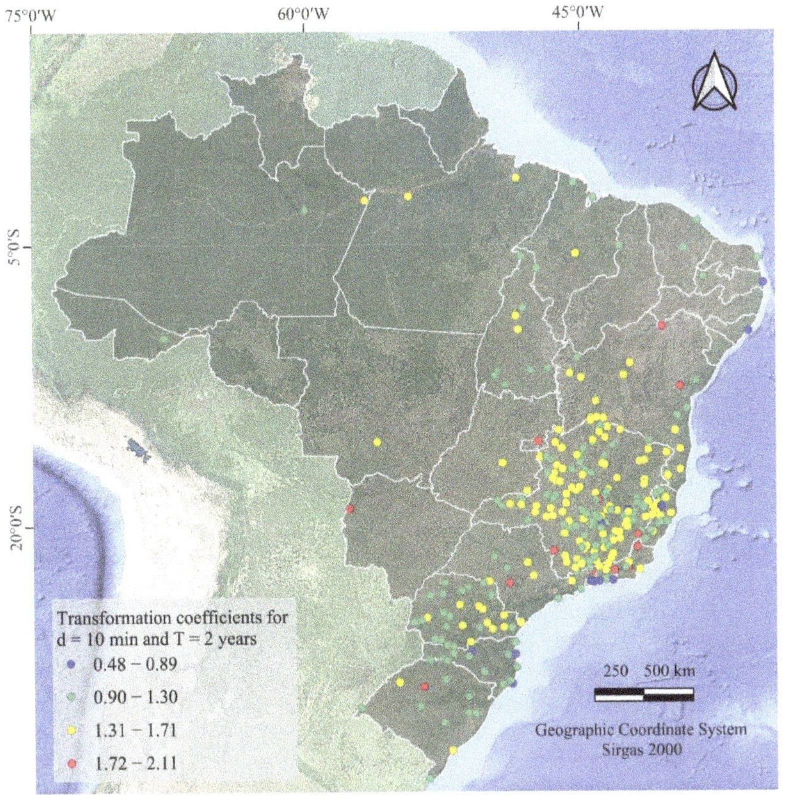

Figure 3. Transformation coefficients for converting maximum daily precipitation into average maximum rainfall intensity with a 10-min duration and a 2-year return period ($tc_{10,2}$).

3.2. *Application of TMDP*

3.2.1. $h_{day,T}$ Values

Regarding the PDFs used to establish $h_{day,T}$, in 35% of the adjustments, GEV was the most suitable, followed by LN2 (33%), GUM (14%), G (10%), and LN3 (8%). As in the present study, GEV has also been showing better performance in other Brazilian locations [22,34,35], proving to be highly regarded in studies related to heavy rainfall.

Figure 4 shows, as an example, only the magnitudes of the maximum daily precipitation $h_{day,T}$ related to the 10-year return period. $h_{day,T}$ of smaller magnitudes, for all combinations of d and T is found mainly in the Northeast Region, in the North of Minas Gerais, and in the Central-West Region. The Brazilian Northeast has semi-arid regions caused by cyclonic air vortices due to the interplateau relief (which hinders the circulation of moist air masses), high temperatures, and low air humidity [36], justifying the lower values. The North of Minas Gerais and the Central-West region have low rainfall rates and higher temperatures, with semi-arid and sub-humid regions [37,38].

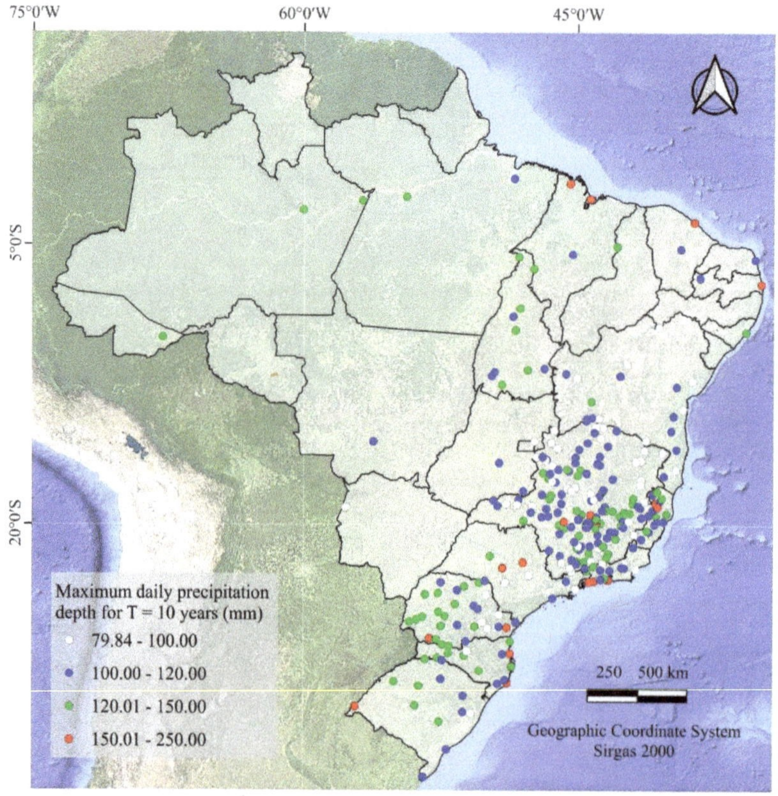

Figure 4. Spatialization of the maximum probable daily precipitation depth for a 10-year return period.

Despite the absence of a well-defined spatial pattern, the largest $h_{day,T}$ are mainly in much of the coastal strip in the Southeast Region and the extreme south (Figure 4). The orographic effect in mountain ranges with orientation parallel to the coast, such as the Serra do Mar, allows the occurrence of orographic effects in the coastal strip facing the Atlantic Ocean, increasing the probability of extreme events associated with ocean humidity [35]. The coastal region of the Northeast presents larger rainfall volumes [39], with rainfall better distributed spatially in the states to the north of the region, such as Piauí and Maranhão [36], and the precipitation is related to the intertropical convergence zone, anticyclones, and frontal systems.

In the Southeast region, high magnitudes of $h_{day,T}$ happen close to large urban centers such as São Paulo and Belo Horizonte, causing increasingly frequent flooding [40]. In the metropolis of São Paulo, extreme rainfall events are linked to the heat island effect and the local influx of sea breezes [41]. Belo Horizonte suffered great losses caused by the extreme rainfall of 2020, which was attributed to climate change induced by human actions [42].

The extreme south of the country is influenced by mesoscale convective events, formed in part of southern Brazil during the spring, summer, and autumn months, and cold fronts, which are responsible for the occurrence of strong storms, with gusts of wind and highly concentrated rainfall [43].

Furthermore, climate change studies project that rainfall will be increasingly concentrated in the rainy season, and the dry season will increase in regions with well-defined seasons (rainy and dry) [40]. The reduction in projected rainfall occurs mainly in the semi-arid region and its surroundings, reaching a reduction of up to 12% [44]. The increase in maximum daily precipitation is observed in the coastal strip and has already reached around 30% in São Paulo [40,45], due to the associated effects of climate change, urbanization [40], growth of the urban heat island, and air pollution [45].

3.2.2. TMDP and RDD Performances

From the 301 stations analyzed, only 293 (Figures 2 and 3) were able to adjust the IDF equation using the assessed TMDP method. In 8 stations, the adjustment was not performed for one of the following reasons: some data series used to adjust the IDF equation were short; the supporting $tc_{d,T}$ was not representative in predicting maximum rainfall at these stations, making it impossible to adjust the IDF curve. Alternative methods to the use of sub-daily series do not replace the need for a density of rainfall measurement stations with consistent series [16]. The stations without equation adjustment are mostly in regions with low or medium station density.

The equivalence hypothesis (1:1 relationship) between the results of the reference equations and those adjusted by the TMDP and RDD methods was rejected for the vast majority of stations (98%). Therefore, the equations adjusted by the methods are statistically different from the original IDF equations. This pattern was also observed in Minas Gerais when applying RDD [16].

Figure 5 shows the scatter plots between the maximum rainfall intensity values observed and estimated using the IDF equations established by the two methods. The stations are grouped as a whole (Figure 5a,b) and within each station density class: HD (Figure 5c,d), MD (Figure 5e,f), and LD (Figure 5g,h). It can be observed that the slope line is closer to unity for TMDP in all situations (general analysis and by region of application). In the HD (Figure 5c,d) and LD (Figure 5g,h) regions, these were practically equal to unity for the TMDP method, demonstrating the superiority of TMDP over RDD. The data dispersion (R^2) was similar for both methods.

The results of the indices of performance are shown in the notched box plots in Figure 6. Overall, the two methods (TMDP and RDD) perform similarly. The *RMSE* and *MPE* were very close among the alternative methods, with notched overlapping for most regions (except for the *RMSE* of the MD region). The *RMSE*, for instance, varied from 1 to 90 mm h^{-1}, with an approximate average of 17 mm h^{-1}. The TMDP performed slightly better than the RDD, according to the *KGE*, due to values closer to unity. Comparatively, the TMDP method had better overall performance. Also, in HD regions, the TMDP, compared to the RDD, the notches of the two box plots do not overlap, showing higher values for the TMDP. Regarding the regional behavior, both performed very similarly in the LD region, which is considered critical regarding the need for information on heavy rainfall. The better performance of TMDP in the HD region indicates a greater spatial dependence of the method's data for the transformation of maximum daily precipitation. The RDD had a slightly better performance in the MD region, possibly because the relations of the precipitation depths were more suitable for disaggregating the maximum annual daily precipitation layer. The relations of precipitation depths of different durations vary less across seasons. Also, the dc used in the RDD was established using mostly data from the Southeast region, especially from São Paulo [12], which may justify its better performance in the MD region.

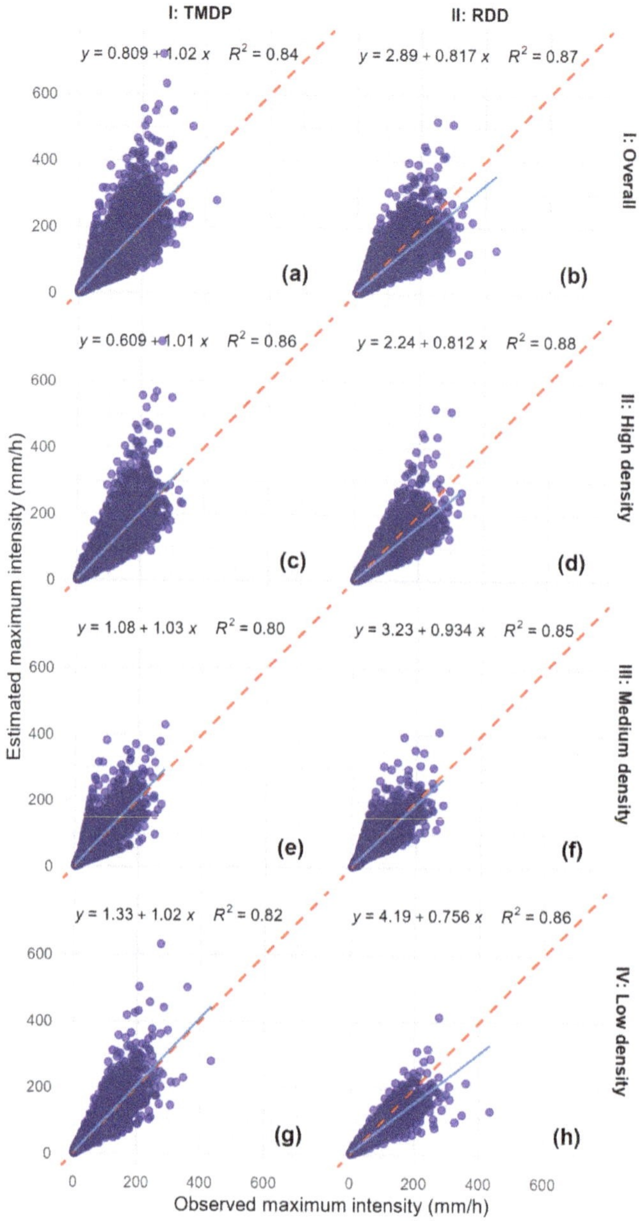

Figure 5. Scatter plot of rainfall intensity obtained by applying the original IDF equations and those established through TMDP (panels **a**,**c**,**e**,**g**) and RDD (panels **b**,**d**,**f**,**h**), considering the station groups: all stations (**a**,**b**), stations in the region with high density—HD (panels **c**,**d**), stations in the region with medium density—MD (panels **e**,**f**), and stations in the region with low density—LD (panels **g**,**h**).

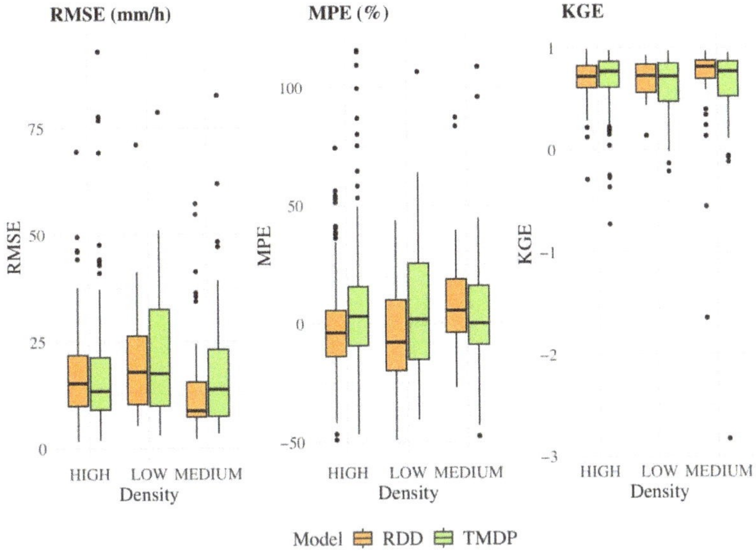

Figure 6. Results of the indices of performance, on average, for TMDP and RDD models.

The spatial analysis of the *MPE* (Figure 7) confirms the similarity between the TMDP and the RDD, with values ranging from −51 to 120%. MPE magnitudes between 6 and 70% were also found in other studies by [13,16,46] when evaluating the equations established with the RDD. Both methods overestimated and underestimated maximum intensities compared to the original IDF. The RDD method's errors are usually negative, therefore representing a tendency towards underestimation of this method. In the TMDP, overestimation tendencies were predominant, as shown by the *MPE* (Figure 6) and the regressions (Figure 5). The underestimation of $\hat{I_{d,T}}$, mostly due to RDD, is a restriction that must be addressed when using equations established by this method. This is because the design rainfall will be reduced and, consequently, the safety of large-scale hydraulic works, such as structures for flood control and/or water storage, will be reduced. This can be accomplished by adopting a longer return period [13].

Although subtle, the better performance of the proposed methodology (TMDP), especially in places with a high density of data on heavy rainfall (HD region), is likely due to a greater "possibility" that both the support and interest stations have similar patterns of heavy rainfall as a result of their proximity. In a way, this corroborates the initial premise of the TMDP method better representing the regional reality of heavy rainfall with the use of *tc*. In the LD region, the distance among the stations is greater and the possibility of alternative stations is reduced; consequently, this subjects the region to great differences in the characteristics of heavy rainfall in each station. Considering the great spatial variability of heavy rainfall, and its isolated eventuality [20,21,47], predicting this climate variable becomes quite hard with low information density and information that is very spatially distant from each other.

Another reason for TMDP performing better than RDD is the return period considered when establishing *tc*. Studies show that applying RDD without considering T can lead to worse estimates of $\hat{I_{d,T}}$ [14,16]. Figure 8 presents, for a specific return period, the relationship between rainfall intensities and their duration changes for different locations. When RDD is used to establish the equations, this relationship is kept the same regardless of the location, resulting in IDF curves that differ only in the magnitude of the events, that is, they are parallel to each other. Applying TMDP minimizes this error, resulting in curves that maintain the local specificities of the behavior of rainfall intensities as a function of their duration.

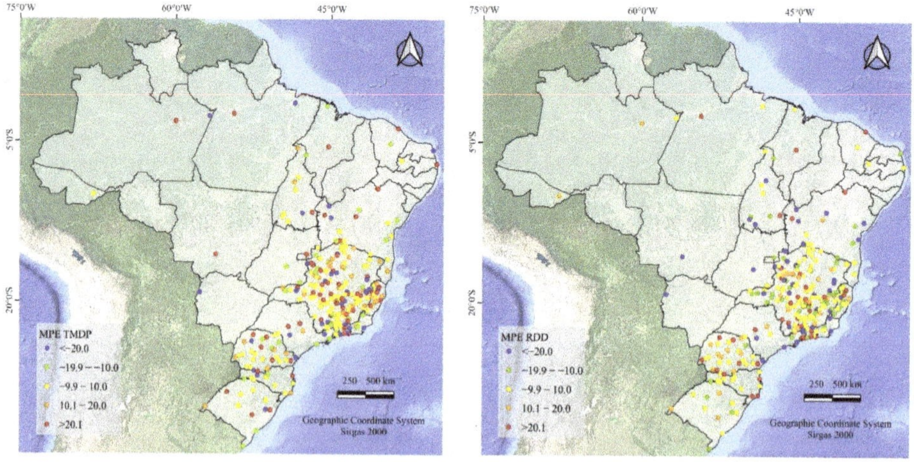

Figure 7. Mean Percentage Error (MPE) for estimating rainfall intensities caused by the application of the TMDP and RDD methods.

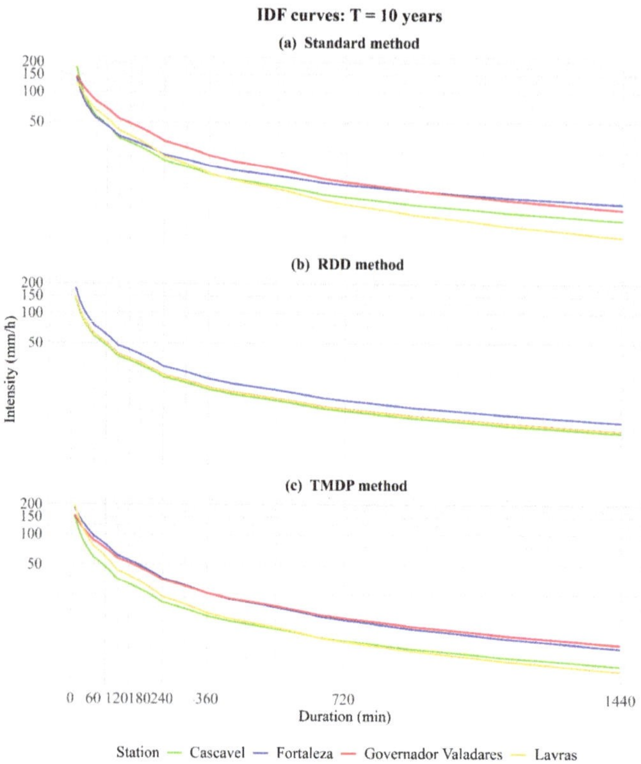

Figure 8. IDF curves for 5 different locations, and 10-year return periods: original equations (**a**), established through TMDP (**b**), and established through RDD (**c**). The intensity (y-axis) is presented on a logarithmic scale.

Among other advantages of the TMDP method, the *tc* directly transforms the maximum daily precipitation into maximum rainfall intensity associated with sub-daily durations, i.e., direct and easy application. The transformation coefficients are also not applied using the cascade effect of RDD, which potentially reduces estimation errors.

Even though small, the proposed TMDP method showed improvements compared to the RDD method. Furthermore, for this work to be even more applicable, and to corroborate the reliability of its application, it is recommended to continually define hydrologically homogeneous areas for each *tc*, since they have similar extreme rainfall patterns across seasons [8,14]. The estimation of extreme rainfall events plays a pivotal role in the design and safety assessment of hydraulic structures. Traditional IDF equations derived from long-term pluviographic records are often considered the most reliable source of input data for such purposes [11]. The estimation of IDF equations through alternative methods poses inherent challenges, as statistical equivalence between models derived from pluviographic records and those obtained from alternative approaches is rarely observed. Such dissonance stems from the limited availability of rainfall data and the complex dynamics governing extreme precipitation processes, particularly due to the pronounced spatial and temporal variability [16,22,40]. While these alternative approaches expand spatial coverage, the TMDP often tends to overestimate rainfall intensities when compared to local observational data. From a conservative engineering perspective, such overestimation may be perceived as a safety-enhancing factor. Nevertheless, systematic overestimation can lead to unnecessarily oversized infrastructure, resulting in inflated construction and maintenance costs, as well as potential inefficiencies in resource allocation. This raises the question of proportionality between the safety margin and the economic feasibility of the project. One pragmatic solution proposed in the literature is the adjustment of the design return period when using alternative estimation methods. The implementation of this adjustment requires a prior and rigorous assessment of potential systematic errors associated with the proposed estimation methods, as such biases directly influence the reliability and applicability of the adjusted design parameters [13].

Regionalization based on hydrometeorological information, for example, can help to identify regions with similar characteristics of intense rainfall [48], which can reduce statistical errors. Future studies testing the proposed method may benefit from incorporating cluster analysis that extends beyond the mere geographical proximity of rain gauges, thereby evaluating the potential similarity in terms of extreme rainfall or broader climatic characteristics. Although the ideal scenario for such an approach involves the use of time series that are temporally consistent and comparable in length, this is not possible in the Brazilian scenario. This condition is essential to mitigate the risk of identifying dissimilarities that stem solely from temporal inconsistencies rather than genuine differences in rainfall behavior. Addressing these considerations could enhance the robustness and applicability of the method across diverse climatic and spatial contexts. Anyway, some studies have demonstrated a similarity in the magnitudes of extreme rainfall events associated with geographical proximity, which may explain the promising performance of the proposed method [28,41].

4. Conclusions

The method proposed in this work, entitled Transformation of the maximum daily precipitation layer (TMDP), was shown to be capable of transforming the maximum daily precipitation layer into maximum sub-daily rainfall intensities. To this end, it was based on the use of known intense rainfall equations and the maximum daily precipitation depths to establish transformation coefficients associated with different durations and return periods ($tc_{d,T}$).

By applying TMDP to Brazil, it was possible to adjust 301 $tc_{d,T}$ obtained from rainfall stations with a known IDF equation. The performance of TMDP was evaluated by establishing IDF equations using the closest station as a support station for locations that already have an IDF. Its performance was also compared to the disaggregation method Rainfall Depth-Duration (RDD), the most widely used in Brazil.

The results showed that TMDP and RDD showed similar errors in the estimation of the average maximum rainfall intensity, ranging from 1 to 90 mm h^{-1}, and averages of 17 mm h^{-1}. TMDP was better in the global analysis and, especially, in the regions with the highest density of IDF equations.

Author Contributions: Conceptualization, R.A.C. and L.T.d.A.; methodology, L.T.d.A., R.A.C., M.C.A. and I.P.T.; validation, L.T.d.A. and I.P.T.; formal analysis, L.T.d.A., R.A.C., M.C.A. and I.P.T.; writing—original draft preparation, L.T.d.A., R.A.C., M.C.A. and I.P.T.; writing—review and editing, L.T.d.A., R.A.C., M.C.A. and I.P.T.; visualization, L.T.d.A. and I.P.T.; supervision, R.A.C.; funding acquisition, R.A.C. All authors have read and agreed to the published version of the manuscript.

Funding: This research was funded by the Coordenação de Aperfeiçoamento de Pessoal de Nível Superior—CAPES [grant number 001]; the Conselho Nacional de Desenvolvimento Científico e Tecnológico—CNPq [grant number 305262/2021–1]; and the Fundação de Amparo à Pesquisa e Inovação do Espírito Santo–FAPES [grant number 996/2022].

Data Availability Statement: The original contributions presented in this study are included in the article. Further inquiries can be directed to the corresponding author.

Conflicts of Interest: The authors declare no conflicts of interest.

References

1. Moura Cardoso do Vale, T.; Helena Constantino Spyrides, M.; De Melo Barbosa Andrade, L.; Guedes Bezerra, B.; Evangelista da Silva, P. Subsistence Agriculture Productivity and Climate Extreme Events. *Atmosphere* **2020**, *11*, 1287. [CrossRef]
2. Wang, J.; Zhang, Y.; Li, K.; Zhang, Z.; Chen, C. Gully Internal Erosion Triggered by a Prolonged Heavy Rainfall Event in the Tableland Region of China's Loess Plateau. *Int. Soil Water Conserv. Res.* **2023**, *11*, 610–621. [CrossRef]
3. Chhetri, T.B.; Dhital, Y.P.; Tandong, Y.; Devkota, L.P.; Dawadi, B. Observations of Heavy Rainfall and Extreme Flood Events over Banke-Bardiya Districts of Nepal in 2016–2017. *Prog. Disaster Sci.* **2020**, *6*, 100074. [CrossRef]
4. Ma, S.; Shao, X.; Xu, C. Landslides Triggered by the 2016 Heavy Rainfall Event in Sanming, Fujian Province: Distribution Pattern Analysis and Spatio-Temporal Susceptibility Assessment. *Remote Sens.* **2023**, *15*, 2738. [CrossRef]
5. Prokop, P.; Wiejaczka, Ł.; Sarkar, S.; Bryndal, T.; Bucała-Hrabia, A.; Kroczak, R.; Soja, R.; Płaczkowska, E. Morphological and Sedimentological Responses of Small Stream Channels to Extreme Rainfall and Land Use in the Darjeeling Himalayas. *CATENA* **2020**, *188*, 104444. [CrossRef]
6. Fekete, A.; Sandholz, S. Here Comes the Flood, but Not Failure? Lessons to Learn after the Heavy Rain and Pluvial Floods in Germany 2021. *Water* **2021**, *13*, 3016. [CrossRef]
7. Martel, J.-L.; Brissette, F.P.; Lucas-Picher, P.; Troin, M.; Arsenault, R. Climate Change and Rainfall Intensity–Duration–Frequency Curves: Overview of Science and Guidelines for Adaptation. *J. Hydrol. Eng.* **2021**, *26*, 03121001. [CrossRef]
8. Yang, Z.; Burn, D.H. Automatic Feature Selection and Weighting for the Formation of Homogeneous Groups for Regional IDF Estimation. *J. Hydrol.* **2019**, *575*, 292–307. [CrossRef]
9. Koutsoyiannis, D.; Kozonis, D.; Manetas, A. A Mathematical Framework for Studying Rainfall Intensity-Duration-Frequency Relationships. *J. Hydrol.* **1998**, *206*, 118–135. [CrossRef]
10. World Meteorological Organization. *Guide to Hydrological Practices: Hydrology from Measurement to Hydrological Information (WMO-No. 168)*; World Meteorological Organization: Geneva, Switzerland, 2008; ISBN 978-92-63-10168-6.
11. Torres, I.P.; Cecílio, R.A.; de Almeida, L.T.; Abreu, M.C.; da Silva, D.D.; Zanetti, S.S.; Xavier, A.C. Rainfall Intensity–Duration–Frequency Curves Dataset for Brazil. *Data* **2025**, *10*, 17. [CrossRef]
12. COMPANHIA DE TECNOLOGIA DE SANEAMENTO AMBIENTAL. *Drenagem Urbana: Manual de Projeto*, 3rd ed.; CETESB: São Paulo, Brazil, 1986.

13. Carvalho Abreu, M.; Thebit de Almeida, L.; Bernardes Silva, F.; de Souza Fraga, M.; Bastos Lyra, G.; Souza, A.; da Silva, D.D.; Avelino Cecílio, R. Disaggregation Coefficients for Obtaining Rainfall Intensity-Duration-Frequency Curves: Concepts, Models, Errors and Trends in Minas Gerais, Brazil. *Urban Water J.* **2023**, *20*, 1647–1660. [CrossRef]
14. Passos, J.B.d.M.C.; da Silva, D.D.; Lima, R.P.C. Daily Rainfall Disaggregation Coefficients for the Doce River Basin, Brazil: Regional Applicability and the Return Period Influence. *Eng. Agrícola* **2021**, *41*, 223–234. [CrossRef]
15. de Bodas Terassi, P.M.; Pontes, P.R.M.; Xavier, A.C.F.; Cavalcante, R.B.L.; de Oliveira Serrão, E.A.; Sobral, B.S.; de Oliveira-Júnior, J.F.; de Melo, A.M.Q.; Baratto, J. A Comprehensive Analysis of Regional Disaggregation Coefficients and Intensity-Duration-Frequency Curves for the Itacaiúnas Watershed in the Eastern Brazilian Amazon. *Theor. Appl. Climatol.* **2023**, *154*, 863–880. [CrossRef]
16. Abreu, M.C.; Cecílio, R.A.; Pruski, F.F.; de Almeida, L.T.; dos Santos, G.R.; Zanetti, S.S.; Pereira, S.B.; Silva, D.D. Daily Rainfall Disaggregation to Estimate the Intensity-Duration-Frequency Relationship in Minas Gerais State, Brazil. *Braz. Arch. Biol. Technol.* **2022**, *65*, e20210694. [CrossRef]
17. Damé, R.d.C.F.; Teixeira-Gandra, C.F.A.; Villela, F.A.; dos Santos, J.P.; Winkler, A.S. Analysis of the Relationship Intensity, Duration, Frequency of Disaggregated Daily Rainfall in Southern Rio Grande Do Sul, Brazil. *Eng. Agrícola* **2014**, *34*, 660–670. [CrossRef]
18. Back, Á.J.; Pola, A.C. Relações Entre Precipitações Intensas de Diferentes Durações de Videira, Santa Catarina. *Tecnol. e Ambient.* **2016**, *22*, 125–135. [CrossRef]
19. Dorneles, V.R.; Damé, R.d.C.F.; Teixeira-Gandra, C.F.A.; Méllo, L.B.; Ramirez, M.A.A.; Manke, E.B. Intensity-Duration-Frequency Relationships of Rainfall through the Technique of Disaggregation of Daily Rainfall. *Rev. Bras. Eng. Agrícola e Ambient.* **2019**, *23*, 506–510. [CrossRef]
20. dos Anjos, R.S.; Nóbrega, R.S.; de Araújo, F.E.; Rocha Filho, G.B. Spacial Distribution of Rain Types in Pernambuco with the Usage of Remote Sensing. *J. Hyperspectr. Remote Sens.* **2016**, *6*, 154–163. [CrossRef]
21. Szwed, M. Variability of Precipitation in Poland under Climate Change. *Theor. Appl. Climatol.* **2019**, *135*, 1003–1015. [CrossRef]
22. Abreu, M.C.; Cecílio, R.A.; Pruski, F.F.; dos Santos, G.R.; De Almeida, L.T.; Zanetti, S.S. Critérios Para Escolha de Distribuições de Probabilidades Em Estudos de Eventos Extremos de Precipitação. *Rev. Bras. Meteorol.* **2018**, *33*, 601–613. [CrossRef]
23. de Mello, C.R.; Viola, M.R.; de Mello, J.M.; Silva, A.M. Continuidade Espacial de Chuvas Intensas No Estado de Minas Gerais. *Ciência e Agrotecnol.* **2008**, *32*, 532–539.
24. Back, Á.J.; Henn, A.; Oliveira, J.L.R. Heavy Rainfall Equations for Santa Catarina, Brazil. *Rev. Bras. Ciência do Solo* **2011**, *35*, 2127–2134. [CrossRef]
25. Denardin, J.E.; Freitas, P.L. de Características Fundamentais Da Chuva No Brasil. *Pesqui. Agropecuária Bras.* **1982**, *10*, 1409–1416.
26. Fendrich, R. *Chuvas Intensas Para Obras de Drenagem (No Estado Do Paraná)*, 1st ed.; Champagnat: Curitiba, Brazil, 1998.
27. Freitas, A.J.; da Silva, D.D.; Pruski, F.F.; Pinto, F.A.; Pereira, S.B.; Gomes Filho, R.R.; Teixeira, A.F.; Baena, L.G.N.; Mello, L.T.A.; Novaes, L.F.d. *Equações de Chuvas Intensas No Estado de Minas Gerais*, 1st ed.; Companhia de Saneamento de Minas Gerais: Belo Horizonte, Brazil, 2001.
28. da Silva, D.D.; Pinto, F.R.L.; Pruski, F.F.; Pinto, F.A. Estimativa e Espacialização Dos Parâmetros Da Equação de Intensidade-Duração-Freqüência Da Precipitação Para o Rio de Janeiro e o Espírito Santo. *Eng. Agrícola* **1999**, *19*, 11–21.
29. da Silva, D.D.; Gomes Filho, R.R.; Pruski, F.F.; Pereira, S.B.; Novaes, L.F. de Chuvas Intensas No Estado Da Bahia. *Rev. Bras. Eng. Agrícola e Ambient.* **2002**, *6*, 362–367. [CrossRef]
30. da Silva, D.D.; Pereira, S.B.; Pruski, F.F.; Gomes Filho, R.R.; Lana, Â.M.Q.; Baena, L.G.N. Equações de Intensidade-Duração-Freqüência Da Precipitação Pluvial Para o Estado de Tocantins. *Eng. na Agric.* **2003**, *11*, 7–14.
31. Agência Nacional de Águas Sistema de Informação Hidrológicas—Hidroweb. Available online: https://www.snirh.gov.br/hidroweb/ (accessed on 3 February 2020).
32. Zambrano-Bigiarini, M. hydroGOF: Goodness-of-Fit Functions for Comparison of Simulated and Observed Hydrological Time Series; R Package Version 0.3-10. 2017. Available online: https://cran.r-project.org/ (accessed on 28 August 2021).
33. R Core Team. *R: A Language and Environment for Statistical Computing*; R Foundation for Statistical Computing: Vienna, Austria, 2018.
34. Junqueira Júnior, J.A.; de Mello, C.R.; Alves, G.J. Eventos Extremos de Precipitação No Alto Rio Grande, MG: Análise Probabilística. *Rev. Bras. Eng. Agrícola e Ambient.* **2015**, *19*, 301–308. [CrossRef]
35. Lima, A.O.; Lyra, G.B.; Abreu, M.C.; Oliveira-Júnior, J.F.; Zeri, M.; Cunha-Zeri, G. Extreme Rainfall Events over Rio de Janeiro State, Brazil: Characterization Using Probability Distribution Functions and Clustering Analysis. *Atmos. Res.* **2021**, *247*, 105221. [CrossRef]
36. Silva, C.B.; Oliveira, L.F.C. Relação Intensidade-Duração-Frequência de Chuvas Extremas Na Região Nordeste Do Brasil. *Rev. Bras. Climatol.* **2017**, *20*, 267–283. [CrossRef]

37. Martins, F.B.; Gonzaga, G.; dos Santos, D.F.; Reboita, M.S. Classificação Climática de Köppen e de Thornthwaite Para Minas Gerais: Cenário Atual e Projeções Futuras. *Rev. Bras. Climatol.* **2018**, *1*, 129–156. [CrossRef]
38. Reboita, M.S.; Rodrigues, M.; Silva, L.F.; Alves, M.A. Aspectos Climáticos Do Estado de Minas Gerais. *Rev. Bras. Climatol.* **2015**, *17*, 209–229. [CrossRef]
39. Back, Á.J. Alternative Model of Intense Rainfall Equation Obtained from Daily Rainfall Disaggregation. *RBRH* **2020**, *25*, e2. [CrossRef]
40. Marengo, J.; Valverde, M.; Obregon, G. Observed and Projected Changes in Rainfall Extremes in the Metropolitan Area of São Paulo. *Clim. Res.* **2013**, *57*, 61–72. [CrossRef]
41. Vemado, F.; Pereira Filho, A.J. Severe Weather Caused by Heat Island and Sea Breeze Effects in the Metropolitan Area of São Paulo, Brazil. *Adv. Meteorol.* **2016**, *2016*, 8364134. [CrossRef]
42. Dalagnol, R.; Gramcianinov, C.B.; Crespo, N.M.; Luiz, R.; Chiquetto, J.B.; Marques, M.T.A.; Neto, G.D.; de Abreu, R.C.; Li, S.; Lott, F.C.; et al. Extreme Rainfall and Its Impacts in the Brazilian Minas Gerais State in January 2020: Can We Blame Climate Change? *Clim. Resil. Sustain.* **2022**, *1*, e15. [CrossRef]
43. Sanches, F.; Verdum, R.; Fisch, G.; Gass, S.L.B.; Rocha, V.M. Extreme Rainfall Events in the Southwest of Rio Grande Do Sul (Brazil) and Its Association with the Sandization Process. *Am. J. Clim. Chang.* **2019**, *08*, 441–453. [CrossRef]
44. Pousa, R.; Costa, M.H.; Pimenta, F.M.; Fontes, V.C.; de Brito, V.F.A.; Castro, M. Climate Change and Intense Irrigation Growth in Western Bahia, Brazil: The Urgent Need for Hydroclimatic Monitoring. *Water* **2019**, *11*, 933. [CrossRef]
45. Silva Dias, M.A.F.; Dias, J.; Carvalho, L.M.V.; Freitas, E.D.; Silva Dias, P.L. Changes in Extreme Daily Rainfall for São Paulo, Brazil. *Clim. Change* **2013**, *116*, 705–722. [CrossRef]
46. Back, Á.J.; Oliveira, J.L.R.; Henn, A. Relações Entre Precipitações Intensas de Diferentes Durações Para Desagregação Da Chuva Diária Em Santa Catarina. *Rev. Bras. Eng. Agrícola e Ambient.* **2012**, *16*, 391–398. [CrossRef]
47. Manke, E.B.; Teixeira-Gandra, C.F.A.; Damé, R.d.C.F.; Nunes, A.B.; Neta, M.C.C.C.; Karsburg, R.M. Seasonal Intensity-Duration-Frequency Relationships for Pelotas, Rio Grande Do Sul, Brazil. *Rev. Bras. Eng. Agrícola e Ambient.* **2022**, *26*, 85–90. [CrossRef]
48. Goyal, M.K.; Shivam, G.; Sarma, A.K. Spatial Homogeneity of Extreme Precipitation Indices Using Fuzzy Clustering over Northeast India. *Nat. Hazards* **2019**, *98*, 559–574. [CrossRef]

hydrology MDPI

Article

Assessment of Satellite Precipitation Products in an Andean Catchment: Ambato River Basin, Ecuador

Pablo Arechúa-Mazón, César Cisneros-Vaca *, Julia Calahorrano-González and Mery Manzano-Cepeda

Facultad de Ingeniería, Universidad Nacional de Chimborazo (UNACH), Riobamba 060110, Ecuador; pablo.arechua@unach.edu.ec (P.A.-M.); jcalahorrano@unach.edu.ec (J.C.-G.); mmanzano@unach.edu.ec (M.M.-C.)
* Correspondence: ramiro.cisneros@unach.edu.ec

Abstract

Accurate precipitation data are essential for hydrological planning in mountainous regions with sparse opportunities for observation, such as the Ambato River basin in Ecuador. In this study, CHIRPS and IMERG satellite precipitation products were compared against six automatic rain gauges from 2014 to 2023, using both categorical metrics (to assess daily rainfall detection skill) and continuous validation (to evaluate rainfall amount), complemented by bias decomposition and spatiotemporal analysis. Our results show that IMERG demonstrated higher skill in detecting daily rainfall, while CHIRPS delivered a more stable performance during dry conditions, with fewer false alarms. Both products capture the main seasonal precipitation patterns but differ in bias behavior: CHIRPS tends to under-estimate daily rainfall less, whereas IMERG provides more reliable volumetric estimates overall. These findings suggest that IMERG may be best suited for flood risk and hydrological modelling, while CHIRPS could be preferred for drought monitoring and climatological studies in Andean catchments.

Keywords: satellite precipitation products; rainfall validation; bias decomposition; Andean catchments

1. Introduction

Climate change has significantly altered rainfall regimes at local, regional, and global scales [1]. These alterations manifest in the distribution and intensity of precipitation, directly affecting ecosystem processes, increasing environmental vulnerability, and impacting human activities [2,3]. Accurate and reliable precipitation estimation, as a key meteorological variable, is therefore essential for understanding and predicting these climate variations [4].

Historically, ground-based meteorological stations have been the sole source of precipitation data for studies and forecasting. However, the scarcity of monitoring stations, limited accessibility, inconsistent data quality, lack of extreme event records, and low spatial and temporal coverage have constrained the availability of reliable data for integrated water resources management [5–8].

Currently, the availability and continuity of satellite-based precipitation records are transforming the way scientific and societal issues related to this variable are addressed, due to their rapid and easy accessibility through various data portals [9]. Several climate datasets have been developed at different spatiotemporal scales based on in situ observations. For instance, the Global Historical Climatology Network (GHCN) is an integrated database with approximately 31,000 stations and observations spanning the entire 20th century [10]. Satellite-based algorithms have enabled significant advances in estimating climate variables such as precipitation, temperature, and albedo, especially in data-scarce regions. However, these are indirect estimates that rely on cloud-top properties, utilizing infrared (IR) algorithms, and cloud liquid and ice content, utilizing polar microwave (PM) algorithms [11]

Quantifying the uncertainty and accuracy associated with each estimate permits users to infer the reliability of satellite precipitation products (SPPs) and improve their applicability across disciplines. Errors are assessed through validation studies that compare satellite-based precipitation estimates with ground truth, using rain gauges or radar observations [12].

Integrated Multi-satellite Retrievals for GPM (IMERG) [13] and Climate Hazards Group InfraRed Precipitation with Station data (CHIRPS) are two widely used SPPs [14]. These tools have proven valuable for supplementing sparse ground station networks, especially in remote or poorly monitored areas, even though each product has advantages and limitations depending on its spatial and temporal resolution. Thus, combining the strengths of each product can enhance their overall performances [15,16]. CHIRPS is a high-resolution, quasi-global precipitation dataset spanning over 35 years, developed by the Climate Hazards Group at the University of California, Santa Barbara. It merges infrared satellite data with ground-based observations to generate accurate precipitation estimates in regions with limited station coverage [14]. IMERG, on the other hand, is a global precipitation dataset produced through the integration of multiple satellite sensors, developed as part of the Global Precipitation Measurement (GPM) mission, which is a collaboration between NASA and the Japan Aerospace Exploration Agency (JAXA) [13].

These products have been evaluated in various studies across different regions and scales. In China, daily and annual precipitation from nine satellite and reanalysis datasets were assessed between 2000 and 2020. The results indicated that CHIRPS performed best in the arid and semi-arid areas of the study region [17].

In South America, Benitez et al. [18] assessed the ability of satellite data to estimate rainfall across the Southeastern South America (SESA) region, covering Argentina, Uruguay, Brazil, and Paraguay. They compared four satellite products—IMERG V.06 Final Run, PERSIANN, PERSIANN CCS-CDR, and PDIR-NOW—against observations from 118 meteorological stations from 2001 to 2020. The results showed that IMERG and CCS-CDR better captured precipitation patterns at annual and seasonal scales. All products had difficulties estimating winter and summer rainfall, and accuracy was generally higher in humid regions and lower in dry zones. Although IMERG showed the highest accuracy (with correlation coefficients up to 0.95), all products struggled to capture extreme events and data-sparse areas.

In Bolivia, Mattos et al. [19] compared CHIRPS and IMERG satellite estimates with rain gauge measurements from 2002 to 2020 across three distinct topographic zones—highlands, valleys, and lowlands. IMERG performed better in detecting precipitation, particularly in the highlands, whereas CHIRPS provided more accurate precipitation measurements across all regions, with lower random errors and relative biases below 10%. IMERG, however, tended to overestimate rainfall, reaching up to 75% overestimation in the Bolivian Altiplano [19].

In Perú, Salas et al. [20] evaluated the performance of GPM IMERG_F v6 by comparing monthly satellite rainfall estimates with data from eight SENAMHI stations in the Madre de Dios basin. Using statistical techniques such as Nash–Sutcliffe Efficiency (NSE) and Root Mean Square Error (RMSE), the authors found that, while satellite data are generally useful, they can exhibit over- or underestimation depending on the season, with an average RMSE of 146.48 mm. In some cases, such as the Iñapari station, NSE values fell below 0.65, indicating poor model performance in specific contexts. The study also highlights the benefits of combining satellite data with data from other sources to optimize water management and evaluate climate phenomena in areas with limited infrastructure.

In Ecuador, Ballari et al. [21] validated monthly precipitation estimates from the TRMM 3B43 product using 14 rain gauges across Ecuador's coastal plains, the Andes, and the Amazon. The study underlined the importance of rainfall monitoring for water availability, especially in mountainous areas with sparse measurements. Results showed that TRMM 3B43 captured precipitation seasonality well, with better performance in coastal and Amazon regions than in the Andes. The authors recommended further calibration and bias correction for satellite data in ungauged catchments.

Recently, Huber et al. [22] compared gridded global datasets of temperature and precipitation (IMERG, CHIRPS, ERA5, GLDAS) with data from a local automatic weather station network initiated in 2014. The findings emphasize that, while satellite products can help fill data gaps, their performance varies depending on location and precipitation type. IMERG and CHIRPS were found to be more effective under certain conditions, yet the lack of in situ data in regions such as the Amazon limits the accuracy of satellite estimates. Expanding the ground station network is therefore essential in improving validation and supporting water planning and climate monitoring in Ecuador.

The above studies demonstrate that, while satellite precipitation products are valuable tools for rainfall estimation, their performances depend on factors such as topography, station density and distribution, climate conditions, cloud characteristics, and temporal resolution. Therefore, evaluating their performances in different geographical and hydrological contexts remains essential.

The Ambato River Basin, located in Ecuador's Sierra region, is part of the upper Pastaza River system. It plays a critical role in agricultural production in Tungurahua Province and is home to approximately 75% of the province's population. The basin faces high water demand and contributes significantly to downstream pollution, affecting neighboring catchments such as the Patate River and tributaries of the Pastaza [23]. In 2013, the provincial government of Tungurahua deployed a network of automatic weather stations to monitor key meteorological variables and provide reliable data for risk prevention and water planning. However, complex topography leads to high climatic variability, and the existing stations are not sufficient to fully capture local precipitation dynamics [24,25].

In this context, the present study aims to evaluate the accuracy of satellite precipitation products CHIRPS and IMERG in the Ambato River Basin. The goal is to assess detection capacity using categorical and continuous statistics, as well as to calculate accuracy indicators through bias decomposition metrics to characterize the spatiotemporal rainfall patterns in the basin. This will contribute to improving the quality of available data for integrated water resources management and enhance our understanding of SPP applicability in high Andean catchments.

2. Materials and Methods

2.1. Study Area

The Ambato River Basin is located in the Western Cordillera of the central Ecuadorian Andes, within the administrative territory of Ambato, Tungurahua Province. It forms part of the upper Pastaza River basin system, which ultimately drains into the Amazon River. Covering an approximate area of ~130,000 ha, this hydrological unit accounts for 38.4% of the total surface area of Tungurahua Province. It is bounded to the north by the Cutuchi

River Basin; to the south by the Pachanlica River Basin; to the east by the Cutuchi, Patate Rivers; and to the west by the Babahoyo and Yaguachi sub-basins [26] (Figure 1).

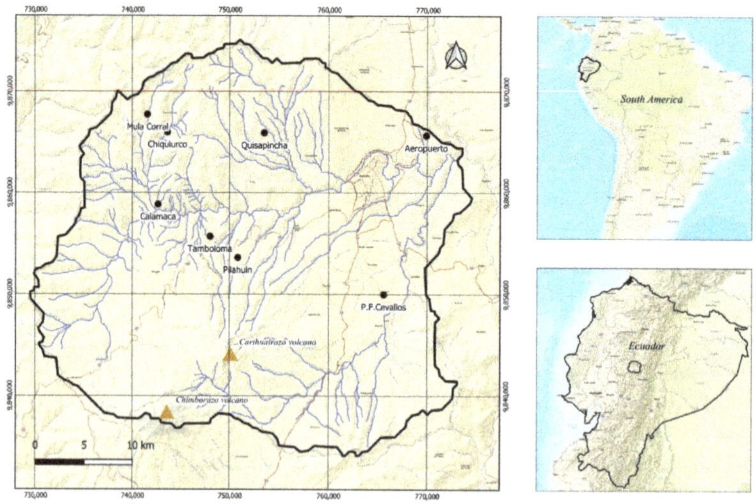

Figure 1. The map shows the geographical location of the Ambato River Basin within Tungurahua Province, in the central Ecuadorian Andes. Black dots represent ground rainfall gauging stations; orange triangles represent main volcano locations.

Climatically, the Ambato River Basin exhibits a predominantly temperate and dry mountain climate, with the presence of multiple microclimates being due to abrupt altitudinal gradients and varied topographic exposure [27]. According to WorldClim v2.1 [28], the basin receives relatively low annual precipitation, ranging from 600 to 1200 mm/year, and has mean annual temperatures between 8 °C and 14 °C, depending on elevation. At higher altitudes, conditions become markedly colder, with permanent snow cover on nearby peaks such as Carihuairazo and Chimborazo, especially over 4800 m a.s.l. Compared to adjacent regions, this basin is drier than the Amazonian foothills (>3000 mm/year) and the coastal plains (~1500 mm/year), highlighting the rain-shadow effect of the Andean cordillera. The temperature regime is relatively stable throughout the year, but the basin shows high interannual variability in precipitation.

Land cover in the basin is distributed as follows: 49% natural ecosystems, 44% agricultural land, 4% areas undergoing erosion processes, 1% water bodies, and 2% human settlements [26].

The hydrological cycle in the basin begins at elevations above 3600 m above sea level, in the páramo ecosystems of the Casahuala, Pilishurco, and Chimborazo Wildlife Production Reserve. These areas function as water regulators and give rise to 127 rivers, 237 streams, 80 lakes, and 244 creeks. The basin supplies water to five of the nine cantons (a second-level political division) of Tungurahua Province, with an estimated population of 320,315 inhabitants and an average population density of 130.9 inhabitants/km^2. The predominant economic activities are agriculture and livestock farming, particularly in rural zones and transitional areas between the páramos and urban regions [29].

2.2. Data Sources

2.2.1. Rain Gauges Data

Daily precipitation records from eight automatic meteorological stations within the Ambato River basin were obtained from the Decentralized Autonomous Government of Tungurahua Province (Figure 1) for 2014–2023.

For the core validation over the full 2014–2023 period, six stations with multi-year records of sufficient length and continuity were used. Tamboloma and Pilahuin stations were not retained because their records are limited to 2019–2023 and contain substantial gaps, which prevents the consistent calculation of validation metrics across the entire study period. Both excluded stations are located near the basin center, close to Calamaca station (Figure 1), so their omission does not create a data gap in the south part of the catchment. However, the southern sector (coincident with the Chimborazo and Carihuairazo volcanic slopes) lacks rain gauges; this coverage limitation is acknowledged.

Following Veiga et al. [30], the six gauge series underwent quality control (QC), including gross-error checks, fixed-range checks, and temporal continuity tests to ensure internal consistency. The characteristics of the six retained stations are reported in Table 1. After QC, these series are internally consistent and suitable for the analyses presented here.

Table 1. Characteristics of the meteorological stations in the Ambato River basin. Coordinates are referred to UTM zone 17 S. Completeness values are computed over 2014–2023. 'Years used' counts the years with ≥80% daily data completeness, which are the years included in the annual statistics of mean annual precipitation (MAP) and standard deviation (SD).

Name	Location			Completeness		Years Used	MAP	SD
	Coord X	Coord Y	Altitude	Daily (%)	Monthly (%)	(y)	(mm/y)	(mm/y)
Chiquiurco	743,682	9,866,064	3763	98.6	80.8	10	1108	88.1
Aeropuerto Ambato	769,929	9,865,679	2590	74.5	56.7	7	587	298
P. Fermín Cevallos	765,641	9,849,972	2910	95.0	80.0	9	478	83.7
Calamaca	742,705	9,858,860	3483	90.3	75.8	8	757	122
Mula Corral	741,602	9,867,738	3890	85.6	65.8	9	958	130
Quisapincha	753,559	9,865,921	3670	96.6	85.8	10	988	185

2.2.2. Satellite Data

This study uses two satellite datasets that have been proved to have a better performance for the conditions in Ecuador: CHIRPS and IMERG. The Climate Hazards Group InfraRed Precipitation with Station Data (CHIRPS) is a quasi-global rainfall dataset developed by the Climate Hazards Group at the University of California, Santa Barbara. It offers high-resolution precipitation estimates (0.05° spatial resolution, ~5 km) from 1981 to the present by combining satellite infrared data with in situ observations from weather stations [14].

The CHIRPS data generation process involves two main steps. First, satellite-derived infrared precipitation (IRP) pentads are calculated by assessing the proportion of time that cold cloud tops (<235 K) are detected in infrared imagery during each five-day period. These frequencies are then converted into rainfall amounts (in mm) through a calibrated regression using TRMM 3B42 pentad precipitation data. To account for climatic variability, the IRP pentads are normalized against their long-term means (1981–2012), resulting in unitless anomaly values indicating deviations from average precipitation (i.e., below-normal, near-normal, or above-normal conditions). These normalized values are multiplied by corresponding pentads from the CHPclim monthly climatology to produce gridded, bias-corrected rainfall estimates known as CHIRP. In the second step, CHIRP estimates are merged with ground-based station data to generate the final CHIRPS product, which improves accuracy across complex terrains and regions with sparse monitoring networks [14].

The Integrated Multi-satellite Retrievals for GPM (IMERG) dataset, developed by the NASA Goddard Space Flight Center's Precipitation Processing System, provides high-resolution, satellite-based precipitation estimates through the integration of passive microwave and infrared sensor data with ground-based rain gauge observations [31]. The IMERG algorithm performs inter-calibration, data merging, and temporal interpolation to generate seamless precipitation records.

IMERG offers near-global coverage, spanning latitudes between 60° N and 60° S, with a spatial resolution of approximately 10 km. Its temporal resolution ranges from 30 min intervals to monthly accumulations, making it suitable for a wide range of hydrological and climatological applications, particularly in data-scarce regions or areas with low-quality observational networks [31].

In this study, daily and monthly precipitation data from both the CHIRPS and IMERG datasets were retrieved using the Google Earth Engine platform [32]. For CHIRPS, "UCSB-CHG/CHIRPS/DAILY" image collection was filtered to extract daily precipitation values corresponding to the geographic coordinates of the meteorological stations located in the Ambato River basin. Similarly, for IMERG, the GPM L3/IMERG V07 collection (IMERG-Final run) was used to obtain daily and monthly precipitation estimates, which were spatially matched to the same set of stations across the basin.

2.3. Evaluation Methods

CHIRPS and IMERG were evaluated at two temporal scales: daily, to assess rainfall daily detection, and monthly, to assess accumulated amounts. Satellite values were extracted from the native grid cell containing each gauge (point-to-pixel pairing). Daily accumulations for gauges and SPEs were aligned to local time (UTC-05:00). Days with missing data in either source were excluded pairwise; no gap infilling was applied. A rainy day was defined as gauge precipitation $\geq$1.0 mm day^{-1} [33], used for all categorical metrics. This denotes daily rain occurrence and does not correspond to a hydrologically coherent storm; daily aggregation can merge or split sub-daily storms. Monthly totals are calendar aggregates suitable for climatological characterization and volumetric accuracy, but catchment response periods need not align with calendar months

To ensure hydrological relevance, the evaluation targets both daily rainy-day detection and rainfall amounts. The daily rainy-day detection skill was assessed with the following measures: Probability of Detection (*POD*), False Alarm Ratio (*FAR*), Frequency Bias Index (*FBI*), and Heidke Skill Score (*HSS*); these capture misses, false alarms, systematic over/under-detection, and chance-corrected categorical skill. These are key attributes for flood early warnings and daily modelling. Quantitative agreement was evaluated with Mean Absolute Error (*MAE*), Root Mean Square Error (*RMSE*), Index of Agreement (*d*), and bias; these relate to error magnitude, pattern fidelity, and systematic deviation that influence water balance closure and drought assessments. To diagnose error sources, bias was decomposed into hit, miss, and false components (*HB*, *MB*, *FB*), which are informative for the mountainous Andean terrain where orographic and seasonal effects can differentially affect peak flow simulations, baseflow conditions, and soil moisture recharges.

2.3.1. Categorical Validation Statistics

To evaluate the ability of satellite-based precipitation estimates (SPEs) to detect rainfall, categorical validation statistics were applied based on daily comparisons with ground-based rain gauge measurements [19,33]. Categorical metrics were computed on the intersection of days with valid gauge, IMERG, and CHIRPS records at each station; daily rainy-day detections were classified as occurring when daily precipitation exceeded 1 mm per day [33], for both gauge and SPPs. Missing values are not imputed. Four categorical metrics were employed to assess performance: Probability of Detection (*POD*), False Alarm Ratio (*FAR*), Frequency Bias Index (*FBI*), and Heidke Skill Score (*HSS*); each of these provides a distinct measure of agreement between IMERG and CHIRPS estimates and ground observations.

The *POD* metric quantifies the proportion of actual rainfall events that were correctly identified by the satellite product, with values approaching 1 indicating better detection performance. Its calculation involves two key variables: *A* (hits, where both the SPE and gauge detect precipitation) and *C* (misses, where precipitation is recorded by the gauge but not detected by the SPE), as shown in Equation (1). *POD* values range from 0 to 1, with 1 representing perfect detection.

$$POD = \frac{A}{A+C} \quad (1)$$

In contrast to *POD*, the False Alarm Ratio (FAR) quantifies the proportion of precipitation events falsely detected by the satellite-based estimates, that is, cases where the SPE reported rainfall, but no precipitation was recorded by the ground-based rain gauges. This metric incorporates variable *B*, representing false alarms, defined as instances where precipitation was estimated by the SPE but not observed in situ (Equation (2)). *FAR* values range from 0 to 1, with lower values indicating better performance and 0 being the ideal outcome.

$$FAR = \frac{B}{A+B} \quad (2)$$

The Frequency Bias Index (*FBI*) assesses the tendency of satellite-based precipitation estimates to either overestimate or underestimate the occurrence of daily precipitation when compared to ground-based observations. It is calculated using both hits (*A*) and false alarms (*B*), as shown in Equation (3). FBI values range from 0 to ∞, with an ideal value of 1. *A* value below 1 suggests that the SPE underestimates precipitation frequency, while a value above 1 indicates an overestimation relative to rain gauge records.

$$FBI = \frac{A+B}{A+C} \quad (3)$$

The Heidke Skill Score (HSS) is a categorical metric that evaluates the overall accuracy of precipitation estimates, correcting for the influence of random chance (Equation (4)). It incorporates all four elements of the contingency table [34]: hits (*A*), false alarms (*B*), misses (*C*), and correct negatives (*D*). Here, *D* corresponds to days with no precipitation detected by either the SPE or the ground-based rain gauge.

HSS values range from $-\infty$ to 1. *A* score below 0 indicates that random chance outperforms the SPE, a value of 0 implies no predictive skill, and a value of 1 represents perfect agreement between the satellite-based estimates and ground observations in identifying precipitation days.

$$HSS = \frac{2(AD-BC)}{[(A+C)(C+D)+(A+B)(B+D)]} \quad (4)$$

2.3.2. Continuous Validation Statistics

To assess the quantitative agreement between satellite-based precipitation estimates (SPEs) and ground-based observations, a set of continuous validation statistics was computed. These included the Mean Absolute Error (MAE), Root Mean Square Error (RMSE), Index of Agreement (*d*), and bias, as defined in Equations (5)–(8). All metrics were calculated separately for daily and monthly datasets.

MAE and *RMSE* quantify the magnitude of error, with ideal values tending toward 0. In contrast, *d* and Bias evaluate relative agreement and systematic deviation, respectively, with optimal values equal to 1.

$$MAE = \left(\frac{1}{N}\right)\sum_{i=1}^{N}|S_i - E_i| \tag{5}$$

$$RMSE = \sqrt{\left(\frac{1}{N}\right)\sum_{i=1}^{N}(S_i - E_i)^2} \tag{6}$$

$$d = 1 - \frac{\sum(S - E)^2}{\sum(|S - \overline{E}| + |E - \overline{E}|)^2} \tag{7}$$

$$Bias = \frac{\sum S_i}{\sum E_i} \tag{8}$$

where S_i is the satellite-derived precipitation at time *i*, E_i is the rain gauge observation at time *i*, $\overline{E}$ means the observed precipitation over the comparison period, and N is the total number of paired observations.

2.3.3. Bias Decomposition Analysis

To better characterize the sources of error in satellite-based precipitation estimates, a bias decomposition analysis was conducted using three complementary metrics: Hit Bias (*HB*), Miss Bias (*MB*), and False Bias (*FB*). These indicators quantify the relative contributions of different types of disagreement between SPE products and ground-based rainfall observations.

HB represents the systematic difference in precipitation volumes during coincident detection events, that is, when both SPE and rain gauges register rainfall. This metric captures whether the SPE tends to overestimate or underestimate rainfall intensity and is calculated using Equation (9). Values close to zero indicate strong agreement, whereas positive values denote overestimation and negative values indicate underestimation.

$$HB = \left(\sum(S - E) \ / \ \sum E\right) \times 100 \quad for: \ S > 0 \ and \ E > 0 \tag{9}$$

The Miss Bias (*MB*) quantifies undetected precipitation events—those observed by ground stations but not captured by the SPE. It is computed as shown in Equation (10) and reflects the proportion of missed rainfall relative to the total observed precipitation. MB values are always negative, and greater absolute values indicate larger under-detection.

$$MB = \left(-\sum E_1 \ / \ \sum E\right) \times 100 \quad for: \ S = 0 \ and \ E > 0 \tag{10}$$

The False Bias (*FB*) measures the magnitude of precipitation estimated by the SPE in the absence of corresponding observations from rain gauges. It highlights false positive detections and is calculated according to Equation (11). As with the other metrics, values closer to zero indicate better performance.

$$FB = \left(\sum S \ / \ \sum E\right) \times 100 \quad for: \ S > 0 \ and \ E = 0 \tag{11}$$

2.4. Analyzing Interannual Spatiotemporal Precipitation Patterns

To analyze spatiotemporal rainfall patterns at the grid-cell level, annual precipitation averages over the study period were mapped using IMERG and CHIRPS datasets. The spatial variability of precipitation was assessed by calculating the coefficient of variation (CV) using Equation (12):

$$CV = \sigma_P / \mu_P \tag{12}$$

where σ_P is the standard deviation of precipitation and μ_P is the mean precipitation. The resulting CV maps provide insights into spatial differences in precipitation variability within the study area. Higher CV values indicate greater relative variability across the grid cells, while lower values suggest more uniform precipitation patterns.

3. Results

3.1. Daily Rainfall Detection

Across the basin, IMERG detects daily rainfall more consistently than CHIRPS at all gauges. POD values for IMERG were around 0.4 at every station (often >0.4), whereas CHIRPS remains <0.3 in most cases; FAR is comparable between products (≈0.28; Table 2).

Table 2. Detection skill scores for daily rainfall estimates from CHIRPS and IMERG across the six selected stations. The variable *n* common number of valid gauge–IMERG–CHIRPS days (2014–2023).

		Daily Categorical Validation—IMERG			
Station	***n***	**POD**	**FAR**	**FBI**	**HSS**
Chiquiurco	3599	0.47	0.37	0.74	0.17
Aeropuerto Ambato	2721	0.39	0.52	0.81	0.2
P.F. Cevallos	3468	0.47	0.55	1.02	0.23
Calamaca	3297	0.46	0.49	0.89	0.13
Mula Corral	3126	0.45	0.39	0.74	0.15
Quisapincha	3526	0.38	0.33	0.57	0.18
Mean		0.44	0.44	0.8	0.18
		Daily Categorical Validation—CHIRPS			
Station	***n***	**POD**	**FAR**	**FBI**	**HSS**
Chiquiurco	3599	0.23	0.35	0.35	0.1
Aeropuerto Ambato	2721	0.26	0.53	0.56	0.14
P.F. Cevallos	3468	0.3	0.55	0.68	0.16
Calamaca	3297	0.23	0.44	0.42	0.1
Mula Corral	3126	0.21	0.36	0.33	0.09
Quisapincha	3526	0.24	0.35	0.36	0.1
Mean		0.24	0.43	0.45	0.12

Spatially, the strongest agreement for IMERG occurs at the eastern station (P. Fermín Cevallos) and the northwestern station (Chiquiurco). In the northwestern sector (Mula Corral, Chiquiurco), CHIRPS shows the weakest daily rainfall detection, while IMERG still outperforms CHIRPS though with more moderate skill than at the eastern/northeastern sites. The northern station (Quisapincha) exhibits intermediate performance, with IMERG consistently ahead of CHIRPS.

The frequency bias further indicates systematic under-detection by CHIRPS (FBI < 1 across most stations), whereas IMERG is closer to the ideal (FBI $\rightarrow$ 1); for example, CHIRPS attains its highest FBI at P.F. Cevallos (~0.68) yet remains below unity. HSS follows the same pattern, with IMERG higher at nearly all gauges. Overall, these daily occurrence results point to basin-wide superiority of IMERG and limited categorical skill for CHIRPS, particularly in the northwest (Figure 1; Table 2).

3.2. Daily Precipitation Totals (Continuous Validation)

At the daily scale, IMERG and CHIRPS show broadly comparable performance, with IMERG exhibiting a slight advantage in bias and index of agreement (d), and marginally higher MAE than CHIRPS (Table 3). IMERG tends to overestimate daily totals on average, whereas CHIRPS generally underestimates them.

Spatially, the largest daily discrepancies for both products occur at the northwestern station Chiquiurco, reflected in elevated MAE and RMSE. In the north sector (Quisapincha), performance is intermediate; IMERG achieves the highest d observed across the network at Quisapincha ($\approx$0.65). The central western station Calamaca and the eastern station P.F. Cevallos show comparatively better agreement patterns, consistent with the station-wise means.

Table 3. Continuous validation statistics at the daily time scale. The variable n indicates the number of paired data points considered in the comparison.

		Daily Continuous Validation—IMERG			
Station	n	**MAE**	**RMSE**	**d**	**Bias**
Chiquiurco	3599	3.33	5.99	0.62	0.86
Aeropuerto Ambato	2721	2.55	6.09	0.48	1.26
P.F. Cevallos	3468	2.41	5.74	0.54	1.71
Calamaca	3297	2.94	5.73	0.55	1.25
Mula Corral	3126	3.16	5.81	0.62	0.95
Quisapincha	3526	2.94	5.33	0.65	0.76
Mean		2.89	5.78	0.58	1.13
		Daily Continuous Validation—CHIRPS			
Station	n	**MAE**	**RMSE**	**d**	**Bias**
Chiquiurco	3599	3.5	6.24	0.47	0.52
Aeropuerto Ambato	2721	2.35	4.88	0.43	0.88
P.F. Cevallos	3468	2.97	6.88	0.38	1.83
Calamaca	3297	2.89	5.6	0.43	0.83
Mula Corral	3126	3.02	5.38	0.46	0.43
Quisapincha	3526	3.1	5.41	0.55	0.58
Mean		2.97	5.73	0.45	0.84

Bias results indicate contrasting tendencies: IMERG averages bias $\approx$ 1.13, i.e., an overestimation of ~13% in daily totals, while CHIRPS averages bias $\approx$ 0.84, i.e., an underestimation of ~16%. Station-wise, IMERG is close to unity at Mula Corral and Chiquiurco ($\approx$0.95 and $\approx$0.86, northwest), whereas CHIRPS shows pronounced underestimation across sites; at Mula Corral (northwest) its bias is ~0.43

IMERG provides slightly higher daily agreement (higher d) but with a tendency to overestimate amounts, while CHIRPS yields slightly lower error magnitudes on average (RMSE) yet systematically underestimates daily totals; this is particularly evident at the northwestern station Chiquiurco and Mula Corral.

3.3. Daily-Scale Bias Decomposition

The bias decomposition analysis at daily time scale for the IMERG and CHIRPS products was conducted to evaluate their detection performance, as well as the magnitude of undetected precipitation and false detections. The overall bias was disaggregated into three components: Hit Bias (HB), Miss Bias (MB), and False Bias (FB) (Table 4).

Network-wide, CHIRPS exhibits a lower HB on average (+14.2%) than IMERG (+18.5%), indicating less overestimation on coincident rainy days. In contrast, MB is substantially less negative for IMERG (≈−43.3%) than for CHIRPS (≈−62.9%), implying that IMERG misses a smaller fraction of gauge-observed daily rainfall. For FB, CHIRPS is slightly lower on average (38.25%) than IMERG (39.17%), indicating fewer false-positive rainfall detections.

At the station scale, patterns align with basin-relative location. On the eastern side (P.F. Cevallos), both products exhibit their largest Hit Bias (HB; CHIRPS 51.42%, IMERG 39.63%) together with elevated False Bias (FB; CHIRPS 92.99%, IMERG 69.73%), indicating strong overestimation on coincident wet days and a high frequency of false-positive rainfall detections. In the northwestern sector (Mula Corral, Chiquiurco), CHIRPS shows the most negative Miss Bias (MB; down to −67.97% at Mula Corral and −66.95% at Chiquiurco), evidencing substantial missed rainfall, while FB remains low (≈15%). IMERG in this sector reduces the magnitude of MB (≈−43% at both stations), with modest positive HB (7–13%) and low-to-moderate FB (≈20%). In the northeast (Quisapincha, Ambato Airport), IMERG attains its lowest FB at Quisapincha (16.33%) but records its most negative MB within its range at Ambato Airport (−49.36%); CHIRPS also underestimates at both sites (MB −62.91% and −62.5%, respectively), with FB of 17.99% and 49.42%. In the central west (Calamaca), IMERG presents one of its least negative MB values (−41.3%), consistent with comparatively better daily detection there; CHIRPS shows moderate HB (15.27%), moderate FB (37.42%), and strongly negative MB (−64.33%).

Table 4. Bias decomposition statistics at the daily time scale. The variable *n* common number of valid gauge–IMERG–CHIRPS days (2014–2023).

		Bias Decomposition IMERG		
Station	*n*	**HB**	**MB**	**FB**
Chiquiurco	3599	7.82	−43.5	20.99
Aeropuerto Ambato	2721	20.35	−49.36	57.79
P.F. Cevallos	3468	39.63	−35.14	69.73
Calamaca	3297	21.61	−41.31	45.64
Mula Corral	3126	13.83	−43.5	24.54
Quisapincha	3526	7.45	−46.99	16.33
Mean		18.45	−43.3	39.17
		Bias Decomposition CHIRPS		
Station	*n*	**HB**	**MB**	**FB**
Chiquiurco	3599	5.62	−66.95	16.2
Aeropuerto Ambato	2721	7.66	−62.5	49.42
P.F. Cevallos	3468	51.42	−53.01	92.99
Calamaca	3297	15.27	−64.33	37.42
Mula Corral	3126	−1.07	−67.97	15.47
Quisapincha	3526	6.24	−62.91	17.99
Mean		14.19	−62.94	38.25

Overall, the decomposition reveals complementary behaviors. IMERG reduces the fraction of missed rainfall (less-negative MB) across the network but tends to overestimate on coincident wet days (higher HB); and, on average, it incurs slightly higher FB than CHIRPS. CHIRPS, in turn, produces fewer product-only wet days (lower FB) and lower HB, yet misses a larger share of gauge-observed rainfall (more-negative MB), particularly in the northwest. These station-wise contrasts are consistent with the categorical findings and help explain the product-specific tendencies reported elsewhere in Section 3 (Table 4).

3.4. Monthly Precipitation Totals (Continuous Validation)

At the monthly scale, interpreted here as calendar aggregates for climatological/volumetric evaluation, IMERG generally shows tighter agreement with gauges than CHIRPS, with higher index of agreement (*d*) and, on average, lower or comparable MAE/RMSE (Table 5). Bias indicates opposing tendencies: IMERG overestimates monthly totals by ~15% on average (bias ≈ 1.15), whereas CHIRPS underestimates by ~15% (bias ≈ 0.85).

Spatially, the largest monthly discrepancies for both products occur at the northwestern station, Chiquiurco (elevated MAE/RMSE). In the north sector, IMERG attains the highest *d* across the network at Quisapincha (≈0.65), while CHIRPS reaches its best d at Ambato Airport (≈0.55). An outlier appears at the eastern station, P.F. Cevallos, where both products show positive bias (IMERG ≈ 1.72, CHIRPS ≈ 1.79).

The gauge–satellite scatter plots (Figure 2) corroborate these patterns: IMERG (blue) points cluster more tightly around the 1:1 line (red dashed line), whereas CHIRPS (black) points more often fall below 1:1, reflecting underestimation; Ambato Airport is a notable exception where CHIRPS aligns relatively well with observations.

Table 5. Continuous validation statistics at the monthly time scale. The variable *n* indicates the number of paired data points considered in the comparison.

		Monthly Continuous Validation—IMERG			
Station	*n*	**MAE**	**RMSE**	**d**	**Bias**
Chiquiurco	97	40.20	54.99	0.58	0.87
Aeropuerto Ambato	69	35.29	50.94	0.38	1.32
P.F. Cevallos	96	38.96	50.31	0.55	1.72
Calamaca	91	39.41	52.45	0.55	1.28
Mula Corral	79	38.42	52.24	0.57	0.95
Quisapincha	103	36.46	44.03	0.65	0.75
Mean		38.12	50.83	0.55	1.15
		Monthly Continuous Validation—CHIRPS			
Station	*n*	**MAE**	**RMSE**	**d**	**Bias**
Chiquiurco	97	52.45	65.60	0.42	0.53
Aeropuerto Ambato	69	24.23	33.16	0.55	0.93
P.F. Cevallos	96	40.64	51.33	0.51	1.79
Calamaca	91	34.85	44.82	0.47	0.87
Mula Corral	79	55.87	66.52	0.39	0.41
Quisapincha	103	44.13	54.26	0.50	0.55
Mean		42.03	52.62	0.47	0.85

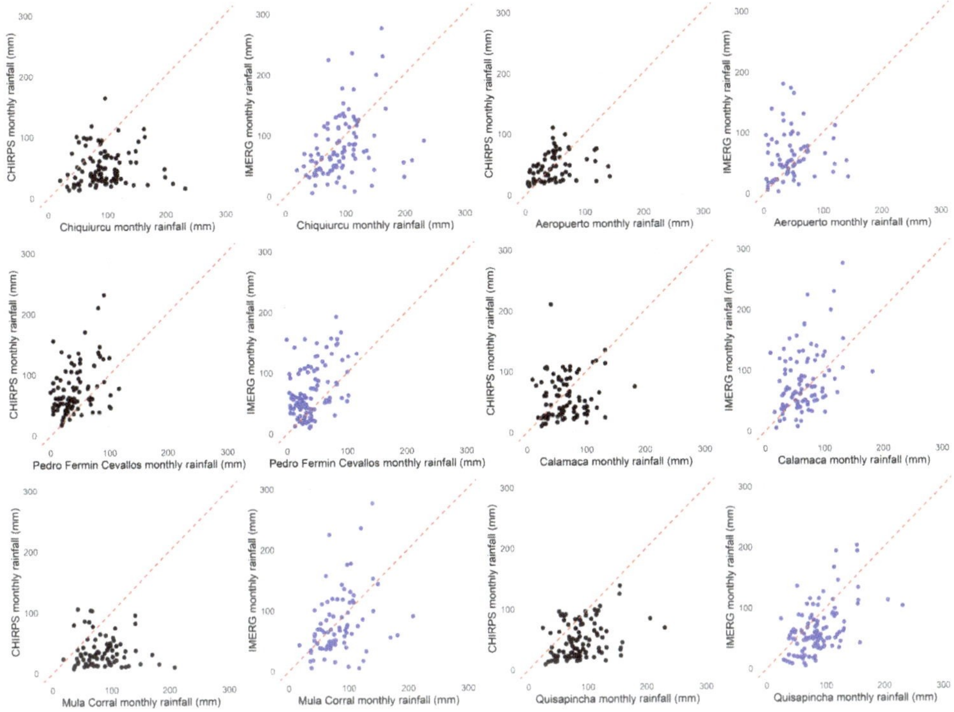

Figure 2. Scatter plots of monthly precipitation estimate from CHIRPS (black) and IMERG (blue) versus meteorological station measurements for the period 2014–2023 in the Ambato River basin, Ecuador.

Notably, IMERG showed stronger concordance with observations at the Quisapincha, Calamaca, and Mula Corral stations, which aligns with the results from the continuous validation metrics (Table 4). Conversely, CHIRPS tends to underestimate monthly precipitation values when compared to gauge data, as evidenced by most scatter points lying below the 1:1 line. An exception to this pattern is observed at the Ambato Airport station, where CHIRPS demonstrates relatively better agreement, consistent with the continuous validation results: bias = 0.93 and d = 0.55.

Monthly climatology data (2014–2023; Figure 3) indicate that both products reproduce the seasonal cycle only at two stations—Calamaca (central west) and Quisapincha (north)—and show limited skill at the remaining four gauges. Consistent with the metrics, IMERG tends to lie above the observed monthly totals, while CHIRPS tends to lie below them.

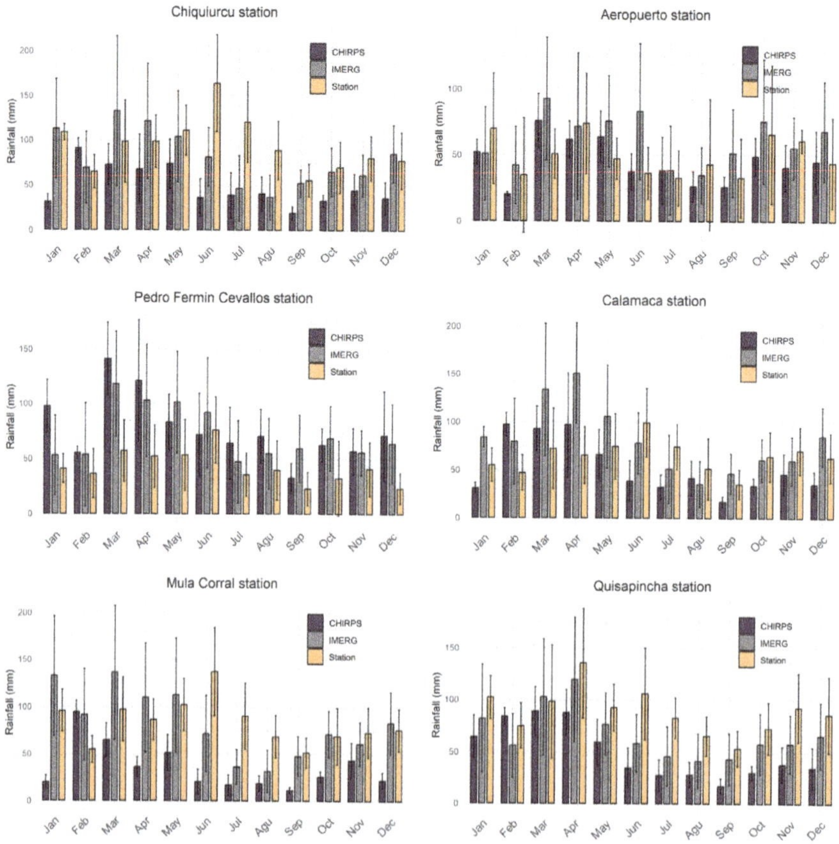

Figure 3. Comparison of long-term mean monthly precipitation estimates from the satellite-based rainfall products (IMERG and CHIRPS) for the period 2014–2023 across six meteorological stations in the Ambato River basin, Ecuador. Vertical bars represent the standard deviation of monthly precipitation.

3.5. Interannual Spatiotemporal Precipitation Patterns

To investigate the spatiotemporal behavior of rainfall across the study area, annual mean precipitation was calculated for both satellite-based products: IMERG and CHIRPS (Figure 4). IMERG estimates ranged from 788 mm to 1577 mm per year (Figure 4a), while CHIRPS estimates varied between 333 mm and 1187 mm annually (Figure 4c).

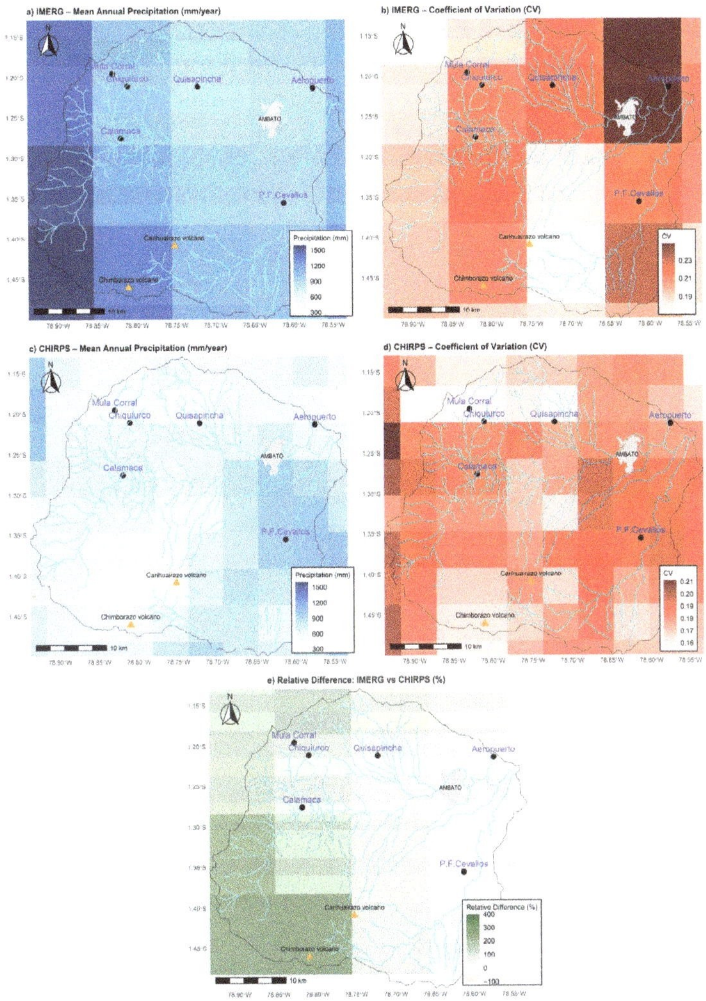

Figure 4. Spatiotemporal distribution of annual mean precipitation and coefficient of variation for the period 2014–2023. Panels (**a**,**b**) correspond to IMERG data; panels (**c**,**d**) correspond to CHIRPS data. Panel (**e**) corresponds to the relative difference (IMERG-CHIRPS) in percentage (%). Black dots represent ground rainfall gauging stations; orange triangles represent main volcano locations.

The coefficient of variation (CV), used to assess interannual variability, further revealed spatial differences between the two products. For IMERG, the highest CV reached 25% (Figure 4b), mainly concentrated in the northeastern sector of the Ambato River basin. The rest of the basin exhibited relatively low CV values, suggesting stable annual precipitation over the study period. In the case of CHIRPS, the maximum CV was slightly lower at 21% (Figure 4d), located to the northwest of the catchment, outside the basin boundaries. Despite these differences, both products revealed similar spatial patterns of interannual variability within the core area of the basin.

Within the study area, three distinct rainfall regimes can be observed. The first corresponds to the inter-Andean corridor spanning the northeastern to southeastern sectors of the catchment. In this corridor, IMERG estimates show a gradual increase in annual precipitation from around 800 mm in the northeast to 1000 mm in the southeast of the catchment. CHIRPS, in contrast, reports lower values ranging from approximately 500 mm to 900 mm.

The second region encompasses the eastern slopes of the Western Andes in the northwest part of the basin, an area of higher elevations with páramo ecosystems. IMERG indicates mean annual precipitation of around 1000 mm, whereas the CHIRPS estimates are significantly lower, near 600 mm.

The third region lies in the southwest sector along the eastern flank of the Western Andes. This area overlaps substantially with the Chimborazo and Carihuairazo volcanoes. This is the largest inter-product discrepancy in the basin (Figure 4e), with IMERG exceeding CHIRPS by ~900 mm.

These spatial contrasts further support the patterns revealed in the validation analyses, highlighting the consistent overestimation by IMERG and the underestimation by CHIRPS across various sectors of the basin.

4. Discussion

Satellite precipitation products such as IMERG and CHIRPS have become indispensable tools for hydrological analysis in regions with sparse ground-based data, particularly in mountainous and forested catchments. However, their performance varies depending on topographic complexity, data processing algorithms, and the temporal resolution of evaluation. For instance, in a study conducted across Bolivia's diverse terrain, IMERG showed superior skill in detecting daily rainfall (POD > 0.5) compared to CHIRPS, especially in high-altitude areas, but suffered from a high false alarm ratio and rainfall overestimation of up to 75% in the highlands [19]. Similarly, in the Madre de Dios basin in Peru, Salas-Choquehuanca et al. [20] reported that IMERG_F v6 data presented notable deviations from observed values, with RMSEs averaging 146 mm and Nash–Sutcliffe Efficiency (NSE) values often below acceptable thresholds (< 0.65). Although the Pearson correlation coefficients (r > 0.8) suggested a generally consistent trend, systematic over- and underestimations persisted across the stations. These results collectively emphasize that, while IMERG can capture general rainfall patterns, it tends to overestimate intensities and may underperform at finer scales or in specific microclimates.

In the Ambato River catchment, characterized by steep topography, the categorical validation metrics (POD, FAR, FBI, HSS) confirmed the superior performance of IMERG over CHIRPS in detecting daily rainfall. IMERG achieved higher POD and HSS values, reflecting its greater ability to correctly identify precipitation occurrences, while both products showed similar FAR values. This aligns with findings from Benítez et al. [18], who highlighted IMERG's capability to detect rainy days (POD > 0.7) and its high agreement with observed precipitation in the wetter regions of southeastern South America. However, as also observed in Bolivia [19], our analysis revealed that IMERG exhibited a higher False Bias Index (FBI), suggesting the presence of false rainfall detections not supported by gauge observations.

In terms of continuous validation at a daily scale, both products produced comparable results for MAE and RMSE. Nevertheless, IMERG showed a slight advantage in MAE and presented a lower bias index, suggesting less systematic deviation from observed rainfall. Interestingly, IMERG tended to overestimate precipitation, whereas CHIRPS consistently underestimated it—a pattern also reported by [19,20]. At the monthly scale, this behavior became more pronounced, with IMERG consistently providing higher rainfall totals. The scatterplot analysis of the monthly estimates confirmed this trend, as IMERG's predictions clustered more closely around the 1:1 line, whereas CHIRPS values tended to fall below it, indicative of underestimation. One factor that may have influenced the monthly validation results is the selection of only those months with complete datasets, whereas daily validation included time series that may have come from months with missing data.

In this context, the findings of López-Bermeo et al. [35] offer relevant evidence, as they conducted a robust validation of CHIRPS using 75 rain gauge stations across the diverse topography of Antioquia, Colombia. Their results revealed that CHIRPS generally performs well at annual and interannual scales, particularly in Andean regions, but tends to overestimate precipitation in most stations and underestimates in warmer, lowland subregions. These biases become more pronounced at the daily scale, where CHIRPS exhibited limited accuracy. This supports our current analysis, which also shows that, while CHIRPS captures the general precipitation patterns, its reliability varies by subregion and time scale, and local overestimation or underestimation must be accounted for when interpreting results against station data.

Quispe et al. [36] assessed the performance of GPM IMERG products (early, late, and final versions) across the Lake Titicaca Basin and found that accuracy improved with coarser temporal scales, with the monthly IMERG-F (final) product consistently outperforming its early and late versions. This finding aligns with our results, in which IMERG showed stronger agreement with rain gauge data, particularly in monthly accumulations.

The bias decomposition analysis further clarified the strengths and limitations of each product. CHIRPS exhibited better performance in terms of Hit Bias (HB), meaning that, on average, its estimations more closely matched the number of actual rainy days. However, IMERG had a better Miss Bias (MB), with fewer undetected precipitation events compared to CHIRPS. Regarding False Bias (FB), CHIRPS again outperformed IMERG, supporting previous findings that IMERG's overestimation is primarily driven by spurious rainfall detections, particularly in regions of complex terrain and variable surface emissivity [19].

Spatial analysis revealed further discrepancies. CHIRPS produced lower annual average precipitation estimates than IMERG, especially in the southwestern region of the basin, where CHIRPS suggested values below 400 mm, while IMERG exceeded 1300 mm. This contrast is critical, as prior studies by Hunink et al. [27] suggest that annual rainfall in this area should be closer to 1000 mm. Therefore, while IMERG may overestimate rainfall amounts, its estimates may still offer a closer approximation to reality in certain high-precipitation subzones.

The implications of these findings are particularly relevant for hydrological modeling, flood forecasting, and water resource planning in the Ambato River basin and similar Andean catchments. The tendency of CHIRPS to underestimate precipitation may lead to underestimation of runoff and water availability in rainfall-runoff models, potentially compromising water allocation and drought risk assessments. Conversely, the overestimation by IMERG—especially in the highland zones—could inflate simulated streamflow values, influencing flood hazard assessments and infrastructure design. Given these trade-offs, the selection of an appropriate product should be context-specific: CHIRPS may be more suitable for long-term climatological analyses and drought studies, where minimizing false rainfall is crucial, while IMERG might be better suited for applications requiring accurate detection of rainfall occurrence, such as early warning systems and daily hydrological modeling. Integrating bias-correction techniques or combining both products using ensemble or fusion methods may further enhance their applicability in operational settings.

Nonetheless, these results should be interpreted considering certain limitations that may affect their generalizability and accuracy. Although six automatic rain gauge stations were used, their spatial distribution may be insufficient for fully representing the heterogeneity of rainfall patterns in such a topographically complex Andean basin. This limited density may reduce the precision of ground-based validation, especially in areas with strong orographic influence. In addition, although a strict quality control process was applied to ensure the reliability of the data, missing records from some stations were not gap-filled. This decision was taken to preserve the integrity of the validation analysis, avoiding artificial bias. While this approach favors methodological rigor, it may also restrict the completeness of the comparison with satellite estimates. Despite these constraints, the study offers valuable insight for hydrometeorological applications in data-scarce regions, where ground-based networks are sparse and satellite precipitation products serve as essential tools. Future research could further enhance validation robustness by integrating complementary data sources and expanding the observational network.

The present comparative assessment of CHIRPS and IMERG reveals that, while both products exhibit valuable capabilities, they also present significant limitations that must be accounted for in hydrometeorological applications. IMERG outperforms CHIRPS in rainfall detection and correlation with observed data, but suffers from higher false detections and systematic overestimation in certain zones. CHIRPS provides more conservative and spatially coherent estimates, though it underrepresents precipitation in high-rainfall regions. These findings reinforce the importance of regional validation of SPEs and support the use of multi-source strategies to improve rainfall estimation accuracy in topographically complex and data-scarce Andean environments.

5. Conclusions

This study provides a comprehensive evaluation of the performance of CHIRPS and IMERG satellite precipitation products against rain gauge observations in the Ambato River catchment, a mountainous Andean basin characterized by steep topography and variable climate conditions. The findings demonstrate that both products capture general precipitation patterns but differ markedly in their accuracy, bias structure, and suitability depending on the temporal scale and intended hydrometeorological application.

IMERG outperformed CHIRPS in rainfall daily rainy-day detection (POD = 0.44 vs. 0.24), and monthly Mean Absolute Error (MAE = 38.12 vs. 42.03), making it more reliable for rainfall-driven hydrological applications. These results suggest that IMERG is more reliable for daily hydrological modeling, flood forecasting, and early warning systems, especially when accurate detection of rainfall occurrence and intensity is required. However, its tendency to overestimate rainfall and produce false alarms (FAR = 0.44) may compromise accuracy in high-resolution or risk-sensitive applications.

Conversely, at a daily time scale, CHIRPS showed better performance in Hit Bias (HB) and False Bias (FB), along with more conservative rainfall estimates, making it potentially more suitable for long-term climatological assessments or drought monitoring, where underestimations are more acceptable than false detections. Its stability under dry conditions further supports this use case.

These results highlight the importance of regional validation before operational use of satellite precipitation estimates. For hydrological modeling, drought assessment, or flood forecasting in the Andes, product selection should be context-specific. Future research should explore bias-correction techniques and multi-sensor integration to improve rainfall representation in complex terrains. Ultimately, while satellite products are indispensable in data-scarce regions, their limitations must be explicitly accounted for to ensure accurate and reliable hydrometeorological analyses.

Author Contributions: Conceptualization, P.A.-M., C.C.-V., J.C.-G. and M.M.-C.; Data Curation, P.A.-M. and C.C.-V.; Formal Analysis, P.A.-M., C.C.-V. and J.C.-G.; Funding Acquisition, P.A.-M., C.C.-V., J.C.-G. and M.M.-C.; Investigation, P.A.-M., C.C.-V., J.C.-G. and M.M.-C.; Methodology, P.A.-M., C.C.-V., J.C.-G. and M.M.-C.; Project Administration, C.C.-V. and J.C.-G.; Resources, C.C.-V. and J.C.-G.; Software, P.A.-M. and C.C.-V.; Supervision, C.C.-V., J.C.-G. and M.M.-C.; Validation, P.A.-M., C.C.-V. and M.M.-C.; Visualization, P.A.-M. and C.C.-V.; Writing—Original Draft Preparation, P.A.-M., C.C.-V. and J.C.-G.; Writing—Review and Editing, P.A.-M., C.C.-V., J.C.-G. and M.M.-C. All authors have read and agreed to the published version of the manuscript.

Funding: This work was supported by Universidad Nacional de Chimborazo-Ecuador (UNACH).

Data Availability Statement: The raw data supporting the conclusions of this article will be made available by the authors on request.

Acknowledgments: The authors would like to express their gratitude to the Honorable Gobierno Provincial de Tungurahua (HGPT) for providing the time series data from the meteorological stations. The authors have reviewed and edited the output and take full responsibility for the content of this publication.

Conflicts of Interest: The authors declare no conflicts of interest.

References

1. Intergovernmental Panel on Climate Change. *Climate Change 2021–The Physical Science Basis: Working Group I Contribution to the Sixth Assessment Report of the Intergovernmental Panel on Climate Change*; Cambridge University Press: Cambridge, UK, 2023.
2. Dai, A. Increasing Drought under Global Warming in Observations and Models. *Nat. Clim. Change* **2013**, *3*, 52–58. [CrossRef]
3. Fu, Y.; Wu, Q. Recent Emerging Shifts in Precipitation Intensity and Frequency in the Global Tropics Observed by Satellite Precipitation Data Sets. *Geophys. Res. Lett.* **2024**, *51*, e2023GL107916. [CrossRef]
4. Maggioni, V.; Meyers, P.C.; Robinson, M.D. A Review of Merged High-Resolution Satellite Precipitation Product Accuracy during the Tropical Rainfall Measuring Mission (TRMM) Era. *J. Hydrometeorol.* **2016**, *17*, 1101–1117. [CrossRef]
5. Mekonnen, M.M.; Hoekstra, A.Y. Sustainability: Four Billion People Facing Severe Water Scarcity. *Sci. Adv.* **2016**, *2*, e1500323. [CrossRef] [PubMed]
6. Célleri, R.; Feyen, J. The Hydrology of Tropical Andean Ecosystems: Importance, Knowledge Status, and Perspectives. *Mt. Res. Dev.* **2009**, *29*, 350–355. [CrossRef]
7. Dalla Torre, D.; Di Marco, N.; Menapace, A.; Avesani, D.; Righetti, M.; Majone, B. Suitability of ERA5-Land Reanalysis Dataset for Hydrological Modelling in the Alpine Region. *J. Hydrol. Reg. Stud.* **2024**, *52*, 101718. [CrossRef]
8. Kusre, B.C.; Phawa, R. Evaluation of Gridded and Satellite Precipitation Product over Hilly Terrain of Meghalaya (India) for Water Management. *Sustain. Water Resour. Manag.* **2023**, *9*, 62. [CrossRef]
9. Kucera, P.A.; Ebert, E.E.; Turk, F.J.; Levizzani, V.; Kirschbaum, D.; Tapiador, F.J.; Loew, A.; Borsche, M. Precipitation from Space: Advancing Earth System Science. *Bull. Am. Meteorol. Soc.* **2013**, *94*, 365–375. [CrossRef]
10. Sun, Q.; Miao, C.; Duan, Q.; Ashouri, H.; Sorooshian, S.; Hsu, K.L. A Review of Global Precipitation Data Sets: Data Sources, Estimation, and Intercomparisons. *Rev. Geophys.* **2018**, *56*, 79–107. [CrossRef]
11. Kidd, C.; Levizzani, V. Status of Satellite Precipitation Retrievals. *Hydrol. Earth Syst. Sci.* **2011**, *15*, 1109–1116. [CrossRef]
12. Lober, C.; Fayne, J.; Hashemi, H.; Smith, L.C. Bias Correction of 20 Years of IMERG Satellite Precipitation Data over Canada and Alaska. *J. Hydrol. Reg. Stud.* **2023**, *47*, 101386. [CrossRef]
13. Hou, A.Y.; Kakar, R.K.; Neeck, S.; Azarbarzin, A.A.; Kummerow, C.D.; Kojima, M.; Oki, R.; Nakamura, K.; Iguchi, T. The Global Precipitation Measurement Mission. *Bull. Am. Meteorol. Soc.* **2014**, *95*, 701–722. [CrossRef]
14. Funk, C.; Peterson, P.; Landsfeld, M.; Pedreros, D.; Verdin, J.; Shukla, S.; Husak, G.; Rowland, J.; Harrison, L.; Hoell, A.; et al. The Climate Hazards Infrared Precipitation with Stations—A New Environmental Record for Monitoring Extremes. *Sci. Data* **2015**, *2*, 150066. [CrossRef]
15. Popovych, V.F.; Dunaieva, I.A. Assessment of the GPM IMERG and CHIRPS Precipitation Estimations for the Steppe Part of the Crimea. *Meteorol. Hydrol. Water Manag.* **2021**, *9*. [CrossRef]
16. Li, L.; Chen, X.; Ma, Y.; Zhao, W.; Zuo, H.; Liu, Y.; Cao, D.; Xu, X. Implications for Validation of IMERG Satellite Precipitation in a Complex Mountainous Region. *Remote Sens.* **2023**, *15*, 4380. [CrossRef]
17. Yang, L.; Shi, Z.; Liu, R.; Xing, M. Evaluating the Performance of Global Precipitation Products for Precipitation and Extreme Precipitation in Arid and Semiarid China. *Int. J. Appl. Earth Obs. Geoinf.* **2024**, *130*, 103888. [CrossRef]

18. Benítez, V.D.; Forgioni, F.P.; Lovino, M.A.; Sgroi, L.; Doyle, M.E.; Müller, G.V. Capability of Satellite Data to Estimate Observed Precipitation in Southeastern South America. *Int. J. Climatol.* **2024**, *44*, 792–811. [CrossRef]
19. Mattos Gutierrez, S.R.; Fenta, A.A.; Meshesha, T.M.; Belay, A.S. Unveiling the Accuracy of New-Generation Satellite Rainfall Estimates across Bolivia's Complex Terrain. *Remote Sens.* **2024**, *16*, 2211. [CrossRef]
20. Salas-Choquehuanca, A.J.; Rosales-Solórzano, E.R.; Correa-Cuba, O.; Taquire-Arroyo, A.F.; Asención-Cordova, R.M. Análisis de Datos de Precipitacion Satelital GPM IMERG En La Cuenca Madre de Dios, Perú. *El Ceprosimad* **2023**, *11*, 6–16. [CrossRef]
21. Ballari, D.; Castro, E.; Campozano, L. Validation of Satellite Precipitation (TRMM 3B43) in Ecuadorian Coastal Plains, Andean Highlands and Amazonian Rainforest. *Int. Arch. Photogramm. Remote Sens. Spat. Inf. Sci.* **2016**, *XLI-B8*, 305–311. [CrossRef]
22. Huber Magoffin, R.; Hales, R.C.; Erazo, B.; Nelson, E.J.; Larco, K.; Miskin, T.J. Evaluating the Performance of Satellite Derived Temperature and Precipitation Datasets in Ecuador. *Remote Sens.* **2023**, *15*, 5713. [CrossRef]
23. OEA. *Proyecto de Manejo y Conservacion, Cuenca Alta Del Rio Pastanza. Resumen Ejecutivo*; OEA: Portland, Oregon, 1991.
24. Uttarwar, S.B.; Napoli, A.; Avesani, D.; Majone, B. Elevation-Driven Biases in Seasonal Weather Forecasts: Insights from the Alpine Region. *Phys. Chem. Earth Parts A/B/C* **2025**, *139*, 103957. [CrossRef]
25. Ma, J.; Li, H.; Wang, J.; Hao, X.; Shao, D.; Lei, H. Reducing the Statistical Distribution Error in Gridded Precipitation Data for the Tibetan Plateau. *J. Hydrometeorol.* **2020**, *21*, 2641–2654. [CrossRef]
26. Gobierno Provincial de Tungurahua. *Plan Provincial de Riego de Tungurahua 2014–2029*; Gobierno Provincial de Tungurahua: Ambato, Ecuador, 2014.
27. Hunink, J.E.; Immerzeel, W.W.; Droogers, P. *Análisis de Patrones Espaciales de Precipitación En La Provincia de Tungurahua*; FutureWater: Wageningen, The Netherlands, 2013.
28. Fick, S.E.; Hijmans, R.J. WorldClim 2: New 1-Km Spatial Resolution Climate Surfaces for Global Land Areas. *Int. J. Climatol.* **2017**, *37*, 4302–4315. [CrossRef]
29. Salomón, M.; Guaman Ríos, C.; Rubio, C.; Galárraga, R.; Abraham, E. Indicadores de uso del agua en una zona de los Andes centrales de Ecuador. Estudio de la cuenca del Río Ambato. *Ecosistemas* **2008**, *17*, 72–85.
30. Veiga, H.; Herrera, N.; Skansi, M.; Podestá, G. *Descripción de Controles de Calidad de Datos Climáticos Diarios Implementados Por El Centro Regional Del Clima Para El Sur de América Del Sur*; Serie Reportes Técnicos–Reporte Técnico CRC–SAS; Regional Climate Centre for Southern South America (CRC-SAS): Suva, Fiji, 2014.
31. Huffman, G.J.; Stocker, E.F.; Bolvin, D.T.; Nelkin, E.J.; Tan, J. *GPM IMERG Final Precipitation L3 Half Hourly 0.1 Degree × 0.1 Degree V06*; Goddard Earth Sciences Data and Information Services Center (GES DISC): Greenbelt, MD, USA, 2019.
32. Gorelick, N.; Hancher, M.; Dixon, M.; Ilyushchenko, S.; Thau, D.; Moore, R. Google Earth Engine: Planetary-Scale Geospatial Analysis for Everyone. *Remote Sens. Environ.* **2017**, *202*, 18–27. [CrossRef]
33. Rahmati Ziveh, A.; Bakhtar, A.; Shayeghi, A.; Kalantari, Z.; Bavani, A.M.; Ghajarnia, N. Spatio-Temporal Performance Evaluation of 14 Global Precipitation Estimation Products across River Basins in Southwest Iran. *J. Hydrol. Reg. Stud.* **2022**, *44*, 101269. [CrossRef]
34. Mason, I.B. Binary Events. In *Forecast Verification: A Practitioner's Guide in Atmospheric Science*; Jolliffe, I., Stephenson, D., Eds.; John Wiley & Sons: Chichester, UK, 2003.
35. López-Bermeo, C.; Montoya, R.D.; Caro-Lopera, F.J.; Díaz-García, J.A. Validation of the Accuracy of the CHIRPS Precipitation Dataset at Representing Climate Variability in a Tropical Mountainous Region of South America. *Phys. Chem. Earth Parts A/B/C* **2022**, *127*, 103184. [CrossRef]
36. Quispe, L.A.; Paxi, E.; Lujano, E. Evaluation of GPM IMERG Performance Over the Lake Titicaca Basin at Different Time Scales. *Environ. Sci. Proc.* **2023**, *25*, 65. [CrossRef]

MDPI AG
Grosspeteranlage 5
4052 Basel
Switzerland
Tel.: +41 61 683 77 34

Hydrology Editorial Office
E-mail: hydrology@mdpi.com
www.mdpi.com/journal/hydrology

www.ingramcontent.com/pod-product-compliance
Lightning Source LLC
Chambersburg PA
CBHW040745250926
51850CB00007BB/1411
9783725857210